Contents

Chapter 1

Introduction

1.1 Commodore VIC-20

The **VIC-20** (Germany: **VC-20**;*[3] Japan: **VIC-1001**) is an 8-bit home computer which was sold by Commodore Business Machines. The VIC-20 was announced in 1980,*[4] roughly three years after Commodore's first personal computer, the PET. The VIC-20 was the first computer of any description to sell one million units.*[5]

1.1.1 History

Origin, marketing

The VIC-20 was intended to be more economical than the PET computer. It was equipped with 5 KB of static RAM and used the same MOS 6502 CPU as the PET. The VIC-20's video chip, the MOS Technology VIC, was a general-purpose color video chip designed by Al Charpentier in 1977 and intended for use in inexpensive display terminals and game consoles, but Commodore could not find a market for the chip. As the Apple II gained momentum with the advent of VisiCalc in 1979, Jack Tramiel wanted a product that would compete in the same segment, to be presented at the January 1980 CES. For this reason Chuck Peddle and Bill Seiler started to design a computer named *TOI* (The Other Intellect).

The TOI computer failed to materialize, mostly because it required an 80-column character display which in turn required the MOS Technology 6564 chip. However, the chip could not be used in the TOI since it required very expensive static RAM to operate fast enough. In the meantime, freshman engineer Robert Yannes at MOS Technology (then a part of Commodore) had designed a computer in his home dubbed the *MicroPET* and finished a prototype with some help from Al Charpentier and Charles Winterble. With the TOI unfinished, when Jack Tramiel was confronted with the MicroPET prototype, he immediately said he wanted it to be finished and ordered it to be mass-produced following a limited demonstration at the CES.

The prototype produced by Yannes had very few of the features required for a real computer, so Robert Russell at Commodore headquarters had to co-ordinate and finish large parts of the design under the codename *Vixen*. The parts contributed by Russell included a port of the operating system (kernel and BASIC interpreter) taken from John Feagans design for the Commodore PET, a character set with the characteristic PETSCII, an Atari 2600-compatible joystick interface, and a ROM cartridge port. The serial IEEE 488-derivative interface (which could use far cheaper cabling than a real IEEE-488 as was used on the PET)*[6] was designed by Glen Stark. Some features, like the memory add-in board, were designed by Bill Seiler. Altogether, the VIC 20 development team consisted of five people, who referred to themselves as the VIC Commandos.*[7] According to one of the development team, Neil Harris. "[W]e couldn't get any cooperation from the rest of the company who thought we were jokers because we were working late, about an hour after everyone else had left the building. We'd swipe whatever equipment we needed to get our jobs done. There was no other way to get the work done! [...] they'd discover it was missing and they would just order more stuff from the warehouse, so everybody had what they needed to do their work."*[7] At the time, Commodore had an oversupply of 1 kbit×4 SRAM chips, so Tramiel decided that these should be used in the new computer. The end result was arguably closer to the *PET* or *TOI* computers than to Yannes' prototype, albeit with a 22-column VIC chip instead of the custom chips designed for the more ambitious computers.

In April 1980 at a meeting of general managers outside London, Jack Tramiel declared that he wanted a low-cost color computer. When most of the GMs argued against it, he said: "The Japanese are coming, so we will become the Japanese." This was in keeping with Tramiel's philosophy which was to make "computers for the masses, not the classes". The concept was championed at the meeting by Michael Tomczyk, newly hired marketing strategist and assistant to the president, Tony Tokai, General Manager of Commodore-Japan, and Kit Spencer, the UK's top marketing executive. Then, the project was given to Commodore

Japan; an engineering team led by Yash Terakura created the VIC-1001 for the Japanese market. The VIC-20 was marketed in Japan as VIC-1001 before VIC-20 was introduced to the US.

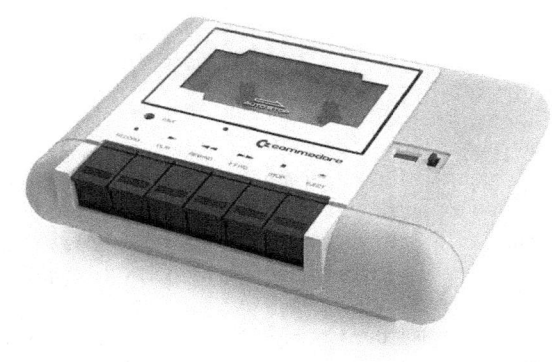

The Commodore 1530 C2N-B Datasette provided inexpensive external storage for the VIC-20

When they returned to California from that meeting, Tomczyk wrote a 30-page memo detailing recommendations for the new computer, and presented it to Tramiel. Recommendations included programmable function keys (inspired by competing Japanese computers),*[8] full-size typewriter-style keys, and built-in RS-232. Tomczyk insisted on "user-friendliness" as the prime directive for the new computer, to engineer Yash Terakura (who was also a friend),*[8] and proposed a retail price of US$299.95. He recruited a marketing team and a small group of computer enthusiasts, and worked closely with colleagues in the UK and Japan to create colorful packaging, user manuals, and the first wave of software programs (mostly games and home applications).

Scott Adams was contracted to provide a series of text adventure games. With help from a Commodore engineer who came to Longwood, Florida to assist in the effort, five of Adams's Adventure International game series were ported to the VIC. They got around the limited memory of VIC-20 by having the 16 KB games reside in a ROM cartridge instead of being loaded into main memory via cassette as they were on the TRS-80 and other machines. The first production run of the five cartridges generated over $1,500,000 in sales for Commodore.

While the PET was sold through authorized dealers, the VIC-20 primarily sold at retail—especially discount and toy stores, where it could compete more directly with game consoles. It was the first computer to be sold in K-Mart. Commodore took out advertisements featuring actor William Shatner (of *Star Trek* fame) as its spokesman, asking: "Why buy just a video game?" Television personality Henry Morgan (best known as a panelist on the TV game show *I've Got a Secret*) became the commentator in a series of Commodore product ads.

The VIC-20 had 5 KB of RAM, of which only 3.5 KB remained available on startup (exactly 3583 bytes). This is roughly equivalent to the words and spaces on one sheet of typing paper, meeting a design goal of the machine. The computer was expandable up to 40 KB with an add-on memory cartridge (a maximum of 27.5 KB was usable for BASIC). Although the VIC-20 was criticized in print as being underpowered, the strategy worked.

In 1981, Tomczyk contracted with an outside engineering group to develop a direct-connect modem-on-a-cartridge (the VICModem), which at US$99 became the first modem priced under US$100. The VICModem was also the first modem to sell over 1 million units. VICModem was packaged with US$197.50 worth of free telecomputing services from The Source, CompuServe and Dow Jones. Tomczyk also created an entity called the Commodore Information Network to enable users to exchange information and take some of the pressure off of Customer Support inquiries, which were straining Commodore's lean organization. In 1982, this network accounted for the largest traffic on CompuServe.

Decline

In 1982 the VIC-20 was the best-selling computer of the year, with 800,000 machines sold. One million had been sold by the end of the year and at one point, 9000 units a day were being produced. That summer, Commodore unveiled the Commodore 64, a more advanced machine with 64 KB of RAM and considerably improved sound and graphics capabilities. Sales were slow at first due to reliability problems and lack of software, but by the middle of 1983, the latter had turned into a flood and VIC-20 sales abruptly plunged. It was quietly discontinued in January 1985.*[9] Perhaps the last new commercially available VIC-20 peripheral was the VIC-Talker, a speech synthesizer; *Ahoy!* in January 1986 wrote when discussing it, "Believe it or not, a new VIC accessory ... We were as surprised as you".*[10]

Applications

Before the VIC-20's release, a Commodore executive promised that the forthcoming computer would have "enough additional documentation to enable an experienced programmer/hobbyist to get inside and let his imagina-

Software cartridge

tion work".[11] Because of its small memory and low-resolution display compared to some other computers of the time, the VIC-20 was primarily used for educational software and games. However, productivity applications such as home finance programs, spreadsheets, and communication terminal programs were also made for the machine. Its high accessibility to the general public meant that many software developers-to-be cut their teeth on the VIC-20, being introduced to BASIC programming or assembly language.

Several computer magazines sold on newsstands, such as *Compute!*, *Family Computing*, *RUN*, *Ahoy!*, and the CBM-produced *Commodore Power Play*, offered programming tips and type-in programs for the VIC-20. Many VIC users learned to program by entering, studying, running, and modifying these type-ins.

The ease of programming the VIC and availability of an inexpensive modem combined to give the VIC a sizable library of public domain and freeware software, although much smaller than that of the C64. This software was distributed via online services such as CompuServe, BBSs, as well as offline by mail order and by user groups.

The VIC's low cost led to it being used by the Fort Pierce, Florida Utilities Authority to measure the input and output of two of their generators and display the results on monitors throughout the plant. The utility was able to purchase multiple VIC and C-64 systems for the cost of one IBM PC Compatible system.[12]

As for commercial software offerings, an estimated 300 titles were available on cartridge, and another 500+ titles were available on tape. By comparison, the Atari 2600—the most popular of the video game consoles at the time—had a library of about 900 titles near the end of its production life (although many titles were extremely similar). Most

cartridge games were ready to play as soon as the VIC-20 was turned on, as opposed to games on tape which required a time-consuming loading process. Titles on cartridge included *Gorf*, *Cosmic Cruncher*, *Sargon II Chess*, and many others. A handful of disk applications were released for the VIC-20.

1.1.2 Technical specifications

Basic features

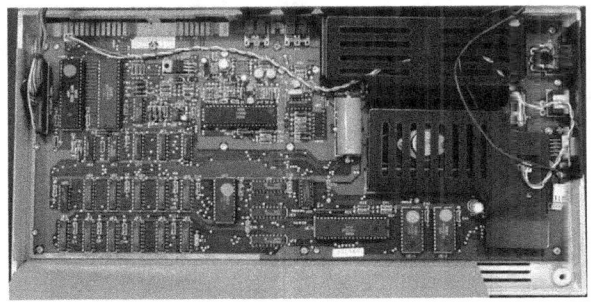

VIC-20 Mainboard

The VIC-20 had proprietary connectors for program/expansion cartridges and a tape drive (PET-standard Datassette). It came with 5 KB RAM, but 1.5 KB of this was used by the system for various things, like the video display (which had a rather unusual 22×23 char/line screen layout), and other dynamic aspects of the ROM-resident BASIC interpreter and KERNAL (a low-level operating system). Thus, only 3583 bytes of BASIC program memory for code and variables was actually available to the user of an unexpanded machine.

The computer also had a single DE-9 game controller port, compatible with the digital joysticks and paddles used with Atari 2600 videogame consoles[13] (the use of a standard port ensured ample supply of Atari-manufactured and other third-party joysticks; Commodore itself offered an Atari-protocol joystick under the Commodore brand) a serial bus (a serial version of the PET's IEEE-488 bus) for daisy chaining disk drives and printers; a TTL-level "user port" with both RS-232 and Centronics signals (most frequently used as RS-232, for connecting a modem[14]).

Importantly, like most video game consoles and many computers at the time the VIC had a ROM cartridge port to allow for plug-in cartridges with games and other software as well as for adding memory to the machine. Port expander boxes were available from Commodore and other vendors to allow more than one cartridge to be attached at a time. Cartridge software ranged from 4 - 16 KB in size, although the latter was uncommon due to its cost and only larger software houses produced 16 KB cartridges.

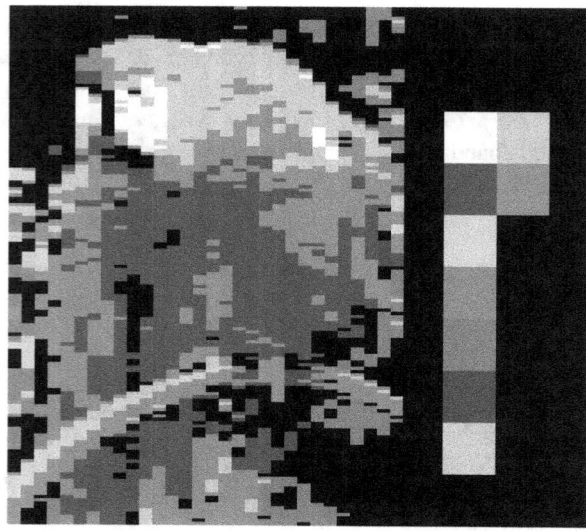

16-color (multicolor) capability

The graphics capabilities of the VIC chip (6560/6561) were limited but flexible. At startup the screen showed 176×184 pixels, with a fixed-colour border to the edges of the screen; since an NTSC or PAL screen has a 4:3 width-to-height ratio, each VIC pixel was much wider than it was high. The screen normally showed 22 columns and 23 rows of 8-by-8-pixel characters; it was possible to increase these dimensions up to 27 columns, but the characters would soon run out the sides of the monitor at about 25 columns. Like on the PET, 256 different characters could be displayed at a time, normally taken from one of the two character generators in ROM (one for upper-case letters and simple graphics, the other for mixed-case—non-English characters were not provided). Normally, the VIC-20 was operated in high-resolution mode whereby each character was 8×8 pixels in size and used one color. A lower-resolution multicolor mode could also be used with 4×8 characters and three colors each, but it was not used as often due to its extreme blockiness.

The VIC chip did not support a true bitmap mode, but programmers could define their own custom character set. It was possible to get a fully addressable screen, although slightly smaller than normal, by filling the screen with a sequence of different double-height characters, then turning on the pixels selectively inside the RAM-based character definitions. The Super Expander cartridge added BASIC commands supporting such a graphics mode using a resolution of 160×160 pixels. It was also possible to fill a larger area of the screen with addressable graphics using a more dynamic allocation scheme, if the contents were sparse or repetitive enough. This was used, for instance, by the game Omega Race. The VIC chip did not support sprites.

The VIC chip had readable scan-line counters but could not generate interrupts based on the scan position (as the VIC-II chip could). However, the two VIA timer chips could be tricked into generating interrupts at specific screen locations, by setting up the timers after a position has been established by repetitive reading of the scan-line counter, and letting them run the exact number of cycles that pass by during one full screen update. Thus it was possible, but difficult, to e.g. mix graphics with text above or below it, or to have two different background and border colors, or to use more than 200 characters for the pseudo-high-resolution mode. The VIC chip could also process a light pen signal (a light pen input was provided on the DE-9 joystick connector) but few of those ever appeared on the market.

The VIC chip had three pulse wave sound generators. Each had a range of three octaves, and the generators were located on the scale about an octave apart, giving a total range of about five octaves. In addition, there was a white noise generator. There was only one volume control, and the output was in mono.

The VIC chip output composite video; Commodore did not include an RF modulator inside the computer's case because of FCC regulations. It could either be attached to a dedicated monitor or a TV set using the external modulator included with the computer.

Memory expansion

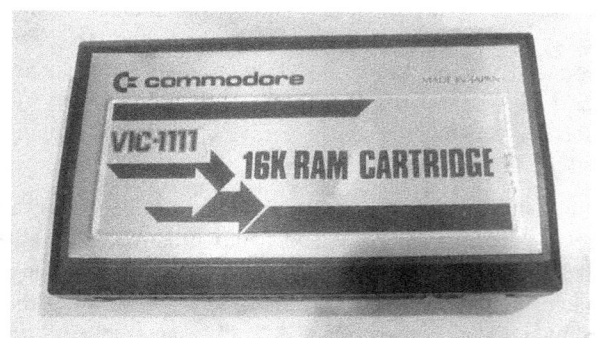

A 16 KB RAM expansion cartridge

The VIC-20's RAM was expandable through the cartridge port. RAM cartridges were available in several sizes: 3 KB (with or without an included BASIC extension ROM), 8 KB, 16 KB, 32 KB and 64 KB, the latter two only from third-party vendors. The internal memory map was dramatically reorganized with the addition of each size cartridge, leading to the situation that some programs would only work if the right amount of memory was present (to cater for this, the 32 KB cartridges had switches, and the 64 KB cartridges had software setups, allowing the RAM to be enabled in user-selected sections).

The most visible part of memory that was reorganized with differing expansion memory configurations was the video memory (with text and/or graphics display data). This was because the video chip could only use the built-in memory for its display data, and at the same time free memory had to remain contiguous for the BASIC interpreter to be able to use it. An unexpanded VIC had 1 KB of system memory, followed by a 3 KB "hole", then 4 KB of contiguous user memory up to address 8191. The 3 KB cartridge would fill the "hole", so on unexpanded and +3K VICs the video area was placed at the top of user memory (8 KB - 512 bytes). If an 8 KB or 16 KB cartridge was added instead, this memory appeared at addresses above 8 KB; the video memory was then placed at the start of user memory at 4 KB, just above the "hole", to provide the maximum amount of contiguous user memory.

The 32 KB cartridges allowed adding up to 24 KB to the BASIC user memory; together with the 3.5 KB built-in user memory, this gave a maximum of 27.5 KB for BASIC programs and variables. The extra 8 KB could usually be used in one of two ways, set by switches:

1. Either it could be mapped into the address space reserved for ROM cartridges, which sat "behind" the I/O register space and thus was not contiguous with the rest of the RAM. This allowed running many cartridge-based games from disk or tape and was thus very useful for software pirates; especially if the RAM expansion allowed switching off writing to its memory after the game was loaded, so that the memory behaved exactly like ROM.

2. Or, 3 KB of the 8 KB could be mapped into the same memory "hole" that the 3 KB cartridge used, letting 5 KB lie fallow. These 3 KB were contiguous with the rest of RAM, but couldn't be used to expand BASIC space to more than 27.5 KB, because the display data would have had to be moved to cartridge RAM, which was not possible.

Some 64 KB expansion cartridges allowed the user to copy ROM images to RAM. The more advanced versions even contained an 80-character video chip and a patched BASIC interpreter which gave access to 48 KB of the memory and to the 80-column video mode. As the latter type of cartridges, marketed primarily in Germany, were not released until late 1984—two years after the appearance of the more capable C64—they went by mostly unnoticed.

*[16]

1.1.3 Reception

While noting the small screen size and RAM, *BYTE* stated that the VIC 20 was "unexcelled as low-cost, consumer-oriented computer. Even with some of its limitations . . it makes an impressive showing against ... the Apple II, the Radio Shack TRS-80, and the Atari 800". It praised the price ("Looking at a picture ... might cause you to think $600 would be a fair price ... But it does not cost $600—the VIC 20 retails for $299.95"), keyboard ("the equal of any personal-computer keyboard in both appearance and performance. This is a remarkable achievement, almost unbelievable considering the price of the entire unit"), graphics, documentation, and ease of software development with the KERNAL.*[17]

1.1.4 Notes

The VIC-20 could be hooked into external electronic circuitry, using parts available from parts outlets like RadioShack and Maplin. Interfaces were designed to use the joystick port, the so-called "user port", or the memory expansion–cartridge port, which exposed various analog to digital, memory bus, and other internal I/O circuits to the experimenter. The BASIC language could then be used (using the PEEK and POKE commands) to perform data acquisition from temperature sensors, control robotic stepper motors, etc. The VIC-20 did not originally have a disk drive, with only a relatively high cost, but extremely reliable digital tape storage system (using audio cassette tapes); the VIC-1540 disk drive was released in 1981. Many experimenters built adapters that allowed any conventional audio cassette recorder to be used for program and data storage (since these were generally cheaper than Commodore's own Datasette recorder, though only as reliable as other manufacturers analog cassette storage solutions).

As on other Commodore 8-bit systems, certain system functions could be accessed by the SYS command. For example, even though the VIC had no hardware reset button, SYS 64802 would cause the computer to reset, because memory location 64802 in the standard memory map was the entry point to the VIC's KERNAL reset routine.

The Commodore VIC-20 continues to have a loyal following today. Programmers continue to write demo, utility, and game programs for the machine (most often shared through the Denial community), and also through commercial retro-software developers such as Psytronik. Recent programs compiled in machine language tend to reveal features of the machine that were never utilized during its production years. A common goal of these programs (and the programmers writing them) is to "show off" how many complex program/graphic features (such as scrolling and pseudo-sprites) and/or intense/realistic gameplay that can be packed into the VIC-20's small amount of available RAM and resolution. Recent software releases such as *Frogger '07* (2007 release) and *Berzerk MMX* (2010 release) have gameplay, graphics, and sound (including voice synthesis in Berzerk) that rival the original arcade machines. Even a port of *Doom*, a 1993 game popularized on much more powerful platforms, has become available for the VIC-20 in 2013.*[18]

1.1.5 See also

- VICE, VIC-20 emulator
- List of Commodore VIC-20 games
- CARDCO

1.1.6 References

[1] "MESS VIC20/VC20 (German) PAL". MESS —Multiple Emulator Super System

[2] "Home Video Game Console Sound Chip Round-Up". 090514 gweep.net

[3] The computer was renamed in German-speaking countries because "VIC" would be pronounced similarly to the obscene word *fick*.

[4] Commodore VIC-20 History

[5] OLD-COMPUTERS.COM : The Museum

[6] "RUN Magazine issue 28".

[7] Herzog, Marty (January 1988). "Neil Harris". *Comics Interview* (54) (Fictioneer Books). pp. 41–51.

[8] http://www.michaeltomczyk.com/Tech-Pioneer.php

[9] http://www.commodore.ca/products/vic20/commodore_vic-20.htm

[10] Kevelson, Morton (January 1986). "Speech Synthesizers for the Commodore Computers / Part II". *Ahoy!*. p. 32. Retrieved 2 July 2014.

[11] "Commodore: New Products, New Philosophies". *Kilobaud*. September 1980. pp. 26–28. Retrieved 23 June 2014.

[12] "RUN Magazine Issue 34".

[13] Flynn, Christopher J. (June 1982). "Using Atari Joysticks With Your VIC". *Compute!*. p. 79. Retrieved 6 October 2013.

[14] The Commodore VICModem and later models connected directly to the user port's edge connector. But in order to connect the VIC to industry-standard modems and other RS-232 devices, the user needed to purchase a separate TTL-to-RS232 voltage converter box (standard TTL voltages lie between 0 and 5 V, while RS-232 uses ±12 V).

[15] "VIC 20 / Commodore 64 RS 232" (PDF). commodore.ca. 30 March 2011. Retrieved 2013-05-21.

[16] "VIC-20 memory map (long)". zimmers.net. 19 September 2005. Retrieved 2013-05-20.

[17] Williams, Gregg (May 1981). "The Commodore VIC 20 Microcomputer: A Low-Cost, High-Performance Consumer Computer". *BYTE*. p. 46. Retrieved 18 October 2013.

[18] "DOOM - First person hit on the Commodore VIC-20 / Commodore VC-20". Retrieved 26 August 2015.

1.1.7 Further reading

- Bagnall, Brian (2005). *On The Edge: The Spectacular Rise and Fall of Commodore*. ISBN 0-9738649-0-7. Retrieved 2009-04-20.
- Finkel, A.; Harris, N.; Higginbottom, P.; Tomczyk, M. (1982). *VIC 20 Programmer's reference guide*. Commodore Business Machines, Inc. and Howard W. Sams & Co, Inc. ISBN 0-672-21948-4. Retrieved 2009-04-20.
- Jones, A.J.; Coley, E. A.; Cole, D. G. J. (1983). *Mastering the Vic-20*. Chichester, UK: Ellis Horwood Ltd. and John Wiley & Sons, Inc. ISBN 0-471-88892-3. Retrieved 2009-04-20.
- Tomczyk, Michael S. (1984). *The Home Computer Wars: An Insider's Account of Commodore and Jack Tramiel*. COMPUTE! Publications, Inc. ISBN 0-942386-75-2. Retrieved 2009-04-20.

BYTE in 1983 published a series of technical articles about the VIC-20:

1. Swank, Joel (January 1983). "Exploring the Commodore VIC-20". *BYTE*. p. 222.

2. Swank, Joel (February 1983). "The Enhanced VIC-20 / Part 1: Adding a Reset Switch". *BYTE*. p. 118.

3. Swank, Joel (March 1983). "The Enhanced VIC-20 / Part 2: Adding a 3K-Byte Memory Board". *BYTE*. p. 34.

4. Swank, Joel (April 1983). "The Enhanced VIC-20 / Part 3: Interfacing an MX-80 Printer". *BYTE*. p. 260.

5. Swank, Joel (May 1983). "The Enhanced VIC-20 / Part 4: Connecting Serial RS-232C Peripherals to the VIC's TTL Port". *BYTE*. p. 331.

1.1.8   External links

- Denial – the Commodore VIC-20 Community

- OLD-COMPUTERS.COM online-museum VIC-20 page

- VIC-20 Programmers reference guide and more

- VIC-20 Gamer - Videos of games produced for the Vic 20

- Rick Melick's Commodore VIC-20 Tribute Page

- atarimagazines.com: A 40/80 Character Expansion For The VIC-20 (*Compute!*, November 1982)

- sleepingelephant.com: 40 and 80 column boards

- "The VIC 40-Column Operating System / Turn Your VIC 20 into a PET" (*Ahoy!*, October 1984)

Chapter 2

Related Topics

2.1 CARDCO

CARDCO was a computer peripheral company during the 1980s in Wichita, Kansas, United States. CARDCO was well known in the Commodore 64 and VIC-20 community because of advertisements in numerous issues of Compute! magazine and availability of their products at large retailers.[*][1]

There were severe shortcomings of early Commodore printers, so CARDCO created the Card Print A (C/?A) printer interface that emulated Commodore printers by converting the Commodore-style IEEE-488 serial interface to a Centronics printer port to allow numerous 3rd-party printers to be connected to a Commodore 64 or VIC-20, such as Epson, Okidata, C. Itoh.[*][2] A second model, a version that supported printer graphics was released called the Card Print +G (C/?+G), supported printing Commodore graphic characters using ESC/P escape codes. CARDCO released additional enhancements, including a model with RS-232 output, and shipped a total over two million printer interfaces.

2.1.1 See also

- Commodore 64 peripherals

2.1.2 References

[1] *Compute!* on Internet Archive

[2] "CARDCO Card Print A (C/?A) - Printer Interface For The Commodore 64 and VIC-20". *COMPUTE Magazine* (34): 251. March 1983.

2.1.3 External links

- CARDCO Card Print A (C/?A) Printer Interface: User Manual, Addendum

- CARDCO Card Print +G (C/?+G) Printer Interface: User Manual, Supplement

2.2 Commodore 1540

The **Commodore 1540** (also known as the **VIC-1540**) introduced in 1982[*][1][*][2] is the companion floppy disk drive for the Commodore VIC-20 home computer. It uses single-sided 5¼" floppy disks, on which it stores roughly 170 kB of data utilizing Commodore's GCR data encoding scheme. The launch price in Germany was 1898 DM (approximate 970 EUR). The US-American version is named VIC 1540 and the German version VC 1540.

Because of the low price of both the VIC-20 and the 1540, this combination was the first computer with a disk drive to be offered on the US market for less than $1000 USD, although the combination of the Commodore 64 and 1541 would prove more enduring. The 1540 is an "intelligent peripheral" in that it has its own MOS Technology 6502 CPU (just like its VIC-20 host) and the resident Commodore DOS on board in ROM – contrary to almost all other home computer systems of the time, where the DOS was loaded from a boot floppy and was executed on the computer's CPU.

Due to a timing conflict with the C64's video chip, the C64 doesn't work properly with the 1540. The better-known 1541 is mechanically and nearly electronically identical to the 1540 but has a revised ROM that permits it to work with the C64 by slowing the drive down slightly. However, it is possible to revert the 1541 into 1540 mode with a Commodore BASIC software command (OPEN 15,8,15, "UI-" : CLOSE 15) to permit better speed when used with a VIC-20.

The 1540 is relatively rare. While cheaper than most other drives of the day, it was more expensive than the VIC-20 computer itself, and the disk media was also still relatively pricey. Also, the relatively small memory of the VIC meant that the faster program loading times of the drive did not gain more than a few seconds compared to tape media. Thirdly, almost all commercial software for the VIC-20 was sold on cartridge or cassette tape media, giving low incentive to buy a floppy drive. The C64 followed close on the heels of the VIC-20, quickly discontinuing the 1540. Most 1540s still in existence were modified with a 1541 ROM so it would work with a C64. Unmodified 1540s are now considered collector's items.

2.2.1 References

[1] "Chronology of Personal Computers (1982)". pctime-line.info. 2013-05-18. Retrieved 2013-06-18. Commodore International releases the 1540 Single-Drive Floppy for the VIC-20. [804.17]

[2] "Commodore VIC-20 computer". oldcomputers.net. 2010-11-08. Retrieved 2013-06-18. 1982: Commodore releases the 1540 Single-Drive Floppy for the VIC-20.

2.3 Commodore Datasette

The **Commodore 1530 (C2N) Datasette** (a portmanteau of *data* and *cassette*), was Commodore's dedicated magnetic tape data storage device. Using compact cassettes as the storage medium, it provided inexpensive storage to Commodore's 8-bit home/personal computers, notably the PET, VIC-20, and C64. A physically similar model **Commodore 1531** was made for the Commodore 16 and Plus/4 series computers.

2.3.1 Description and history

The Datasette contained built-in analog to digital converters and audio filters to convert the computer's digital information into analog sound and vice versa (much like a modem does over a telephone line). Connection to the computer was done via a proprietary edge connector (Commodore 1530) or mini-DIN connector (Commodore 1531). The absence of recordable audio signals on this interface made the Datasette and its few clones the only cassette recorders usable with CBM's machines, until aftermarket converters made the use of ordinary recorders possible.

The Datasette was more popular outside than inside the United States. U.S. Gold, which imported American computer games to Britain, often had to wait until they were converted from disk because most British Commodore 64 owners used tape.[1][2] *Computer Gaming World* reported in 1986 that British cassette-based software had failed in the United States because "97% of the Commodore systems in the USA have disk drives";[3] by contrast, MicroProse reported in 1987 that 80% of its 100,000

sales of *Gunship* in the UK were on cassette.[4] In the United States disk drives quickly became standard, despite the Commodore 1541 floppy drive costing roughly five times as much as a Datasette. In most parts of Europe, the Datasette was the medium of choice for several years after its launch, although floppy disk drives were generally available. The inexpensive and widely available audio cassettes made the Datasette a good choice for the budget-aware home computer mass market.

The Datasette loading process

The Datasette was slow albeit extremely reliable,[5][6] transferring data at around 50 bytes per second; even the very slow 1541 was significantly faster. Some years after the Datasette's launch, however, special *turbo tape* software appeared, providing much faster tape operation (loading and saving). Such software was integrated into most commercial prerecorded applications (mostly games), as well as being available separately for loading and saving the users' homemade programs and data. These programs were only widely used in Europe, as the US market had long since moved onto disks.

Datasettes could typically store about 100 kByte per 30 minute side.[7] The use of *turbo tape* and other fast loaders increased this number to roughly 1000 kByte.

2.3.2 Interface

The Datasette has only one connection cable with a PCB edge connector at the computer end. All input/output signals to the datasette are all digital and so all digital to analog and vice versa is handled within the unit. Power is also included in this cable. The pinout is ground, +5 V DC, motor, read, write, key-sense.[8] The sense signal monitors the play, rewind, and fast-forward buttons, but cannot differentiate between them. A mechanical interlock prevented any two of them being pressed at the same time. Unregulated 6.36 V DC[9] is used to power the cassette motor.[10]

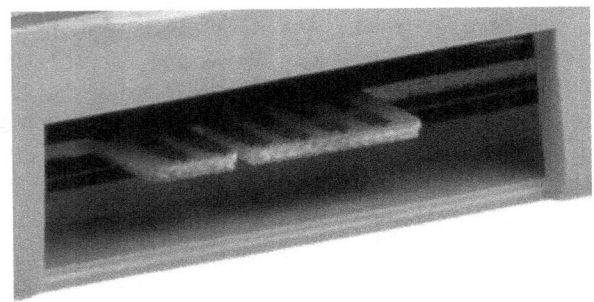

Commodore 64 cassette port

2.3.3   Physical coding

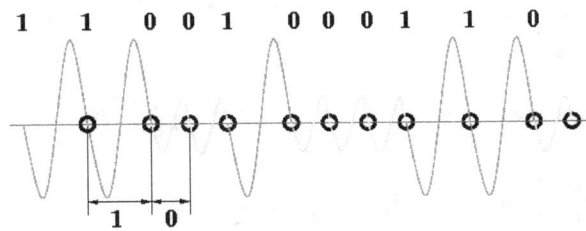

The resulting waveform from storing data

To record physical data, the zero-crossing from positive to negative voltage of the analog signal is measured. The resulting time between these positive to negative crossings is then compared to a threshold to determine whether the time since the last crossing is short (0) or long (1).[*][11] Note the lower amplitude for the shorter periods.

A circuit in the tape unit transforms the analog signal into a logical one or zero, which is then transmitted to the computer via the tape connector. Inside the computer, the first Complex Interface Adapter (6526) in the C64 senses when the signal goes from one to zero. This event is called trigger and causes an interrupt request. This event can be handled by a handler code, or simply discovered by testing bit 4 of location $DC0D. The points that trigger this event are indicated by the black circles in the figure.[*][11]

Inside the tape device the read head signal is fed into an operational amplifier (1) whose output signal is DC-filtered. Op-amp (2) amplifies and feeds an RC-filter. Op-amp (3) amplifies the signal again followed by another DC-filter. Op-amp (4) amplifies the signal into clipping the sine formed signal. The positive and negative rails for all op-amps are wired to +5V DC and GND. The clipped signal therefore fits into the TTL electrical level window of the schmitt trigger step that in turn feeds the digital cassette port.[*][12]

On the PAL version of the C64, the time granularity is 1.014 µs (for NTSC 0.978 µs). For a 300 bit/s data rate and where each bit uses 3284 clock cycles this means 3284

* 1.014 µs = 3330 µs/bit.

Once the bits can be decoded, they are fed into a shift register and are continuously compared to a special bit sequence. This bit sequence can also be seen as a byte. A bit-sequence match means that the stream is byte-synchronized. The first byte to compare with is called *lead-in byte*. If matched, it's compared to the *sync byte* as well.[*][11]

An example: Turbo Tape 64 has a *lead-in byte* $02 (binary 00000010), *sync byte* $09 (binary 00001001) and a following sync sequence of $08, $07, $06, $05, $03, $02, $01.[*][11]

2.3.4   Practical handling

Typical labeling of cassette inlays with the meter reading of the tape drive and the appropriate computer game titles

The way to arrange a "directory" and find software on tapes was accomplished by using an inlay sheet with counter position noted, and program name next to it.

The physical shape of the Commodore Datasette 1530/31 is a weight of 0.7 kg and measurement 19.5 cm wide, 5 cm high and 15 cm deep.

2.3.5   Main models

Used with the PET, VIC-20, C64/128

There are at least four main models of the 1530/C2N Datassette:

- The built-in Datassette in the original PET 2001: black cassette lid, five white keys, no tape counter, no SAVE LED

- Black body original shape model, black cassette lid, five black keys, no tape counter, no SAVE LED

- White body original shape model, black cassette lid, five black keys, with tape counter, no SAVE LED

- White body new shape model, silver cassette lid, six black keys, with tape counter and a red SAVE LED

The first two external models were made as PET peripherals, and styled after the PET 2001 built-in tape drive. The latter two were styled and marketed for the VIC-20 and C64. All 1530s were compatible with all those computers, as well as the C128.

In addition to this, some models came with a small hole above the keys, to allow access to the adjustment screw of the tape head azimuth position. A small screwdriver could thus easily be used to effect the adjustment without disassembling the Datassette's chassis.

Confusingly, the Datassette at various times was sold both as the *C2N DATASETTE UNIT Model 1530* and as the *1530 DATASSETTE UNIT Model C2N*. Note the difference in spelling (one *S* versus two) used on the original product packaging.*[13]

Used with the C16/116 and Plus/4

Similar in physical appearance to the 1530/C2N models is the **Commodore 1531**, made for the Commodore 16 and Plus/4 series computers. This had a Mini-DIN connector in place of the PCB edge connector. This could be used with a C64/128 via an adaptor, which was supplied by Commodore with some units.

- Black/Charcoal body new shape model, silver cassette lid, six light gray keys, with tape counter and a red SAVE LED

2.3.6 Models

- The second, most common version of the 1530 C2N Datassette

- Datassette 1531

- One of the few clones

2.3.7 See also

- Magnetic tape data storage

- Fast loader

- IBM cassette tape

- Kansas City standard

2.3.8 References

[1] Anderson, Chris (June 1985). "On top of the US Goldmine" . *Zzap!64* (interview). pp. 46–48. Retrieved 26 October 2013.

[2] Pountain, Dick (January 1985). "The Amstrad CPC 464" . *BYTE*. p. 401. Retrieved 27 October 2013.

[3] Wagner, Roy (August 1986). "The Commodore Key" . *Computer Gaming World*. p. 28.

[4] Brooks, M. Evan (November 1987). "Titans of the Computer Gaming World / MicroProse" . *Computer Gaming World*. p. 16.

[5] "How TurboTape Works" .

[6] "The Official Book for the Commodore 128" .

[7] "Basic Commodore information" .

[8] pinouts.ru - C64 Cassette pinout, 2012-01-15

[9] "250469 rev.A right" . 100610 zimmers.net

[10] "250469 rev.A left" . 100610 zimmers.net

[11] "How Commodore tapes work" . 091205 wav-prg.sourceforge.net

[12] Datasette service manual model C2N/1530/1531, preliminary, Oct. 1984 PN-314002-02

[13] Bo Zimmerman. "Faster than a speeding South American Grima Slug" . *Commodore Gallery*. Retrieved 20 April 2012.

2.3.9 External links

- Similar Commodore tape drives

- Datasette photos

- Description of tape format with conversion utilities and code

- C2N232 project to build a hardware adaptor/software program to archive Commodore Datasette files to a modern computer.

- DC2N Homepage Digital C2N replacement project.

- Sketchup model of the Commodore Datasette 1530. Sketchup model of the Commodore Datasette 1530.

Technology began work on a video chip named *MOS Technology 6564* intended for the *TOI* computer and had also made some work on another chip, *MOS 6562* intended for a color version of the Commodore PET. Both of these chips failed due to memory timing constraints (both required very fast and thus expensive SRAM, making them unsuitable for mass production). Before finally starting to use the VIC in the VIC-20, chip designer Robert Yannes fed features from the 6562 (a better sound generator) and 6564 (more colors) back to the 6560, so before beginning mass production for the VIC-20 it had been thoroughly revised.

Its features include:

- 16 kB address space for screen, character and color memory (only 5 kB points to RAM on the VIC-20 without a hardware modification)

- 16 colors (the upper 8 can only be used in the global background and auxiliary colors)

- two selectable character sizes (8×8 or 8×16 bits; the pixel width is 1 bit for "hires" characters and 2 bits for "multicolor" characters)

- maximum video resolution depends on the television system (176 × 184 is the standard for the VIC-20 firmware, although at least 224 × 256 is possible on the PAL machine)

- 4 channel sound system (3 square wave + "white" noise + global volume setting)

- on-chip DMA

- two 8-bit A/D converters

- light pen support

Pinout diagram of the 6560 version of the MOS VIC chip. This circuit was packaged in a standard 40-pin DIP casing.

2.4 MOS Technology VIC

The **VIC (Video Interface Chip)**, specifically known as the **MOS Technology 6560** (NTSC version) / **6561** (PAL version), is the integrated circuit chip responsible for generating video graphics and sound in the Commodore VIC-20 home computer. It was originally designed for applications such as low cost CRT terminals, biomedical monitors, control system displays and arcade or home video game consoles.

The chip was designed by Al Charpentier in 1977 but Commodore could not find a market for the chip. In 1979 MOS

Unlike many other video circuits of the era, it does not offer dynamic RAM refresh capabilities. Thus the VIC-20 employed the more expensive static RAM chips; but with the low RAM capacity of that machine, just 5 KB in the main box, that didn't matter too much. Memory expansions for the VIC-20 either used SRAM as well or implemented their own refresh circuit.

The VIC was programmed by manipulating its 16 control registers, memory mapped to the range $9000–$900F in the VIC-20 address space. The on-chip A/D converters were used for dual paddle position readings by the VIC-20, which also used the VIC's light pen facility. The VIC preceded the much more advanced VIC-II, used by the VIC-20's successors, the C64 and C128.

VIC-20s with expansion RAM have their video memory (550 bytes) at $1000 and when it is not present, $1E00. User-defined character sets must be placed within the first

5k of system RAM. The default PETSCII character ROM is at $8000 and each character takes 8 bytes to store. Up to 128 characters may be used at any one time. While the PET had a backslash (\) in its character set, this was replaced on the VIC-20 (and all subsequent Commodore machines) with a British pound sign (£).

Programmable characters are the only way of creating graphics and animation on the VIC as the chip does not have sprites or an all-points-addressable bitmap mode. Of the 16 colors in the palette, eight may be used for the foreground (per the color RAM at $9400) and border while the others are limited to the background and auxiliary multicolors. The MSB of the color RAM is a flag used to indicate if that character is multicolor or high resolution. Due to the extreme blockiness of the former, most VIC-20 games use hires characters.

The VIC does not support scrolling or raster interrupts like on the VIC-II, but the scanline counters could be polled for a specific point on the screen to produce raster effects. This feature was rarely used in games except for a few titles like iMagic's Demon Attack.

Sound programming on the VIC is done by placing a frequency value in one of the four registers at $900A-$900D (they are turned off by writing a zero to them). The first three are square wave generators pitched half an octave from each other and the fourth is for white noise.

2.4.1 Registers

The VIC has 16 read/write registers listed below:

2.4.2 VIC IC list

- MOS Technology 6560 NTSC

- MOS Technology 6561E PAL Ceramic version, used in early VIC-20's

- MOS Technology 6561-101 PAL

2.4.3 References

- Bagnall, Brian (2005). *On The Edge: The Spectacular Rise and Fall of Commodore*. Variant Press. ISBN 0-9738649-0-7.

2.4.4 See also

- Video Display Controller

- List of home computers by video hardware

- MOS VIC-II as used in the Commodore 64

2.4.5 External links

- VIC Chip info from Rick Melick's VIC-20 Tribute Page

- More info, incl register usage

- MOS VIC datasheet (GIF format, zipped)

2.5 PETSCII

PETSCII (*PET Standard Code of Information Interchange*), also known as **CBM ASCII**, is the character set used in Commodore Business Machines (CBM)'s 8-bit home computers, starting with the PET from 1977 and including the VIC-20, C64, CBM-II, Plus/4, C16, C116 and C128. *

2.5.1 History

The character set was largely designed by Leonard Tramiel (the son of Commodore CEO Jack Tramiel) and PET designer Chuck Peddle. The graphic characters of PETSCII were one of the extensions Commodore specified for Commodore BASIC when laying out desired changes to Microsoft's existing 6502 BASIC to Microsoft's Ric Weiland in 1977.*[1] The VIC-20 used the same pixel-for-pixel font as the PET, although the characters appeared wider due to the VIC's 22-column screen. The Commodore 64, however, used a slightly re-designed, heavy upper-case font, essentially a thicker version of the PET's, in order to avoid color artifacts created by the machine's higher resolution screen. The C64's lowercase characters are identical to the lowercase characters in the Atari 800's system font (released several years earlier).

Peddle claims the inclusion of card suit symbols was spurred by the demand that it should be easy to write card games on the PET (as part of the specification list he received).*

2.5.2 Specifications

PETSCII is based on the 1963 version of ASCII (rather than the 1967 version, which most if not all other computer character sets based on ASCII use). Assuming the graphics mode is **unshifted**, PETSCII has only uppercase letters in its powerup state, an up-arrow (↑) instead of a caret (^) in position $5E and a left-arrow (←) instead of an underscore (_) in position $5F. Also, in the VIC-20 and

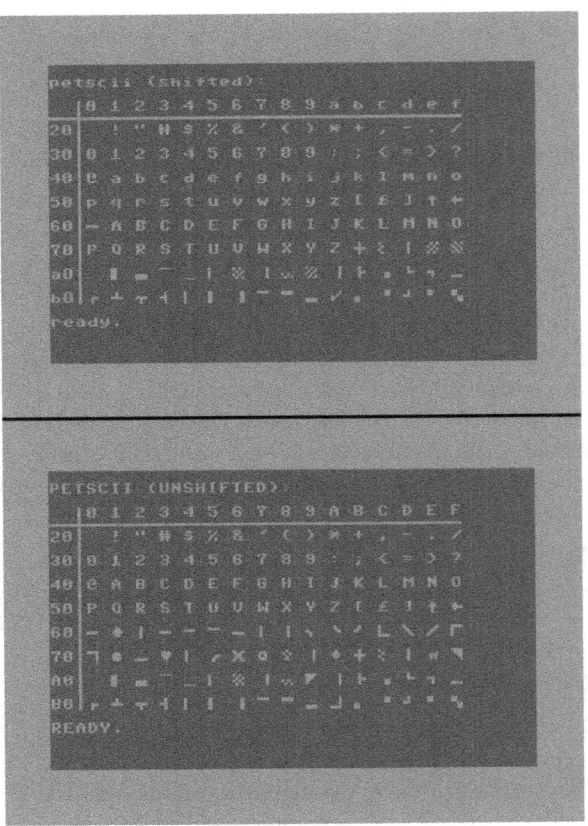

C64 startup screen with shifted and unshifted modes of PETSCII, and the two characters from ASCII-1963.

PETSCII Chart as displayed on the C64 in shifted and unshifted modes. (Not shown are control codes, as well as characters in the $C0-$FF range, which are the standard uppercase keycodes returned from the keyboard, and which are mirrored to the range $60-7F)

C64 version, the backslash (\ ) in position $5C is occupied by a British pound sign (£). In *unshifted mode*, codes $60–$7F and $A0–$FF are allotted to CBM-specific block graphics characters (horizontal and vertical lines, hatches, shades, triangles, circles and card suits). Ranges $00–$1F and $80–$9F have control codes.

The Commodore PET's lack of a programmable bitmap-mode for computer graphics, as well as it having no re-definable character set capability, may be one of the reasons PETSCII was developed; by creatively using the well thought-out block graphics, a higher degree of sophistication in screen graphics is attainable than by using plain ASCII's letter/digit/punctuation characters. In addition to the relatively diverse set of geometrical shapes that can thus be produced, PETSCII allows for several grayscale levels by its provision of differently hatched checkerboard squares/half-squares. Finally, the reverse-video mode (see below) is used to complete the range of graphics characters, in that it provides mirrored half-square blocks.

PETSCII also has a *text mode*, in which lowercase letters occupy the range $41–$5A, and uppercase letters occupy the range $C1–$DA. The text mode is not available at powerup, but must be actuated by holding one of the SHIFT keys and then press and release the *Commodore* key. Regardless of whether the chip has undergone this graphic "shift", there are block graphic characters in the range of $E0-FF. This serves to distinguish PETSCII from those kinds of ASCII that go back no farther than ASCII-1967, so any text transfer between an 8-bit Commodore machine and one that uses 1967-derived ASCII would result in text where uppercase letters appear to be lowercase, and low-ercase letters uppercase. There is no easy Boolean operation to change these cases to the proper case. Thus, as with other computers based on non-standard-ASCII character sets, software conversion is needed when exchanging text files and/or telecommunicating with standard ASCII systems. The other ranges are unchanged in shifted mode; this means that the other characters added in ASCII-1967 be-sides lowercase letters —i.e. the grave accent, curly braces, vertical bar, and tilde —do not exist in PETSCII.

Included in PETSCII are cursor and screen control codes, such as {HOME}, {CLR}, {RVS ON}, and {RVS OFF} (the latter two activating/deactivating reverse-video character display). The control codes appeared in program listings as reverse-video graphic characters, although some computer magazines, in their efforts to provide more clearly readable listings, pretty-printed the codes using their actual names, like the above examples. Such names were commonly enclosed in curly braces in the listings. This prevented ambiguity, since, as mentioned, PETSCII had no curly brace characters. The screen control codes were essentially similar to escape codes for text based computer terminals.

As indicated above, PETSCII provides for shifting between the power-on default (unshifted) uppercase+graphics character set and the alternative (shifted) lower+uppercase set (where the shifted set contains a subset of the block graphic characters of the unshifted set). The shift between modes is done by POKEing location 59468 with the value 14 to select the alternative set or 12 to revert to standard. On C64 the sets are alternated by flipping bit 2 of the byte 53272. On some models of PET this can also be achieved via special control code PRINT CHR$(14) which adjust the line spacing as well as changing the character set; the POKE method is still available and does not alter the line spacing.*[2] Thus, screen editor state changes, rather than the employment of separate ASCII codes, are used to choose between single-case (all capitals) and dual case. In the VIC-20, C64, and later machines (not including the CBM business computers), color codes supplement the other screen control codes. (The colors of the VIC-20 and C64/128 are listed in the VIC-II article.)

2.5.3 Codepage layout

Since not all of the characters encoded by PETSCII are 'graphic' (i.e., control codes) and not all of them have a corresponding Unicode representation, they cannot be portably displayed in a web browser. The following table shows the glyphs for PETSCII graphic characters where there is a corresponding Unicode glyph, and the Unicode replacement character U+FFFD (�) otherwise. Control characters and other non-printing characters are represented by abbreviations for their names. Where a particular code point encodes both a shifted and unshifted character, both characters are shown, with the unshifted character on the left. Row and column headings indicate the hexadecimal digit combinations to produce the eight-bit code value; e.g., the letter *L* is at code value 4C.

Note that the table below is for the Commodore 64. Other Commodore machines used slightly different versions of PETSCII, which used different control characters and in some cases different graphic characters. For example, on the Commodore 128 $07 was the bell control character, and on CBM machines prior to the VIC-20, characters $2C and $6C both produced a comma character, albeit with slightly different semantics.*[3]

The actual character generator ROM used a different set of assignments. For example, to display the characters "@ABC" on screen by directly POKEing the screen memory, one would POKE the decimal values 0, 1, 2, and 3 rather than 64, 65, 66, and 67.

Some PETSCII Codes can't be printed and are only used for Keyboard input (e.g. F1, RUN/STOP).

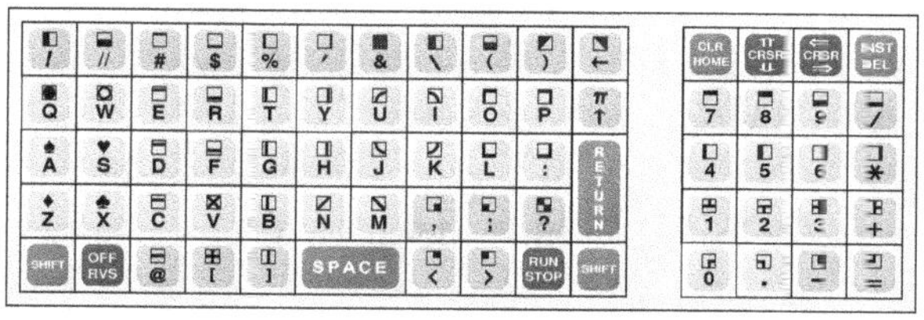

PET 2001 keyboard layout, illustrating PETSCII graphics characters

2.5.4 See also

- ATASCII
- ZX Spectrum character set
- Extended ASCII
- Text semigraphics

2.5.5 Notes

1. ^ The Amiga home/personal computer family uses standard ISO-8859-1.
2. ^ see *On The Edge* by Brian Bagnall, ISBN 0-9738649-0-7, page 43, 54-55.

2.5.6 References

[1] A Conversation with Chuck Peddle, Bil Herd, Jeri Ellsworth - part 3 (2009 videoconference, 06:30)

[2] THE COMMODORE PET COMPUTER / FREQUENTLY ASKED QUESTIONS FILE - VERSION 1.7 (Updated 25 November 2000) BY LARRY ANDERSSON, COMMODORE COLLECTOR AND PET ENTHUSIAST

[3] Commodore Trivia Edition #26 Answers for February 1996

2.5.7   External links

- PETSCII character map, part 1, part 2, part 3 (JPEG)

- An attempt at PETSCII to Unicode mapping, unshifted, shifted

- Commodore 128 PETSCII control characters

- Typography in 8 bits: System fonts

- Online PETSCII-art editor

2.6   PWP

Not to be confused with Professional Women Photographers.

PWP (*Pers' Wastaiset Produktiot*) is a Finnish demogroup founded in 1994. Having originally concentrated on humorous games and storyline-based text mode demos on the low-end PC's, PWP is now one of the leading demogroups on the Commodore VIC-20.

The VIC-20 demos by PWP have been well received by the demomaking community due to their technological excellence and the unusual and entertaining style reminiscent of the older PC-based works. Many of the demos have won competitions at well-respected demo parties such as the Assembly and the *Alternative Party*, often while competing against productions written for far more powerful platforms such as the PC and the Amiga. The single-load demo *Robotic Warrior* has also been the first 8-bit demo ever nominated in the Scene.org Awards (Best 4K Intro, 2004).

2.6.1   Notable releases

PC

- Mazzembly 1997 - a roguelike game satirizing the Assembly demo party

- Helium

- Isi

- Final Isi

VIC-20

- Impossiblator

- Impossiblator 2

- Impossiblator 3

- Robotic Warrior (winner of the Alternative Party 2003 demo compo, later nominated for a Scene.org Award)

- Robotic Liberation (winner of the Assembly 2003 old-school demo compo)

- The Next Level

2.6.2   External links

- The official home page of PWP

- PWP on pouet.net

- More in-depth discussion on some PWP demos

2.7   Stack Light Rifle

The **Stack Light Rifle** is a light gun that was manufactured by Stack Computer Services and created for the ZX Spectrum, Commodore 64, and the Commodore VIC-20. It was released in 1983. The rifle is bundled with three games on tape, High Noon, Shooting Gallery and Grouse Shoot for the Spectrum. Different games were offered for the Commodore 64 and VIC-20 versions (all the games for these two systems were included on one cassette). It retailed for about $60, which is extraordinarily expensive given the fact that most cartridges were $10–20 each. The Stack Light Rifle is differentiated from future light guns as being very realistic looking; future unrealistic light guns such as the NES Zapper and the Sega Light Phaser dealt with controversy due to the guns still being misidentified as real firearms.

The main pistol is attached to 12 feet of cable which ends in a dead-ended ZX81-size connector which plugs into the Spectrum's user port. A barrel, stock and telescopic sight can all be attached to the pistol. The barrel actually facilitated the gun's performance as it filtered out ambient light. These three parts combined to provide a reasonable - if not perfect - degree of accuracy, and allowed the user to effectively use the light gun from the comfort of an armchair. One can extrapolate that the multi-part design was later mimicked on the Sega Menacer.

Variants of the Light Rifle were available for the ZX Spectrum, Commodore VIC-20 and Commodore 64 and all perform the same function. Like the Atari XG-1 light gun, the Stack Light Rifle was treated by the hardware as a light pen. Due to lack of availability of software drivers for the Light Rifle, only the three games that came with the device were available. In April 1985, Sinclair User magazine reported that Stack Computer Services company disappeared.

2.7.1 Technical specifications

The main component of the Stack Light Rifle System is the electronic target pistol that is connected to the computer by a generous length of lead. At the computer end, depending on the version, there is a connector for the appropriate socket or edge connector. On the ZX Spectrum version the connector contains two chips and a couple of simple components to interface the main electronics inside the gun to the computer. To make the pistol more accurate and to turn it into a rifle - it is supplied with a shoulder stock that clips and secures to the rear of the pistol, a barrel and a make-believe telescopic sight.

The electronics inside the pistol consist of a light detector or photo-diode and a small amplifier and buffer. Light coming down the barrel is focused by a small plastic lens onto the photo-diode, and the device is sensitive enough to detect the changes in intensity of the picture. Once boosted by the amplifier, the signal is clipped to provide a digital pulse rather than an analogue waveform and is then fed to the computer via the switch. The screen position that is being scanned at that moment is the position the rifle is pointing at. As the computer receives the pulse from the Light Rifle it compares the value of its scan registers with the screen position of the target and, if a match is found, the played has scored a direct hit.

2.7.2 Supported Games

Commodore 64

- Escape From Alcatraz
- High Noon*[lower-alpha 1]
- Glorious 12th*[lower-alpha 1]
- Gallery*[lower-alpha 1]
- Crowshoot
- Rat's & Cats

VIC-20

- High Noon*[lower-alpha 1]
- Glorious 12th*[lower-alpha 1]
- Gallery*[lower-alpha 1]

ZX Spectrum

- High Noon*[lower-alpha 1]

- Invasion Force by Micromania

[1] bundled by Stack

2.7.3 External links

- Sinclair User Magazine: Issue 37, April 1985
- World of Spectrum Feature on the Stack Light Rifle
- *Stack Light Rifle* at World of Spectrum

2.8 Super Expander

This article is about the VIC-20 expansion cartridge. For the corresponding C64 product, see Super Expander 64.

The **VIC-1211 Super Expander** was a cartridge for the Commodore VIC-20 home computer. It was designed to provide several extensions to the BASIC interpreter on the computer, mostly to help with programming graphics and sound. It also provided 3 kB of extra RAM (of which 136 bytes were used by the cartridge itself). The cartridge was created by Commodore Business Machines (CBM) and released in 1981.

2.8.1 Description

The dialect of BASIC bundled with the VIC-20, Commodore BASIC V2.0, was notorious for its sparse functionality. It didn't even match the features of Commodore's older line of computers, the PET which, at that time, already featured Commodore BASIC version 4.0. As a result it was outdated by the VIC-20's release and seemed quite primitive compared to BASIC dialects available on other microcomputers. To be fair, the decision by Commodore to recycle the old BASIC, and the fact that it could fit in just 16 kB ROM (including the KERNAL), helped keep the VIC-20's price to a minimum and so contributed to its huge success. Plus it was stable and almost entirely bug-free, which could not be said of some competing BASICs.

Nevertheless, not only did "VIC BASIC" lack commands considered fundamental to the BASIC language, such as "else" and "renum", but graphics and sound effects were completely unsupported. To use VIC-20's graphics and sound programmers had to "PEEK and POKE" bytes directly from/to the VIC-20's graphics/sound hardware, the 6560 Video Interface Chip (VIC). This made programming quite tedious and error prone since cryptic memory addresses and codes had to be used constantly, mistakes in these would usually crash the computer instead of giving an error message, and many statements were required to do even simple tasks. Such a thing was death in the tiny RAM and slow interpreted BASIC paradigm of the day.

Programmers could mitigate these problems by using machine code, to an extent, but this in itself was a tedious process with a rather steep learning curve. So to address these shortcomings Commodore created the *Super Expander* cartridge. It provided extra BASIC commands to facilitate using graphics and sound on the VIC-20. It also had commands to read the joystick and lightpen, and unlocked the use of function keys.

2.8.2 Graphics

The VIC-20 did not support high resolution graphics directly. Hi-res graphics were implemented by "painting" the display with characters, and "redefining" the character bitmaps on the fly. This was a complex and long-winded process; implementing it in a BASIC program was virtually useless due to the execution time required to draw anything.

The *Super Expander* took care of all the hard work. It allowed the programmer to draw points, lines, ellipses and arcs, and to paint enclosed regions, with one-line statements. All the VIC-20's 16 colours could be used, although with restrictions due to limitations of the 6560 chip. Display resolution was 160×160 pixels, throttled down from 192×200 allowed by the 6560 chip, in order to permit per-pixel addressability. Multicolor hi-res was supported (with a resolution of 80×160) and could be mixed with normal hi-res.

2.8.3 Sound

The VIC-20's sound capability was fairly simplistic, so programming sound effects using "PEEK and POKE" was not so much of a chore as programming graphics. Even so the *Super Expander* provided a command to play simple tones on the VIC-20's four voice channels, and to control the volume.

Music playback was unsupported on the VIC-20; the usual way of implementing a musical note was to play a tone inside a for loop time delay. In contrast, with *Super Expander* musical scores could played by simply PRINTing a string of characters. (Music strings were distinguished from regular strings using a special reverse-control-character, familiar to anyone who has used colours or cursor controls in VIC-20 programs.) Each of the VIC-20's four voice channels could play their own scores simultaneously, giving harmonious effects which could be striking by the standards of the time.

2.8.4 Other devices

Super Expander provided commands to read the status of the joystick and paddles, and the position of the lightpen. In the case of the joystick, since it was the "digital" or "switch" type, further bit-fiddling was required to decode its position.

2.8.5 Function keys

Ordinarily the VIC-20's function keys could only be used in a program, by scanning for the appropriate key code when reading the keyboard. In the VIC-20's direct mode they were not available to do anything. With the *Super Expander* the function keys could be assigned to execute commands in immediate mode. By default they came pre-programmed with the most common BASIC commands, in a similar fashion to GW-BASIC on the IBM PC. The user could then assign their own commands, or any arbitrary string in fact, to the function keys.

2.8.6 Drawbacks

- Commodore designed the *Super Expander* to map the graphics display to a 1024×1024 coordinate system. Under this scheme, each video pixel was 6.4 (or 12.8) "virtual" pixels in size. This meant that in order to place pixels in exact positions on the screen a further scaling operation had to be coded in. Similar functionality was provided on BASIC 3.5 and 7.0 via the SCALE command, but in these versions of BASIC, scaling was optional and could be done to any arbitrary user-specified size.

- The aspect ratio of the output device (i.e. television set) was not taken into account by the coordinate system. So a circle sized, say, 300×300 would appear elliptical. Similarly, a line drawn from (0,0) to (300,300) would not be displayed as 45°.

- When drawing circle arcs, the starting and ending angles had to be specified in "gradians". In this "metric" angular system there are 400 gradians (also called "grads" or "gons") to the circle, as opposed to the familiar 360 degrees. (One can use gradians on most scientific calculators and even with Microsoft Calculator, but the only profession that makes somewhat regular use of this unit is surveying.) It is a mystery why Commodore chose this obscure and unconventional unit of measure, and in any case, Commodore did not implement it properly, dividing the circle into 100 gradians rather than 400.

- The *Super Expander* had no capability to put a bitmap to the display. This meant arbitrary bitmaps, as might be used in a hi-res game, had to drawn pixel-by-pixel. The slowness of the BASIC interpreter made this unsuitable for applications like arcade-style games. As a result *Super Expander'*s usefulness was really hamstrung, consigning it to shape-centric drawings such as charts and simple pictures, or adventure-style games with static images.

- Programs written using the extra *Super Expander* commands were not portable. A user needed to own the cartridge and have it installed before a program written with the additional commands would run. Loading the program onto an unexpanded VIC-20 gave errors. Therefore the range of software released to take advantage of the *Super Expander'*s capabilities, including type-in programs published in magazines, was very small. This limited the *Super Expander'*s appeal and usefulness.

In spite of the above, the *Super Expander'*s features filled many of the gaps in the VIC-20's programming environment. Additionally, similar microcomputers on the market suffered the same, or equivalent, shortcomings.

2.8.7 See also

- Commodore BASIC

- MOS Technology VIC (aka the MOS Technology 6560 Video Interface Chip)

- Super Expander 64

2.8.8 External links

- ftp.funet.fi: Super Expander manual and demo programs

2.9 VICE

This article is about the software emulator. For the magazine, see VICE (magazine). For other uses, see Vice (disambiguation).

The software program **VICE**, standing for *Versatile Commodore Emulator*, is a free and cross platform emulator for Commodore's 8-bit computers, running on Amiga, Unix, MS-DOS, Win32, Mac OS X, OS/2, Acorn RISC OS, QNX QNX, GP2X GP2X, Dingoo Dingoo A320, Syllable Syllable OS, and BeOS host machines. VICE is free software, released under the GNU General Public Licence.

It's also available for a variety of platforms: for instance VICE for Microsoft Windows (Win32) is known as **WinVICE**, the OS/2 variant is called Vice/2, and the emulator running on BeOS is called BeVICE.

2.9.1 History

As of version 2.1, released December 19, 2008, VICE emulates the Commodore 64, the C128, the VIC-20, the Plus/4, the C64 Direct-to-TV (with its additional video modes) and all the PET models including the CBM-II but excluding the 'non-standard' features of the SuperPET 9000. WinVICE supports digital joysticks via a parallel port driver, and, with a CatWeasel PCI card, is planned to perform hardware SID playback (requires optional SID chip installed in socket).

As of 2004, VICE was one of the most widely used emulators of the Commodore 8-bit microcomputers.[1] It is also one of the few usable Commodore emulators to exist on free *NIX platforms, and one of the first to be distributed under GNU GPL. It is available for most Linux distributions.

2.9.2 Bibliography

- Simon Carless, *Gaming hacks*, O'Reilly Media. 2004, ISBN 0-596-00714-0, pp. 5–8

- Jason Kroll. *Commodore 64 Game Emulation*, Linux Journal, Issue 72, April 2000

2.9.3 See also

- Commodore 64

- CCS64

2.9.4 References

[1] Carless, p. 5

2.9.5 External links

- Official website , with Online manual (HTML)

- VICE knowledge base (preliminary)

- Guide for installation on Linux and Windows

- PRG Starter Simplifies VICE usage

- Windows binaries automated nightly builds

- VICE.js JavaScript port of VICE

- Windows binaries unofficial nightly builds, also SDL and 64-bit binaries available

Chapter 3

Software

3.1 The Automatic Proofreader

The Automatic Proofreader is a series of checksum utilities published by COMPUTE! Publications for its *COMPUTE!* and *COMPUTE!'s Gazette* magazines, and various books. These programs were designed to allow home computer users to easily detect errors on BASIC type-in programs, and worked by displaying a hash value for each line entered that could be compared against the reference value printed in the magazine. Initially published for use with the Commodore 64 and VIC-20, the Proofreader was later made available for the Atari 8-bit family, Apple II family, and IBM PC/PCjr as well.

The line-by-line "real-time" feedback feature was something of a novelty at the time, and represented a significant improvement over earlier checksum utilities, which were typically run only after a user program had been entered, most often giving a single control value for the entire program. The majority of such schemes lacked an error location feature, and, due to quite simplistic checksum algorithms, had serious trouble catching many "minor" typing errors like transposed characters (these errors having the potential of being just as detrimental as "major" ones to the typed-in program's functioning).

3.1.1 Commodore versions

The Automatic Proofreader was first introduced in October 1983 for the Commodore 64 and VIC-20. The same listing was designed to work on both systems. This version of the Proofreader would display a byte-sized numeric value at the top left corner of the screen whenever a program line was entered.

The initial version of the Proofreader, however, had several drawbacks. It was loaded into the cassette buffer (memory area), which was overwritten whenever a program was loaded or saved using the Datassette. This caused difficulties if a cassette user had to resume work on a partially completed listing. A complicated method had to be used to get both the Proofreader and the program listing in memory at the same time. Also, the checksum method used was relatively rudimentary, and did not catch transposition errors, nor did it take whitespace into account.

Because of this, the **New Automatic Proofreader** was introduced in February 1986. This version used a more sophisticated checksum algorithm that could catch transposition errors. It also took spacing into account if they were within quotes (where they were generally significant to the program's operation), while ignoring them outside of quotes (where they were not relevant). Also, the decimal display of the checksum was replaced with a hexadecimal value.

The New Automatic Proofreader was designed to run on any Commodore 8-bit home computer (including the C16/Plus/4 and C128), automatically relocating itself to the bottom of BASIC RAM and moving pointers to hide its presence. It was continuously published until *COMPUTE!'s Gazette* switched over to a disk-only format after the December 1993 issue.

3.1.2 References

The *Compute!'s Gazette* extracts below are stored as JPEG images at Sami Rautiainen's Ancient file library. The *Compute!* extract resides at the Classic Computer Magazine Archive, maintained by Kevin Savetz.

- The Automatic Proofreader – *COMPUTE!'s Gazette*, November 1983, p. 149

- The Automatic Proofreader For VIC, 64, And Atari – *COMPUTE!* March 1984, pp. 60-.

- The New Automatic Proofreader, Part 1/4, Part 2/4, Part 3/4 – *COMPUTE!'s Gazette*, February 1986, pp. 108, 109, 116

3.2 Commodore BASIC

Commodore BASIC, also known as **PET BASIC**, is the dialect of the BASIC programming language used in Commodore International's 8-bit home computer line, stretching from the PET of 1977 to the C128 of 1985. The core was based on 6502 Microsoft BASIC, and as such it shares many characteristics with other 6502 BASICs of the time, such as Applesoft BASIC. Commodore licensed BASIC from Microsoft on a "pay once, no royalties" basis after Jack Tramiel turned down Bill Gates' offer of a $3 per unit fee, stating, "I'm already married," and would pay no more than $25,000 for a perpetual license.*[1]

3.2.1 History

Commodore took the source code of the flat-fee BASIC and further developed it internally for all their other 8-bit home computers. It was not until the Commodore 128 (with V7.0) that a Microsoft copyright notice was displayed. However, Microsoft had built an easter egg into the version 2 or "upgrade" Commodore Basic that proved its provenance: typing the (obscure) command WAIT 6502, 1 would result in Microsoft! appearing on the screen. (The easter egg was well concealed—the message did not show up in any disassembly of the interpreter.)*[2]

The popular Commodore 64 came with BASIC v2.0 in ROM despite the computer being released after the PET/CBM series that had version 4.0 because the 64 was intended as a home computer, while the PET/CBM series were targeted at business and educational use where their built-in programming language was presumed to be more heavily used.

3.2.2 Technical details

A convenient feature of Commodore's ROM-resident BASIC interpreter and KERNAL was the full-screen editor.*[3]*[4] Although Commodore keyboards only featured two cursor keys which alternated direction when the shift key was held, the screen editor allowed users to enter direct commands or to input and edit program lines from anywhere on the screen. If a line was prefixed with a line number, it was tokenized and stored in program memory. Lines not beginning with a number were executed by pressing the RETURN key whenever the cursor happened to be on the line. This marked a significant upgrade in program entry interfaces compared to other common home computer BASICs at the time, which typically used line editors, invoked by a separate EDIT command, or a "copy cursor" that truncated the line at the cursor's position.

It also had the capability of saving named files to any device, including the cassette – a popular storage device in the days of the PET, and one that remained in use throughout the lifespan of the 8-bit Commodores as an inexpensive form of mass storage. Due to the their use as an analog audio medium, cassettes were as ubiquitous during the era as CDs are today. Most systems only supported filenames on diskette, which made saving multiple files on other devices more difficult. The user of one of these other systems had to note the recorder's counter display at the location of the file, but this was inaccurate and prone to error. With the PET (and BASIC 2.0), files from cassettes could be requested by name. The device would search for the filename by reading data sequentially, ignoring any non-matching filenames. The file system was also supported by a powerful record structure that could be loaded or saved to files. Commodore cassette data was recorded digitally, rather than less expensive (and less reliable) analog methods used by other manufacturers. Therefore, the specialized Datasette was required rather than a standard tape recorder. Adapters were available that used an analog to digital converter to allow use of a standard recorder, but these cost only a little less than the Datassette.

Due to Commodore's use of the same BASIC on multiple hardware architectures, the LOAD command was provided with an additional parameter to indicate the memory address where the program should start. A command such as LOAD "*",8 would place the program data at the start of BASIC's area, while LOAD "*",8,1 would place the file at the address from which it was saved. The former was usually used for BASIC programs, since BASIC's location varied between different models. Some Commodore BASIC variants supplied BLOAD and BSAVE commands that worked like their counterparts in Applesoft BASIC, loading or saving bitmaps from specified memory locations.

Like the original Microsoft BASIC interpreter, on which it is based, Commodore BASIC is slower than native machine code. Test results have shown that copying 16 kilobytes from ROM to RAM takes less than a second in machine code, compared to over a minute in BASIC. To execute faster than the interpreter, programmers started using various techniques to speed up execution. One was to store often-used integer values in variables rather than using literal values, as interpreting a variable name was faster than interpreting a literal number. When speed was important, some programmers converted sections of BASIC programs to 6502 or 6510 assembly language which were POKEed into memory from DATA statements at the end of the BASIC program, and executed from BASIC using the SYS command either from direct mode or from the program itself. When the execution speed of machine language was too great, such as for a game or when waiting for user input, programmers could poll by PEEKing selected memory locations (such as $A6 for the C-64, or $D0 for the C-128, denoting size of the keyboard queue) to delay or halt execution.

Commodore BASIC keywords could be abbreviated by entering first an unshifted keypress, and then a shifted keypress of the next letter. This set the high bit, causing the interpreter to stop reading and parse the statement according to a lookup table. This meant the statement up to where the high bit was set was accepted as a substitute for typing the entire command out. However, since all BASIC keywords were stored in memory as single byte tokens, this was a convenience for statement entry rather than an optimization.

In the default uppercase-only character set, shifted characters appear as a graphics symbol; e.g. the command, GOTO, could be abbreviated G{Shift-O} (which resembled GⲦ onscreen). Most such commands were two letters long, but in some cases they were longer. In cases like this, there was an ambiguity, so more unshifted letters of the command were needed, such as GO{Shift-S} (GO♥) being required for GOSUB. Some commands had no abbreviated form, either due to brevity or ambiguity with other commands. For example, the command, INPUT had no abbreviation because its spelling collided with the separate INPUT# keyword, which was located nearer to the beginning of the keyword lookup table. The heavily-used PRINT command had a single ? shortcut, as was common in most BASIC dialects. Abbreviating commands with shifted letters is unique to Commodore BASIC.

By abbreviating keywords, it was possible to fit more code on a single line (line lengths were usually limited to 2 or 4 screen lines, depending on the specific machine). This allowed for a slight saving on the overhead to store otherwise necessary extra program lines, but nothing more. All BASIC commands were tokenized and took up 1 byte (or two, in the case of several commands of BASIC 7 or BASIC 10) in memory no matter which way they were entered. And, such long lines could be difficult to edit. The LIST command displayed the entire command keyword - extending the program line beyond the 2 or 4 screen lines which could be entered into program memory.

Program lines in Commodore BASIC do not require spaces anywhere except between the line number and the statement, e.g., 100 IFA=5THENPRINT"YES":GOTO160, and it was common to write programs with no spacing. This feature was added to conserve memory since the tokenizer never removes any space inserted between keywords: the presence of spaces results in extra 0x20 bytes in the tokenized program which are merely skipped during execution.

The order of execution of Commodore BASIC lines was not determined by line numbering; instead, it followed the order in which the lines were linked in memory.[*][5] Program lines were stored in memory as a singly linked list with a line number, a pointer (containing the address of the beginning of the next program line), and then the tokenized code for the line. While a program was being entered, BASIC would constantly reorder program lines in memory so that the line numbers and pointers were all in ascending order. However, after a program was entered, manually altering the line numbers and pointers with the POKE commands could allow for out-of-order execution or even give each line the same line number. In the early days, when BASIC was used commercially, this was a software protection technique to discourage casual modification of the program.

Variable names were only significant to 2 characters; thus the variable names VARIABLE1, VARIABLE2 and VA all referred to the same variable.

The native number format of Commodore BASIC, like that of its parent MS BASIC, was floating point. Most contemporary BASIC implementations used one byte for the characteristic (exponent) and three bytes for the mantissa. The accuracy of a floating point number using a three-byte mantissa is only about 6.5 decimal digits, and round-off error is common. Commodore, however, used MS BASIC's four-byte mantissa, which made their BASIC much more useful for business.

Although Commodore BASIC supports integer variables (denoted with a percent sign), in practice they only work on array variables and serve the function of conserving memory by limiting array elements to two bytes each. Denoting a normal variable as integer simply causes BASIC to convert it back to floating point, slowing down program execution and wasting memory as each percent sign takes one additional byte to store.

Also akin to MS BASIC, 16-bit signed integers (i.e. in the range −32768 to 32767) were available by postfixing a variable name with a percent symbol. String variables were represented by postfixing the variable name with a dollar sign. Thus, the variables AA$, AA, and AA% would each be understood as distinct.

The Simons' BASIC start-up screen

Many BASIC extensions were released for the Commodore 64, due to the relatively limited capabilities of its native BASIC 2.0. One of the most popular extensions was the DOS Wedge, which was included on the Commodore 1541 Test/Demo Disk. This 1 KB extension to BASIC added a number of disk-related commands, including the ability to read a disk directory without destroying the program in memory. Its features were subsequently incorporated in various third-party extensions, such as the popular Epyx FastLoad cartridge. Other BASIC extensions added additional keywords to make it easier to code sprites, sound, and high-resolution graphics like Simons' BASIC.

Although BASIC 2.0's lack of sound or graphics features was frustrating to many users, some critics argued that it was ultimately beneficial since it forced the user to learn machine language.

The limitations of BASIC 2.0 on the C64 led to use of built-in ROM machine language from BASIC. To load a file to a designated memory location, the filename, drive, and device number would be read by a call: SYS57812"filename",8

The location would be specified in two locations: POKE780,0:POKE781,0:POKE782,192

And the load routine would be called: SYS65493

The disk magazine for the C64, Loadstar was a venue for hobbyist programmers, who shared collections of proto-commands for BASIC, called with the SYS address + offset command.

From a modern programming point of view, the earlier versions of Commodore BASIC presented a host of bad programming traps for the programmer. As most of these issues derived from Microsoft BASIC, virtually every home computer BASIC of the era suffered from similar deficiencies.[*][6] Every line of a Microsoft BASIC program was assigned a line number by the programmer. It was common practice to increment numbers by some value (5, 10 or 100) to make inserting lines during program editing or debugging easier, but bad planning meant that inserting large sections into a program often required restructuring the entire code. A common technique was to start a program at some low line number with an ON...GOSUB jump table, with the body of the program structured into sections starting at a designated line number like 1000, 2000, and so on. If a large section needed to be added, it could just be assigned the next available major line number and inserted to the jump table.

Later BASIC versions on Commodore and other platforms included a DELETE and RENUMBER command, as well as an AUTO line numbering command that would automatically select and insert line numbers according to a selected increment. In addition, all variables are treated as global variables. Clearly defined loops are hard to create, often

causing the programmer to rely on the GOTO command (this was later rectified in BASIC 3.5 with the addition of the DO, LOOP, WHILE, UNTIL, and EXIT commands). Flag variables often needed to be created to perform certain tasks. Earlier BASICs from Commodore also lack debugging commands, meaning that bugs and unused variables are hard to trap.

3.2.3   Use as user interface

In common with other home computers, Commodore's models booted directly into the BASIC interpreter. BASIC's file and programming commands could be entered in direct mode to load and execute software. If program execution was halted using the RUN/STOP key, variable values would be preserved in RAM and PRINTed for debugging. This, along with the advanced screen editor included with Commodore BASIC gave the programming environment a REPL-like feel; programmers could insert and edit program lines at any screen location, interactively building the program.[*][7] This is in contrast to business-oriented operating systems of the time like CP/M or MS-DOS, which typically booted into a command line interface. If a programming language was required on these platforms, it had to be loaded separately.

While some versions of Commodore BASIC included disk-specific DLOAD and DSAVE commands, the version built into the popular Commodore 64 lacked these, requiring the user to specify the disk drive's device number (typically 8 or 9) to the standard LOAD command, which otherwise defaulted to tape. Another omission from the Commodore 64s BASIC 2.0 was a DIRECTORY command to load a disk's contents into screen memory without clearing main memory. On the 64, viewing a disk's contents was implemented as loading a "program" which when listed showed the directory. This had the effect of overwriting the currently-loaded program. Addons like the DOS Wedge overcame this by rendering the directory listing direct to the screen.

3.2.4   Versions and features

A list of CBM BASIC versions in chronological order, with successively added features:

Released versions

- V1.0: PET 2001 with chiclet keyboard and built-in Datassette (original PET)

  - arrays limited to 256 elements
  - PEEK command explicitly disabled over BASIC ROM locations above $C000

- V2.0 (first release): PET 2001 with full-travel keyboard & upgrade ROMs

  - add IEEE-488 support
  - improved the garbage collection*[8]
  - fix array bug
  - Easter egg – entering WAIT6502,1 displays MICROSOFT!

- V4.0: PET/CBM 4000/8000 series (and late version PET 2001s)

  - disk operations: DLOAD,DSAVE,COPY,SCRATCH, etc. (15 in all)
  - disk error-channel variables: DS,DS$
  - greatly improved garbage-collection performance*[9]

- V2.0 (second release, after 4.0): VIC-20; C64

- V4+ : CBM-II series (aka B, P range)

  - memory management: BANK
  - more disk operations: BLOAD, BSAVE,DCLEAR
  - formatted printing: PRINT USING,PUDEF
  - error trapping: DISPOSE
  - alternative branching: ELSE
  - dynamic error handling: TRAP,RESUME,ERR$()
  - flexible DATA read: RESTORE [linenumber]
  - string search function: INSTR

- V3.5: C16/116, Plus/4

  - sound and graphics commands
  - joystick input: JOY
  - decimal ↔ hexadecimal conversion: DEC(),HEX$()
  - structured looping: DO,LOOP,WHILE,UNTIL,EXIT
  - function key assignment: KEY (also direct mode)
  - program entry/editing: AUTO,DELETE,RENUMBER
  - debugging (tracing): TRON, TROFF
  - MLM entry command: MONITOR
  - C(1)16, Plus/4 Easter egg – enter SYS 52650

- V7.0: C128

- more sound and graphics commands, including sprite handling

- built-in sprite editor: SPRDEF

- multi-statement blocks for IF THEN ELSE structures: BEGIN,BEND

- paddle, lightpen input: POT,PEN

- exclusive or function: XOR

- get variable address: POINTER

- text mode windowing: WINDOW

- controlled time delay: SLEEP

- memory management: SWAP,FETCH,STASH,FRE(1)

- used the 128's bank switching to store program code separately from variables. Variable values would be preserved across program executions if the program was started with the GOTO command.

- more disk operations: BOOT,DVERIFY

- CPU speed adjustment: FAST,SLOW (2 vs 1 MHz)

- enter C64 mode: GO64

- undocumented, working: RREG (read CPU registers after a SYS)

- unimplemented commands: OFF,QUIT

- C128 Easter egg – enter SYS 32800,123,45,6

Unreleased versions

- V3.6 : Commodore LCD (unreleased prototype). Almost identical to V7.0, with the following differences:*[10]

  - VOLUME instead of VOL
  - EXIT instead of QUIT
  - FAST,SLOW commands not present
  - Additional command: POPUPS

- V10 : Commodore 65 (unreleased prototype)

  - graphics/video commands: PALETTE,GENLOCK
  - mouse input: MOUSE,RMOUSE
  - text file (SEQ) utility: TYPE
  - program editing: FIND,CHANGE
  - memory management: DMA, FRE(2)
  - unimplemented commands: PAINT,LOCATE,SCALE,WIDTH,SET,VIEWPORTPASTE,CU

3.2.5   Notable extension packages

- Super Expander (VIC-20; delivered on ROM cartridge) (Commodore)

- Super Expander 64 (C64; cartridge) (Commodore)

- Simons' BASIC (C64; cartridge) (Commodore)

- Graphics BASIC (C64; floppy disk) (Hesware)

- BASIC 8 (C128; floppy disk and optional internal ROM chip) (Walrusoft)

3.2.6   References

[1] Stated by Jack Tramiel at the Commodore 64 25th Anniversary Celebration at the Computer History Museum December 10, 2007 .

[2] Bill Gates' Personal Easter Eggs

[3] "Keyboarding and the Screen Editor" .

[4] "Byte July 1983" (PDF).

[5] "Mapping the Commodore 64" .

[6] "Atari BASIC and PET Microsoft BASIC. A BASIC Comparison." .

[7] "An Introduction to the Commodore 64: Adventures in Programming" .

[8] http://www.zimmers.net/anonftp/pub/cbm/firmware/README.txt

[9] http://www.zimmers.net/anonftp/pub/cbm/firmware/README.txt

[10] Contents of the Commodore LCD U103 BASIC ROM

Notes

- Commodore/Microsoft Basic version timeline

- Bill Gates' Personal Easter Eggs in 8 Bit BASIC, pagetable.com

BASIC 2.0

- Angerhausen et al. (1983). *The Anatomy of the Commodore 64* (for the full reference, see the C64 article).

BASIC 3.5

- Gerrard, Peter; Bergin, Kevin (1985). *The Complete COMMODORE 16 ROM Disassembly*. Gerald Duckworth & Co. Ltd. ISBN 0-7156-2004-5.

BASIC 7.0

- Jarvis, Dennis; Springer, Jim D. (1987). *BASIC 7.0 Internals*. Grand Rapids, Michigan: Abacus Software, Inc. ISBN 0-916439-71-2.

BASIC 10.0

- Commodore 65 preliminary documentation (March 1991), with addendum for ROM version 910501. c65manual.txt

3.3   Contiki

This article is about the embedded operating system. For other uses, see Contiki (disambiguation).
Not to be confused with Kontiki.

Contiki is an open source operating system for networked, memory-constrained systems with a particular focus on low-power wireless Internet of Things devices. Examples of where Contiki is used include street lighting systems, sound monitoring for smart cities, radiation monitoring systems, and alarm systems.[*][1] Contiki was created by Adam Dunkels in 2002[*][2] and has been further developed by a world-wide team of developers from Texas Instruments, Atmel, Cisco, ENEA, ETH Zurich, Redwire, RWTH Aachen University, Oxford University, SAP, Sensinode, Swedish Institute of Computer Science, ST Microelectronics, Zolertia, and many others.[*][3] The name *Contiki* comes from Thor Heyerdahl's famous Kon-Tiki raft.

Despite providing multitasking and a built-in TCP/IP stack, Contiki only needs about 10 kilobytes of RAM and 30 kilobytes of ROM.[*][1] A full system, complete with a graphical user interface, needs about 30 kilobytes of RAM.[*][4]

3.3.1   Hardware

Contiki is designed to run on classes of hardware devices that are severely constrained in terms of memory, power, processing power, and communication bandwidth. A typical Contiki system has memory on the order of kilobytes, a power budget on the order of milliwatts, processing speed measured in megahertz, and communication bandwidth on the order of hundreds of kilobits/second. This class of systems includes both various types of embedded systems as well as a number of old 8-bit computers.

3.3.2 Networking

Contiki provides three network mechanisms: the uIP TCP/IP stack,[5] which provides IPv4 networking, the uIPv6 stack,[6] which provides IPv6 networking, and the Rime stack, which is a set of custom lightweight networking protocols designed specifically for low-power wireless networks. The IPv6 stack was contributed by Cisco and was, at the time of release, the smallest IPv6 stack to receive the IPv6 Ready certification.[7] The IPv6 stack also contains the RPL routing protocol for low-power lossy IPv6 networks and the 6LoWPAN header compression and adaptation layer for IEEE 802.15.4 links.

The Rime stack is an alternative network stack that is intended to be used when the overhead of the IPv4 or IPv6 stacks is prohibitive. The Rime stack provides a set of communication primitives for low-power wireless systems. The default primitives are single-hop unicast, single-hop broadcast, multi-hop unicast, network flooding, and address-free data collection. The primitives can be used on their own or combined to form more complex protocols and mechanisms.[8]

3.3.3 Low-power operation

Many Contiki systems are severely power-constrained. Battery operated wireless sensors may need to provide years of unattended operation and with little means to recharge or replace its batteries. Contiki provides a set of mechanisms for reducing the power consumption of the system on which it runs. The default mechanism for attaining low-power operation of the radio is called ContikiMAC.[9] With ContikiMAC, nodes can be running in low-power mode and still be able to receive and relay radio messages.

3.3.4 Simulation

The Contiki system includes a network simulator called Cooja.[10] Cooja simulates networks of Contiki nodes. The nodes may belong to either of three classes: emulated nodes, where the entire hardware of each node is emulated, Cooja nodes, where the Contiki code for the node is compiled for and executed on the simulation host, or Java nodes, where the behavior of the node must be reimplemented as a Java class. A single Cooja simulation may contain a mixture of nodes from either of the three classes. Emulated nodes can also be used to include non-Contiki nodes in a simulated network.

In Contiki 2.6, platforms with the TI MSP430 and Atmel AVR microcontrollers can be emulated.

3.3.5 Programming model

To run efficiently on memory-constrained systems, the Contiki programming model is based on protothreads.[11][12] A protothread is a memory-efficient programming abstraction that shares features of both multi-threading and event-driven programming to attain a low memory overhead of each protothread. The kernel invokes the protothread of a process in response to an internal or external event. Examples of internal events are timers that fire or messages being posted from other processes. Examples of external events are sensors that trigger or incoming packets from a radio neighbor.

Protothreads are cooperatively scheduled. This means that a Contiki process must always explicitly yield control back to the kernel at regular intervals. Contiki processes may use a special protothread construct to block waiting for events while yielding control to the kernel between each event invocation.

3.3.6 Features

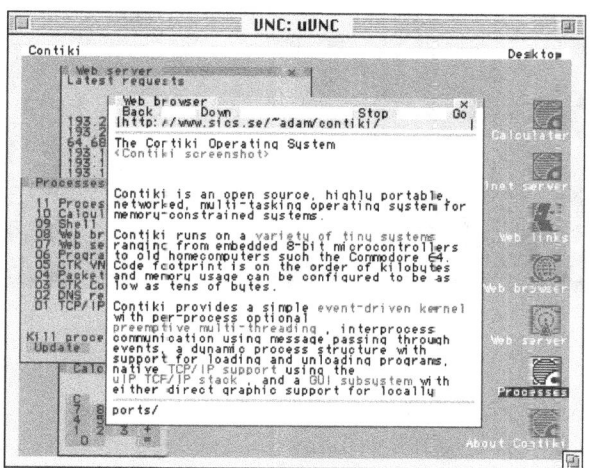

Screenshot of the VNC server running on the Atmel AVR port of Contiki.

Contiki supports per-process optional preemptive multi-threading, inter-process communication using message passing through events, as well as an optional GUI subsystem with either direct graphic support for locally connected terminals or networked virtual display with VNC or over Telnet.

A full installation of Contiki includes the following features:

- Multitasking kernel
- Optional per-application pre-emptive multithreading
- Protothreads

- TCP/IP networking, including IPv6

- Windowing system and GUI

- Networked remote display using Virtual Network Computing

- A web browser (claimed to be the world's smallest)

- Personal web server

- Simple telnet client

- Screensaver

3.3.7 Ports

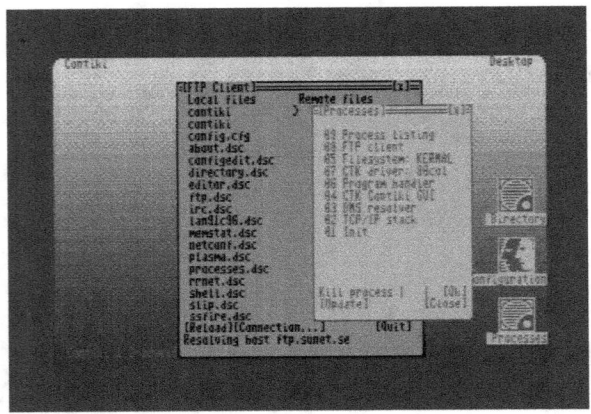

"C64 port" , Contiki *(screenshot).*

The Contiki operating system has been or is being ported to the following systems:

Microcontrollers

- Atmel —ARM, AVR

- Freescale —MC13224

- NXP Semiconductors —LPC1768*[13], LPC2103

- Microchip —dsPIC, PIC32 (PIC32MX795F512L)

- Texas Instruments —MSP430, CC2430, CC2538, CC2630, CC2650

- STMicroelectronics —STM32 W

Computers

- Apple —II series*[lower-alpha 1]

- Atari —8-bit,*[lower-alpha 1] ST, Portfolio

- Casio —Pocket Viewer

- Commodore —PET,*[lower-alpha 1] VIC-20,*[lower-alpha 1] 64,*[lower-alpha 1] 128*[lower-alpha 1]

- Tangerine Computer Systems —Oric*[lower-alpha 1]

- NEC —PC-6001

- Sharp —Wizard

- Intel, AMD, VIA, many others —x86-based Unix-like systems, atop GTK+, or more directly using an X Window System*[14]

Game consoles

- Atari —Jaguar

- Game Park —GP32

- Nintendo —Game Boy, Game Boy Advance, Entertainment System (NES)*[lower-alpha 1]

- NEC —TurboGrafx-16 Entertainment SuperSystem (PC Engine)

3.3.8 See also

- BeRTOS

- ERIKA Enterprise

- SymbOS

- TinyOS

- Wheels (operating system)

3.3.9 Notes

[1] cc65 based development

3.3.10 References

[1] *Contiki OS.*

[2] *Contiki: Bringing IP to Sensor Networks*

[3] "Community" , *Contiki OS.*

[4] Out in the Open: The Little-Known Open Source OS That Rules the Internet of Things

[5] Dunkels, Adam (May 2003), "Full TCP/IP for 8 Bit Architectures" , *Proceedings of the First ACM/Usenix International Conference on Mobile Systems, Applications and Services (MobiSys)*, San Francisco.

[6] Durvy, Mathilde; Abeillé, Julien; Wetterwald, Patrick; O'Flynn, Colin; Leverett, Blake; Gnoske, Eric; Vidales, Michael; Mulligan, Geoff; Tsiftes, Nicolas; Finne, Niclas; Dunkels, Adam (November 2008), "Making sensor networks IPv6 ready" , *Proceedings of the Sixth ACM Conference on Networked Embedded Sensor Systems (SenSys)* (poster session), Raleigh, NC, US: ACM.

[7] *Newsroom*, Cisco, 2008.

[8] Dunkels, Adam; Österlind, Fredrik; He, Zhitao (November 2007), "An adaptive communication architecture for wireless sensor networks" , *Proceedings of the Fifth ACM Conference on Networked Embedded Sensor Systems (SenSys)*, Sydney, AU.

[9] Dunkels, Adam, *The ContikiMAC Radio Duty Cycling Protocol* (PDF).

[10] "Start" , *Contiki OS.*

[11] Dunkels, Adam; Schmidt, Oliver; Voigt, Thiemo; Ali, Muneeb (November 2006), "Protothreads: Simplifying event-driven programming of memory-constrained embedded systems" , *Proceedings of the Fourth ACM Conference on Embedded Networked Sensor Systems (SenSys)*, Boulder, CO, USA Dunkels, A.; Schmidt, O.; Voigt. T.; Ali, M. (2006). "Protothreads" . *Proceedings of the 4th international conference on Embedded networked sensor systems - Sen Sys '06.* p. 29. doi:10.1145/1182807.1182811. ISBN 1595933433. (PDF, Presentation slides).

[12] "Protothread" , *Code*, Google.

[13] http://sourceforge.net/p/contiki/mailman/message/31753844/

[14] Stein, H. *Running Contiki under Windows*, Trix, archived from the original on 2003-12-09.

3.3.11 External links

- Official website
- *C64 Web*: a web site run from an unmodified 1982 built Commodore 64

- "Tools Contiki" , *Hitmen*, AT: C02: unofficial website for historic ports of the 1.x version.

- "Minimal Contiki OS for LPC2103" , *Manishshakya*, NP.

- *Contiki 2.5 config file and disk image generator*, A2 retro systems.

- *Contiki porting on PIC32 (SEED-EYE BOARD)*, IT: SS-SUP.

3.4 Delta Drawing

Delta Drawing Learning Program, later retitled **Delta Drawing Today**, was an early turtle graphics drawing program developed by Computer Access Corporation,[1] and published by Spinnaker Software in 1983.

Delta Drawing was intended for children age 4 to 14. It featured a functional programming language for executing scripted drawing and painting instructions.[2] Spinnaker sought to improve on the educational value of Logo, an earlier educational programming language that could also program turtle graphics.[3]

Power Industries LP of Newton, Massachusetts later acquired Delta Drawing Learning Program and continued its development. They released the new version, Delta Drawing Today v4.0. in 1990,[4] and eventually a Spanish-language edition: Delta Drawing Today Version Español.

3.4.1 References

[1] Staff writer (1983). "Drawing Program for Children" . *Compute!* (34): 251. Retrieved 4 October 2011.

[2] D'Ignazio, Fred (1983). "Turning Logo Upside Down" . *Compute!* (37): 152. Retrieved 4 October 2011.

[3] King, Charles (1983). "Software Review: Welcome to the World of Delta Drawing." . *Hands On!* **6** (2): 15–16.

[4] *THE Journa* (Information Synergy, Inc.) **18** (1): 48. 1990. Missing or empty |title= (help);

3.4.2 Further reading

- Shannon, L.R. (21 January 1992). "A Graphics Program That Isn't Child's Play" . *The New York Times*. Retrieved 5 October 2011.

3.5 Commodore DOS

Commodore DOS, aka **CBM DOS**, is the disk operating system used with Commodore's 8-bit computers. Unlike most other DOS systems, which are loaded from disk into the computer's own RAM and executed there, CBM DOS is executed internally in the drive: the DOS resides in ROM chips inside the drive, and is run there by one or more dedicated MOS 6502 family CPUs. Thus, data transfer between Commodore 8-bit computers and their disk drives more closely resembles a local area network connection than typical disk/host transfers.

3.5.1 CBM DOS versions

At least seven distinctly numbered versions of Commodore DOS are known to exist; the following list gives the version numbers and related disk drives. Unless otherwise noted, drives are 5¼-inch format. The "lp" code designates "low profile" drives. Drives whose model number starts with 15 connect via Commodore's unique serial (TALK/LISTEN) protocols, all others use IEEE-488.

- 1.0 – found in the 2040 and 3040 floppy drives

- 2.0 – found in the 4040 and 3040 floppy drives

- 2.5 – found in the 8050 floppy drives

- 2.6 – found in the 1540, 1541, built-in SX-64 drive, 1551, 2031 (+"lp"), and 4031 floppy drives

- 2.7 – found in the 8050, 8250 (+"lp"), and SFD-1001 floppy drives

- 3.0 – found in the 1570, 1571, and 8280 floppy drives (8280: 8-inch), as well as the 9060 and 9090 hard drives

- 3.1 – found in the built-in 1571 drive of C128D/DCR computers

- 10.0 – found in the 1581 (3½-inch)

Version 2.6 was by far the most commonly used and known DOS version, due to its use in the 1541 as part of C64 systems.

The revised firmware for the 1571 which fixed the relative file bug was also identified as V3.0. Thus it is not possible to tell the two versions apart by version number alone.

3.5.2 Technical overview

1541 Directory and file types

The 1541 Commodore floppy disk can contain up to 144 files in a flat namespace (no subdirectories); the directory is stored on reserved track 18, which is the center track of a 35-track single-sided disk. A file name may be up to 16 bytes in length and theoretically will be unique; by using direct access methods on the directory structure, it is possible to rename a file to that of another—although accessing such files may be difficult or impossible. Files with identical names usually serve no purpose except to inform or visually manage files. One popular trick, used, for example, by The Final Cartridge III, was to add files named "----------------" of type DEL< to the directory, and files could then be rearranged around those lines to form groups. Many game developers, warez group members and demoscene hackers used some more clever custom directory entries as well.

File names may contain a shift+space character ($A0), and if the directory listing is being viewed from BASIC, the portion of the file name beyond the $A0 character will appear to have been separated from the first part of the file name by a quotation mark, causing BASIC to not consider it to be part of the full file name. This feature was used to create directory entries such as SAVE "PROGRAM(shift+space),8,1:",8,1 which will then appear in the directory listing as, for example, 32 "PROGRAM",8,1: PRG. When the user moved the cursor to the beginning of the line and typed the word LOAD over the file size and pressed Enter, BASIC will interpret that as LOAD "PROGRAM",8,1: ..., causing the program to be loaded into memory. Anything after the colon will not be executed, since the LOAD command will execute the other program, never returning to the interpreter.

A null byte embedded in a file name will interrupt the listing after loading by BASIC. If there are three null bytes, that makes it difficult to list through BASIC. Many machine language programmers would experiment with null bytes in an attempt to make it harder for BASIC programmers to access their code and tamper with it.

In BASIC, the directory can be accessed as a non-executable pseudo-BASIC program with LOAD "$0",8 (or LOAD "$1",8 in the case of a dual drive) followed by LIST. The first line has a line number of 0 or 1 (indicating the drive number), showing in reverse video the name and ID of the disk and a shorthand code for the DOS version with which it was created (codes vary only as far as the DOS versions use incompatible disk formats, "2A" is used by most 5.25 inch DOS versions, "3D" by the 3.5 inch 1581). Lines after this have the size of a file (in disk blocks) as their pseudo "line number", followed by the file name in quotes and the three-letter type code. The last line shows the number of unallocated blocks on the disk (again as a pseudo "line number"), followed by the words "BLOCKS FREE."

On the Commodore 64, entering LOAD "$",8,1 will flood the screen with garbage instead of loading the directory into BASIC RAM. This is because the drive assigns the directory a load address of $0401 (1025), which is equivalent to the start of BASIC for the Commodore PET, but corresponds to the default screen memory in the C64 (starting with the second character on the first line of the screen).

Viewing the directory with a LOAD "$",3 command overwrites the BASIC program in memory. The DOS Wedge and various third-party cartridges and extenders such as Epyx FastLoad, Action Replay and The Final Cartridge III allow viewing of the disk directory using special commands that load the directory into screen memory without destroying the current BASIC program. The Commodore 128's BASIC 7.0 includes a DIRECTORY or CATALOG command (assigned on bootup to the F3 key) that performs the same function.

The following file types are supported:

SEQ

> A sequential file is a data file that can be linearly read from start to finish. Many word processors as well as programming text editors used sequential files for data storage. A sequential file is analogous to a flat file in Linux or UNIX, in that it has no specialized internal structure. It is not possible to position to any arbitrary location in a sequential file, as there is no analog of the lseek kernel call found in UNIX-like operating systems.

PRG

> Similar to a SEQ file, a program file has a little-endian-coded 16-bit load address prepended to the actual file content. All machine language and BASIC programs are saved as a program file via the kernal's SAVE call and can be subsequently loaded to memory with the BASIC's LOAD command (or the kernal's LOAD call). It is also possible to explicitly create a program file through DOS commands and then write any arbitrary data into it.

REL

> A relative file is a variation of the sequential file type, in which an indexing mechanism referred to as side-sectors is present to permit record-oriented access. Records may be a maximum of 254 bytes in size and are addressed by a one-based cardinal number, permitting true random access to any part of the file.

USR

> A user-specified file has an internal structure that is identical to that of a sequential file. Commodore's original purpose for this file type was the facilitation of DOS development, as the file content could be copied into a drive buffer for execution by the drive's microprocessor. It is unknown if anyone found a use for the facility. Some applications that use non-standard low-level disk structures save data in USR format, which came to be considered a sort of "leave me alone, don't try to copy or delete" indication to the user. Most notably, GEOS' "VLIR" files show up as USR files.

DEL

> An undocumented internally used file type similar in structure to a sequential file. Creation of this file type must be accomplished by direct manipulation of the disk directory.

The presence of an asterisk (*) prepended to the file type in a directory listing (for example, *SEQ) indicates that the file was not properly closed after writing. When the drive is commanded to close a file that has been opened for writing, the associated buffer is flushed to the disk and the block availability map (BAM) is updated to accurately reflect which blocks have been used. If a program crash or other problem (such as the user removing the disk while a file is open) results in an "orphan file", also referred to as a "poison" or "splat" file, buffers are not flushed and the BAM will not accurately reflect disk usage, putting the disk at risk of corruption. A poison file generally cannot be accessed (but can be opened in "modify" mode), and an attempt to use the DOS scratch command to delete the file may cause filesystem corruption, such as crosslinking. The only practical method of removing one of these files is by opening the file in "modify" mode (and fixing it), or by validating the disk (see the DOS validate command below), the latter which rebuilds the BAM and removes poison file references from the directory. The infamous save-with-replace bug could result in creation of splat files.

*DEL is a special type written into the on-disk directory entry of files that have been deleted. Such files are not shown in a normal directory listing, and their data blocks and directory entries will be reused by files that are subsequently created. Some utility programs allow the "un-deletion" of such files if their data blocks and directory entries haven't not yet been overwritten by other files.

File types with < after them (for example, PRG<) are "locked", and cannot be deleted—they can be opened for reading, however. There is no Commodore DOS command that can explicitly set or clear this status, but many third-party utilities were written to allow this to be done. These utilities generally read the directory using direct-access commands, perform the necessary modifications to the raw data, and then write the changes back to the disk.

File access

Accessing files is primarily an issue for the host computer. The kernal ROM in the computer contains the necessary primitive routines needed to access files, and the BASIC ROM contains a higher level abstraction for file access using BASIC syntax. The components that concern the DOS itself are file name parsing and the secondary address. This section will give an overview of the necessary BASIC commands for the sake of completeness.

Opening a file on a Commodore disk unit entails the processing of a number of parameters that are vaguely analogous to file opening procedures in other environments. Since the DOS is actually running in the drive's controller, the file opening sequence must pass enough information to the drive to assure unambiguous interpretation. A typical statement in BASIC to write to a sequential file would be as follows:

OPEN 3,8,4,"0:ADDRESSBOOK,S,W"

The parameters following the OPEN verb are as follows:

3 This parameter, the *file number*, logically identifies the opened file within the *computer's* operating system and is analogous to a file descriptor in UNIX-like operating systems. It is never sent to the drive and thus is neither known nor used by the drive's own operating system. The file number may be in the range of 1 to 254 inclusive, is assigned by the programmer and must be unique if more than one file is simultaneously opened. Once the file has been opened all program input and output procedures use the file number. In assembly language programs, this value is often referred to as LA (logical address), the abbreviation coming from the mnemonic that refers to the memory location where the file number is stored.

8 This parameter, the *device number*, identifies a specific peripheral attached to the computer. Devices 0 through 3 address the keyboard, tape cassette, RS-232 interface, and video display, respectively, all of which are directly controlled by the kernal ROM. Device numbers 4 and higher address devices attached to the peripheral bus, such as printers or disk drives. In the case of a disk drive, the device number refers to the

unit's controller, not the drive mechanism(s) within the unit. By convention, the first disk drive unit on a system has device number 8, the second drive, if present, 9, etc., up to a maximum of 15. The device number scheme was derived from the IEEE-488 or general purpose interface bus (GPIB) that was used with the Commodore PET/CBM models. In assembly language programs, this value is often referred to as FA or PA (physical address), again from the mnemonic for the memory location where the device number is stored.

4 This parameter, the *secondary address*, which may range from 0 to 15 inclusive, refers to a specific communication channel established with the device's controller and is passed to the device when it is commanded to "talk" or "listen" on the peripheral bus. As with the file number, the secondary address is determined by the programmer and must be unique for the device in question. The range 0 to 14 inclusive is used for passing data to or from the device, whereas 15, referred to as the "command channel", is used to issue commands to the device's controller (such as to rename a disk file), if the device is able to support such an operation. In disk drives, secondary addresses 0 to 14 inclusive are mapped to buffers within the controller, hence establishing communication with a specific file on a specific disk; since as mentioned above the drive does not know about the *file number*, it can only use the *secondary address* to make a difference between several files that are open at the same time. On the other hand, the host operating system is agnostic about the secondary address; it is transmitted to the drive on every access to the file, but not otherwise used by the host. In assembly language programs, this value is often referred to as SA (secondary address).

COMMAND STRING The "0:ADDRESSBOOK,S,W" parameter is officially referred to in Commodore documentation as the *command string* and is interpreted by the controller of the device being accessed. In the case of a disk drive unit, the formal command string structure consists of the drive mechanism number (0:, not to be confused with the device number), filename (ADDRESSBOOK), file type (S, sequential in this example) and access mode (W, opened for writing in this example). In practice, some of these parameters may be omitted. Minimally, only the filename is required if the file is to be opened for reading.

The drive number identifies a drive mechanism attached to a disk unit's controller and is analogous to a logical unit number in a SCSI controller that is capable of controlling multiple mechanisms (e.g., the OMTI SASI controllers that were developed to work with ST-412/ST-506 hard drives in the 1980s). In floppy disk units, the first mechanism is drive 0: and the second is 1:. It is fairly common practice to omit the drive number when communicating with a single drive floppy unit, as 0: is the default in such units, but since omitting the number can trigger a few obscure bugs in the DOS it is not a recommended practice (a colon alone is equivalent to 0: and is enough to avoid those bugs). An exception to this convention is with the Lt. Kernal hard disk subsystem, in which the drive number refers to "logical units" (virtual drives created on a single physical drive), which made syntax such as 4: or 10: necessary if a file to be opened was not on logical unit zero (equivalent to drive mechanism zero in a dual floppy unit).

Files can also be loaded and saved to with LOAD and SAVE commands. File name specifiers can also be used here, for example, SAVE "FILE" ,8 saves the BASIC program to a PRG (program) file and SAVE "0:FILE,SEQ,WRITE" ,8,1 saves the BASIC program to a sequential file. If the secondary address isn't specified or is specified as 0 (e.g. LOAD "FILE" ,8), the file is saved/loaded from the BASIC memory area (which, on the C64, starts by default at $0801). If the secondary address is specified as a non-zero value (e.g. LOAD "FILE" ,8,1), the program is loaded starting from the address specified by the file itself (the PRG header, which is the first two bytes of the file)—this form of command is more common when loading machine code programs.

Load relocation was first introduced on the VIC-20 because this machine could start BASIC RAM in several different locations, depending on the memory expansion that was installed. The older Commodore PET series did not support relocation, so LOAD "FILE" ,8 and LOAD "FILE" ,8,1 would have the same effect: the file would be loaded into the same memory region from which it was saved. Load relocation happens in the host, being an exception to what is said above about the secondary address being used only device-internally.

The command LOAD "*",8,1 will load the first program on the disk starting from the file-specified memory location. This is one of the most popular load commands of the platforms, the method to start majority of commercial software, for example. Of note is the fact that the wildcard expansion will only pick the first catalog name when no other file on that disk has been accessed before; the name of the last-used file is kept in memory, and subsequent LOAD "*",8,1 commands will load that file rather than the first. (However, LOAD "0:*",8,1 or LOAD ":*",8,1 will always load the first file on the disk.)

The directories of disks in two-drive units are accessed as LOAD "$0" , 8 and LOAD "$1" , 8. "0:$" and "1:$" do not access the directory but actual files on one drive or the other that just happen to be named "$". Partial directories can be loaded by adding a colon and a template, for example LOAD "$C:K*=P" ,8 would load a partial directory that shows only the files whose name starts with the letter K and which are of type PRG; all such partial directories still contain the initial disk name line and the final "BLOCKS FREE" line.

The save-with-replace bug

Commodore DOS also offers a "Save-with-Replace" command, which allows a file to be saved over an existing file without the need to first SCRATCH the existing file. This was done by prepending an @ symbol to the file name during the OPEN or SAVE operation - for instance, SAVE "@MY PROGRAM" ,8. For years rumors spread, beginning with the 4040 drive, of a bug in the command's implementation. At first, this was denied by some commentators. Prizes were offered to prove the existence of the bug.[1] By early 1985 *Compute!* magazine advised readers to avoid using the command.[2] That year various authors independently published articles[1][3][4][5] proving that the Save-with-Replace bug was real and including methods by which it could be triggered.

Affected devices included the single-drive 1541 and dual-drive 4040; the 8050 and 8250 did not exhibit the issue.[1] Some commentators suggested the bug could be avoided by always explicitly specifying the 0: drive number when saving,[3] though it was later shown that *any* disk operations without a drive number were sufficient to lead to the bug.[6] The bug stemmed from the fact that the affected DOS implementations were modified versions of the DOS contained in earlier Commodore PET dual drives such as the 8050. This created a "phantom drive 1:" on single-drive systems, resulting in the allocation of an unnecessary buffer under some conditions. Since the Save-with-Replace command used all five drive buffers, and because the method by which the "phantom" buffer was allocated did not meet specifications, this resulted in scrambled data being written to the disk under some conditions.

In September 1986, Philip A. Slaymaker published an article[7] describing in great detail the cause of the bug and providing patches to the 1541 drive ROMs; readers with an EPROM burner could produce their own patched ROMs which could be swapped into the drive. Commodore was made aware of Slaymaker's findings, and while they never issued an official update for the original 1541's ROMs, they did fix the bug in Revision 5 of the 1571 ROMs, and also in the ROMs for the 1541-c and 1541-II drives. Although not supported by Commodore, it is known that the 1541-II firmware (but not that of the 1541-c) can also be used in an original 1541 drive by using EPROMs, which will fix the bug for that drive as well.

Command channel

As previously noted, the Commodore DOS itself is accessed via the "command channel", using syntax like that used to access files. Issuing commands to the DOS and retrieving status and error messages generated in response to commands is accomplished by opening a file to the device using 15 as the secondary address, for example:

OPEN 1,8,15

To retrieve and display the device status, one could code:

OPEN   1,8,15:INPUT#1,E,E$,T,S:PRINT E,E$,T,S:CLOSE 1

In the above example, E will hold the error number (if any, it will be zero if no error exists), E$ will be a terse text description of the error, T will represent the disk track where the error occurred, and S will be the sector on track T to which the error refers. If no error exists, the equivalent of 00,OK,00,00 will be returned in the four variables. Note that INPUT# is a run mode only verb. Also, in programs that issue many disk commands it is customary to open a file to the device's command channel at the start of the program and not close it until the program has finished.

Commodore BASIC versions 4.0 and later provide a pseudo-variable referred to as DS$ that may be used to retrieve drive status in lieu of the above code. This reserved variable is not available on earlier versions of BASIC, so the command channel must be manually read as demonstrated above. Note that immediately after power-on or reset, the DOS revision will be returned. For example, a 1541 will return 73,CBM DOS V2.6 1541,00,00. Error code 73 is common to all drive models and may be used to determine if the drive has been reset to its power-on state.

3.5.3   DOS commands

There are also a command for seeking in RELative type files (RECORD#), several block-level direct-access commands (BLOCK-READ, BLOCK-WRITE, BUFFER-POINTER), block management (BLOCK-ALLOCATE, BLOCK-FREE), drive memory manipulation and execution of program code on the drive's processor (MEMORY-WRITE, MEMORY-READ, MEMORY-EXECUTE, BLOCK-EXECUTE) and user-definable

functions (USER and & commands). Some of the theoretically user-definable functions were rededicated for accessing new functionality in DOS versions after 1.0.

3.5.4   References

[1] Whittern, Charles H. (July 1985). "SAVE with Replace Exposed!!". *The Transactor* **6** (1): 20.

[2] "The Great Commodore Save/Replace Debate". *Compute!*. February 1985. p. 10. Retrieved 6 October 2013.

[3] Slaymaker, P. A. (October 1985). "Save With Replace: Debugged At Last / Part 1". *Compute!*. p. 79. Retrieved 16 October 2013.

[4] Slaymaker, P. A. (November 1985). "Save-With-Replace: Debugged At Last / Part 2". *Compute!*. p. 111. Retrieved 30 October 2013.

[5] Editors. "Save@: Gerry Neufield's Theory on an Old Bug". *Info* № 9, December 1985/January 1986.

[6] Excerpt of e-mail from Philip A. Slaymaker also archived at Groups.Google

[7] Philip A. Slaymaker. "Eliminating SAVE@ and Other 1541 Bugs". *The Transactor* Vol. 7 № O2, September 1986, pp. 33–35.

Notes

- Immers, Richard; Neufeld, Gerald G. (1984). *Inside Commodore DOS. The Complete Guide to the 1541 Disk Operating System*. DATAMOST, Inc & Reston Publishing Company, Inc. (Prentice-Hall). ISBN 0-8359-3091-2.

- Englisch, Lothar; Szczepanowski, Norbert (1984). *The Anatomy of the 1541 Disk Drive*. Grand Rapids, MI: Abacus Software (translated from the original 1983 German edition, Düsseldorf: Data Becker GmbH). ISBN 0-916439-01-1.

- (Finnish) Lundahl, Reijo (1986). *1541-Levyasema*. Amersoft. ISBN 951-35-3206-2

3.6   SpeedScript

SpeedScript is a type-in word processor for various 8-bit home computers of the 1980s. Approximately 5 KB in length, it provided many of the same features as commercial word processing packages of the early 8-bit era, such as PaperClip and Bank Street Writer.

3.6.1 Versions

In April 1983 *Compute!* published **Scriptor**, a word processor written by staff writer Charles Brannon in BASIC and assembly language, as a type-in program for the Atari 8-bit family.[4] In January 1984 version 1.0 of his new word processor SpeedScript appeared in *Compute!'s Gazette* for the Commodore 64 and VIC-20.[1] 1.1 appeared in *Compute!'s Second Book of Commodore 64*, 2.0 on *Gazette Disk* in May 1984, and 3.0 in *Compute!* in March and April 1985.[5][6] Corrections that updated 3 0 to 3.1 appeared in May 1985,[7] and the full version appeared in a book published by Compute!, *SpeedScript: The Word Processor for the Commodore 64 and VIC-20*.[8] A 3.2 update appeared in the December 1985 *Compute!*[9] and January 1986 *Compute! Disk*[2] and again later in the May 1987 *Compute!'s Gazette* issue with three additional utilities.[10]

SpeedScript was later ported to the Atari and the Apple II family in *Compute!* in May[11] and June 1985 respectively.[12][2] SpeedScript was written entirely in assembly language, and Compute! Publications later released book/disk combinations that contained the complete commented source code (as well as the machine language in MLX format) for each platform.[8][13][14]

A version of SpeedScript for MS-DOS was created in 1988 by Randy Thompson and published in book form by Compute! Books.[3] This version was written in Turbo Pascal with portions written in assembly language, and added incremental new features to the word processor such as additional printer commands, full cursor-control (to take advantage of the PC's Home, End, PgUp, and PgDn keys), and a native 80-column mode.

3.6.2 80-column updates

The original versions of SpeedScript were designed for the 40-column Commodore 64 and the 22-column VIC-20. When the Commodore 128 was released, featuring an 80-column display, many users requested an updated version of SpeedScript to take advantage of this new capability. In June 1986, *Compute!'s Gazette* published SpeedScript-80, a short patch for SpeedScript 3.0 or higher, which enabled the use of the VDC's new 80-column capabilities on a Commodore 128 running in 64 mode.[15] However, this did not take advantage of the C128's expanded memory, and a few minor commands were eliminated due to the alterations to the existing code. SpeedScript-80 was enhanced soon after with SpeedScript-80 Revisited, by Bob Kodadek.

A native version for the C128 called SpeedScript 128, also written by Kodadek, was finally released in October 1987. This version eliminated the problems of the patch and took full advantage of the C128's 80-column screen, its ex-

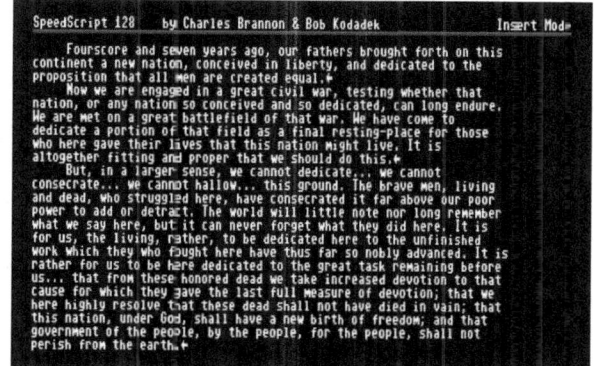

SpeedScript 128

panded memory and the enhanced keyboard.[16] A later update appeared in September 1989, adding full text justification, tab setting, and online help.[17]

In December 1987, *Compute!'s Gazette* published *Instant 80*, a utility for the C64 version of SpeedScript that allowed 80-column document previewing (though not editing) on a standard C64. This was done by using half-width characters on a high-resolution graphics screen.[18]

3.6.3 Utilities

Although SpeedScript did not include a built-in spell checker, additional utilities were soon published. In December 1985, SpeedCheck was published in *Compute!'s Gazette*.[19] This external utility accepted SpeedScript files (as well as those from compatible word processors, such as *PaperClip*) and spell-checked them against a user-defined dictionary. An enhanced 80-column version for the C128, SpeedCheck 128, was published in September 1988.[20]

Another utility, ScriptSave, was developed to provide automatic saving functionality to the Commodore 64 version of SpeedScript 3.0.[21] This program would set up a timer program to save documents to disk, before loading and running SpeedScript itself.

Several additional utilities were published in the May 1987 issue of *Compute!'s Gazette* along with SpeedScript 3.2. ScriptRead[22] was developed to identify and preview SpeedScript documents on a disk, with the ability to scratch any files no longer needed. This was an important addition as on a single-drive system there would be no way to save work if the disk became full. SpeedSearch[23] provided full-text search of all SpeedScript documents on a disk, returning a count of how many times the searched word or phrase was used in each document. Date and Time Stamper[24] introduces a program to the disk drive that adds time stamps to files on disk, then executes SpeedScript.

3.6.4   Reception

In a review of four word processors, *The Transactor* in May 1986 praised SpeedScript as "extremely sophisticated", citing its large text buffer, logical cursor navigation, and undo command. While criticizing it for lacking right justification, the magazine concluded that SpeedScript was not only "an easy winner" among budget-priced word processors, but also "a serious contender even when compared with the higher priced programs".*[25]

SpeedScript was sufficiently popular to receive coverage in some reference works, such as the "Wordprocessing Reference Guide" of Karl Hildon's popular *Inner Space Anthology*.*[26]

3.6.5   Gallery

3.6.6   References

[1] Brannon, Charles (January 1984). "*SpeedScript* Word Processor For Commodore 64 And VIC-20". *COMPUTE!'s Gazette* (Greensboro, North Carolina: COMPUTE! Publications) (7): 38–59. Retrieved 18 February 2015.

[2] Mitchener, Leo (June 1986). "*SpeedScript*'s Lineage". *COMPUTE!* (Letters to the Editor) (73): 11. ISSN 0194-357X. Retrieved 8 November 2013.

[3] Thompson, Randy (1989). *PC SpeedScript*. Radnor, Pennsylvania: COMPUTE! Books. ISBN 0-87455-166-8.

[4] Brannon, Charles (April 1983). "*Scriptor*: An Atari Word Processor". *COMPUTE!* (35): 56–70. ISSN 0194-357X. Retrieved 30 October 2013.

[5] Brannon, Charles (March 1985). "*SpeedScript 3.0*: All Machine Language Word Processor For Commodore 64". *COMPUTE!* (58): 123–133. ISSN 0194-357X. Retrieved 1 March 2015.

[6] Brannon, Charles (April 1985). "*SpeedScript 3.0*: All Machine Language Word Processor For Expanded VIC-20". *COMPUTE!* (59): 100–106. ISSN 0194-357X. Retrieved 1 March 2015.

[7] "Capute!". *COMPUTE!* (Column) (60): 99. May 1985. ISSN 0194-357X. Retrieved 1 March 2015.

[8] Brannon, Charles (1985). *SpeedScript, the Word Processor for the Commodore 64 and VIC-20*. Greensboro, North Carolina: COMPUTE! Publications. ISBN 0-94238-694-9.

[9] Brannon, Charles (December 1985). "*SpeedScript 3.0* Revisited". *COMPUTE!* (67): 90–91. ISSN 0194-357X. Retrieved 1 March 2015.

[10] Brannon, Charles (May 1987). "*SpeedScript 3.2* For The Commodore 64". *COMPUTE!'s Gazette* (47): 54–71. ISSN 0737-3716. Retrieved 1 March 2015.

[11] Brannon, Charles (May 1985). "*SpeedScript 3.0*: All Machine Language Word Processor For Atari". *COMPUTE!* (60): 103–111. ISSN 0194-357X. Retrieved 1 March 2015.

[12] Brannon, Charles; Martin, Kevin (June 1985). "*SpeedScript 3.0*: All Machine Language Word Processor For Apple". *COMPUTE!* (61): 116–123. ISSN 0194-357X. Retrieved 1 March 2015.

[13] Brannon, Charles (1985). *SpeedScript, the Word Processor for Atari Computers*. Greensboro, North Carolina: COMPUTE! Publications. ISBN 0-87455-003-3.

[14] Brannon, Charles; Martin, Kevin (1985). *Speedscript, the Word Processor for Apple Personal Computers*. Greensboro, North Carolina: COMPUTE! Publications. ISBN 0-87455-000-9.

[15] Heimarck, Todd (June 1986). "SpeedScript-80 For The 128". *COMPUTE!'s Gazette* (36): 77–78. ISSN 0737-3716. Retrieved 1 March 2015.

[16] Kodadek, Robert (October 1987). "*SpeedScript 128*". *COMPUTE!'s Gazette* (52): 22–52. ISSN 0737-3716. Retrieved 1 March 2015.

[17] Gruber, Michael (September 1989). "*SpeedScript 128 Plus*". *COMPUTE!'s Gazette* (75): 38–44. ISSN 0737-3716. Retrieved 4 March 2015.

[18] Mackinnon, Glen (December 1987). "*Instant 80*: True 80-Column Preview For SpeedScript". *COMPUTE!'s Gazette* (54): 76. ISSN 0737-3716. Retrieved 1 March 2015.

[19] Cowper, Ottis T. (December 1985). "*SpeedCheck*: An Expandable Spelling Checker For The Commodore 64 And 128". *COMPUTE!'s Gazette* (30): 64–70. ISSN 0737-3716. Retrieved 18 February 2015.

[20] Smith, Larry D (September 1988). "*SpeedCheck 128*: A Spelling Checker For SpeedScript 128". *COMPUTE!'s Gazette* (63): 60–61. ISSN 0737-3716. Retrieved 1 March 2015.

[21] Lambert, J. Blake (May 1985). "ScriptSave: Automatic Disk Saves For Commodore 64 *SpeedScript 3.0*". *COMPUTE!* (60): 84–85. ISSN 0194-357X. Retrieved 1 March 2015.

[22] Childress, Buck (May 1987). "ScriptRead". *COMPUTE!'s Gazette* (47): 77. ISSN 0737-3716. Retrieved 18 February 2015.

[23] St. Clair, Tony (May 1987). "SpeedSearch". *COMPUTE!'s Gazette* (47): 75. ISSN 0737-3716. Retrieved 18 February 2015.

[24] Kodadek, Bob (May 1987). "*SpeedScript* Date and Time Stamper". *COMPUTE!'s Gazette* (47): 76. ISSN 0737-3716. Retrieved 18 February 2015.

[25] Bose, Ranjan (May 1986). "A Comparison of Four Word Processors". *The Transactor* **6** (6): 72–74. ISSN 0827-2530. Retrieved 1 March 2015.

[26] Hildon, Karl J. H. (March 1985). *The Complete Commodore Inner Space Anthology*. Milton, Ontario: Transactor Publishing. pp. 17–19. ISBN 0-9692086-0-X. Retrieved 1 March 2015.

Chapter 4

Games

4.1 List of Commodore VIC-20 games

Here is a **list of games** for the Commodore VIC-20 personal computer system, sorted alphabetically. See lists of video games for other gaming platforms.

Contents :

- Top
- 0–9
- A
- B
- C
- D
- E
- F
- G
- H
- I
- J
- K
- L
- M
- N
- O
- P
- Q
- R
- S
- T
- U
- V
- W
- X
- Y
- Z

4.1.1 0–9

- *3D Maze*
- *3D Silicon Fish*

4.1.2 A

- *Abductor*
- *Adventureland* - Cartridge only
- *Agressor*
- *Airplane*
- *Alien*
- *Alien Attack*
- *Alien Blitz*
- *Alien Raiders*
- *Alien Soccer*
- *Alien Vortex*
- *Alien Wars*

- *Alpha Blaster*
- *Alphoids*
- *Amok!*
- *Ampumapeli*
- *Annihilator*
- *Another Vic in the Wall*
- *Antimatter Splatter*
- *Ape Escape*
- *Apple Bug*
- *Apple Panic*
- *Arcadia*
- *Artillery Duel*
- *Asteroids*
- *Asteroyds*
- *Astro Fighters*
- *Astro Panic*
- *Astro Patrol*
- *Astro-Command*
- *Atlantis*
- *Avalanche (2012)*
- *Avenger (Space Invaders (VIC ROM 1901))*

4.1.3   B

- *Bagdad*
- *Baja 1000*
- *Battlefield*
- *Berzerk (2010)*
- *Bewitched (imagine Software 1983)*
- *Big Bad Wolf*
- *Blip-itz*
- *Blitz (VIC-20 Starter Pack (1982))'*
- *Blitzkrieg*
- *Block Buster*

- *Blockade*
- *Blue Meanies From Outer Space*
- *Blue Star (2008)*
- *The Bomb*
- *Bomber*
- *Bomber Run*
- Bongo
- *Bonking Barrels*
- *Bonzo*
- *Bounceout*
- *Brainstorm*
- *Buck Rogers: Planet of Zoom*
- *Bug Diver*
- *Bunny Hop*
- *Burning Building*

4.1.4   C

- *CB Slot VI*
- *CB Slot VIII*
- *Cannonball Blitz*
- *Canyon Runner*
- *Car Chase*
- *Car Driver*
- *Carling the Spider (2009)*
- *Castle Dungeon*
- *Catcha Snatcha*
- *Caterpilla*
- *Caves*
- *Centipede*
- *Centipod*
- *Chariot Race*
- *Choplifter*
- *Chopper*

- *City Bomber*
- *Clowns*
- *Code Breaker*
- *Computer Adventure*
- *Congo Bongo*
- *Cops 'n' Robbers*
- *Cosmic Cruncher*
- *The Count*
- *Crazy Climber*
- *Crazy Kong*
- *Crossfire*
- *Cyclons*
- *Cosmic Jailbreak*

4.1.5 D

- *Dam Buster*
- *Death House*
- *Death Maze*
- *Deathtrap*
- *Defender*
- *Deflex*
- *Demon Attack*
- *Depth Charge*
- *Devastator*
- *Dig Dug*
- *Dodo Lair*
- *Donkey Kong*
- *Doom (2013)*
- *Downhill*
- *Drag Strip*
- *Dragonfire*
- *Dragon Master*
- *Duck Shoot*
- *Dungeon*

4.1.6 E

- *English Invaders*
- *Explorer*
- *Emits Attack*
- *Escape MCP*

4.1.7 F

- *Football Manager*
- *Forth a computer language on a cartrage*
- *Fourth Encounter*
- *Frantic*
- *Frantic Fisherman*
- *Freeze Factory*
- *Frogger*
- *Frogger '07 (2007)*

4.1.8 G

- *Galaxian*
- *Garden Wars*
- *Galactic Defender*
- *Ghost Hunt*
- *Ghost Manor*
- *Gorf*
- *Grand Prix*
- *Gridrunner*
- *Gridtrap*
- *Gusher (Visions 1983)*

4.1.9 H

- *Hangman*
- *Hareraiser*
- *Harvester*
- *Hellgate*
- *Home Babysitter*
- *Hoppit (VIC-20 Starter Pack (1982))*
- *Hunchback*

4.1.10 I

- *Ice*
- *Impossible Mission*
- *The Improbable War (2010)*
- *Island of Secrets (2009)*

4.1.11 J

- *Jelly Monsters*
- *Jetpac*
- *Jigsaw*
- *Juggler*
- *Jungle Hunt*
- *Jupiter Lander (VIC ROM 1907 (1981))*

4.1.12 K

- *K-Star Patrol*
- *Killer Comet*
- *Krazy Kong (video game)*

4.1.13 L

- *Laserzone*
- *Log Run*
- *Lode Runner*
- *London to Paris Air Race*
- *Lucy Lizard*
- *Lunar Lander*
- *Lunar Leepers* (Sierra on-line cartridge)

4.1.14 M

- *Magical Maldibus*
- *Marble Hunt*
- *Mathematic Missile*
- *Math Hurdler*
- *Matrix*

- *Max*
- *Mayhem (2012)*
- *Menagerie*
- *Metagalactic Llamas Battle at the Edge of Time (LLA-MASOFT 1982)*
- *Metamorphosis*
- *Meteor Maze*
- *Mickey the Brickey*
- *Midnight Drive*
- *Mine Field*
- *Mine Madness*
- *Miner 2049er*
- *Mission Impossible Adventure*
- *Mole Attack*[1]
- *Money Snake*
- *Money Wars*
- *Moon Patrol*
- *Motor Mouse*
- *Motorway*
- *Ms. Pac-Man*
- *Multisound Synthesizer*
- *Mutant Herd*
- *Mystery Fun House*

4.1.15 N

- *New York Blitz*
- *Number Nabber*

4.1.16 O

- *Omega Race* - Cartridge only
- *Othello*

4.1.17   P

- *Pac-Man*
- *Pak Bomber*
- *Pakakuda*
- *Panic*
- *Parachute (2008)*
- *Paratrooper*
- *Pedes & Mutants*
- *The Perils of Willy*
- *Phantom Attack*
- *Pharaohs Curse*
- *Pin Ball*
- *Pinball Wizard*
- *Pirate Adventure*
- *Pistolen Paultje*
- *The Pit*
- *Plague*
- *Poker*
- *Pole Position*
- *Pontoon 21*
- *Pool*
- *Pooyan 20 (2014)*
- *Potholes I*
- *Potholes II*
- *Potholes III*
- *Potholes IV*
- *Potholes V*
- *Problems*
- *Psycho Shopper*
- *PuckMan*
- *Pyramid of Doom*

4.1.18   Q

- *Q*bert*
- *Quest*
- *Quikman (2009)*
- *Quirk*
- *Quizard*
- *Quiz Master*

4.1.19   R

- *RIJ-Test*
- *Rabbit Blitz*
- *Rabbit Chaser*
- *Race*
- *Racefun*
- *Racer* - Cartridge only
- *Radar RatRace (VIC ROM 1910)*
- *Raid on Fort Knox*
- *Rat Hotel*
- *RattenVanger*
- *Realms of Quest I (1991)*
- *Realms of Quest II (2004)*
- *Realms of Quest III (2009)*
- *Realms of Quest IV (2013)*
- *Red Alert*
- *Rescue at Rigel*
- *Return to Fort Knox (2008)*
- *Rhino*
- *Ricochet*
- *River Rescue*
- *Road Driver*
- *Road Race*
- *Road Toad*
- *Roader*

- *Robotron: 2084*
- *Rocket Command*
- *Rockman*
- *Ronnie!*
- *Rubiks Cube*
- *Rugby*
- *Ruimtemonsters*

4.1.20   S

- *Salmon Run*
- *Sargon II Chess*
- *Saucer Shooter*
- *Scare City Motel*
- *Schuifspel*
- *Scrambler*
- *Sea Wolf (VIC ROM 1937)*
- *Secret Mission*
- *Serpentine*
- *Shamus*
- *Shooting Gallery*
- *Ski*
- *Skipping Ball*
- *Sky Diver*
- *Sky Hop*
- *The Sky is Falling*
- *Slap Dab*
- *Slot*
- *Smash*
- *Snake*
- *Snake-Bite*
- *Snooker*
- *Sokkelo*
- *Space Dock*

- *Space Docking*
- *Space Escort*
- *Space Fortress*
- *Space Panic*
- *Space Snake*
- *Space Zap*
- *Speed Boat*
- *Speed-Ski*
- *Spider City*
- *Spiders of Mars*
- *Spike's Peak*
- *Star Battle*
- *Star Chaser*
- *Star Post*
- *Star Trek*
- *Starwars*
- *Starwars II*
- *Stock*
- *Stop Thief*
- *Strange Odyssey*
- *Super Slither*
- *Super Seeker*
- *Super Slot*
- *Super Smash*
- *Swarm*
- *Sword of Fargoal*

4.1.21   T

- *Tank Battalion (2014)*
- *Tank Versus UFO*
- *Tank Wars*
- *Ten Ten (2009)*
- *Thunderflash*

- *Tic-Tac-Toe*

- *Time Bomb*

- *Tooth Invaders*

- *Traxx*

- *Trap-Man*

- *Treasure of the Bat Cave*

- *Turmoil*

- *Type-a-Tune (VIC-20 Starter Pack (1982))*

- *Typing*

4.1.22 U

- *U-Boat*

- *UFO*

- *Ultima: Escape from Mt. Drash*

- *Under Mine*

4.1.23 V

- *Vegas Jackpot*

- *VIC Avenger* - Cartridge only

- *Vic Biorhythms*

- *Vic Calendar*

- *Vic Downs*

- *Vic Panic*

- *Vic Super Lander*

- *Vic-Tac-Toe*

- *VICtoria (2008)*

- *Visible Solar System*

- *Voodoo Castle* - Cartridge only

4.1.24 W

- *WhackE (2009)*

- *Wacky Waiters*

- *Wall Street*

- *Witch Way*

- *Wizard and the Princess*

- *Word Game*

- *Word Match*

- *Wunda Walter*

4.1.25 Z

- *Zac-Man*

4.1.26 References

[1] Komar, Charlene (June 1983). "Mole Attack". *Electronic Games*. p. 70. Retrieved 6 January 2015.

4.2 3D Silicon Fish

3D Silicon Fish is a 1984 computer game for the Commodore VIC-20.[*][1] It was developed in the United Kingdom.

4.2.1 Plot

According to the instruction manual, the game takes place on Earth in the future where inhabitants are dependent on silicon for technology. Unfortunately, the silicon reserves are running out, but another source of silicon has been located in a faraway part of the galaxy. A mercenary going by the name of Sillo is sent to the dangerous area with a vehicle called the "silicon fisher" to gather the silicon and transport the cargo back to Earth.

4.2.2 Gameplay

The game consists of a maze located above a stream that flows from the top to the bottom of the screen. The player navigates through the maze with the silicon fisher vehicle while looking for silicon that floats along the river. Collecting the silicon is done by positioning the vehicle towards the direction where the silicon is heading and throwing out a net to catch it. Every so often, an enemy called a "Kryllon" materializes in the air and explodes after a short time span. If the player is within range of the explosion, they will be killed.[*][2]

As each level is completed, the speed of the game quickens along with more randomly-appearing Kryllons. The colour of the levels change as the game goes on, and the music grows faster as the pace of the game accelerates.[*][3]

4.2.3 References

[1] "My Trip to 'Retrovision 2010'". Frank Gasking. 2 October 2010. Retrieved 27 May 2014.

[2] "3D Silicon Fish for Commodore 64". MobyGames. Retrieved 23 May 2014.

[3] Retro Gamer Team (20 May 2009). "3D Silicon Fish Review". Retro Gamer. Retrieved 30 May 2014.

4.2.4 External links

- 3D Silicon Fish at MobyGames

4.3 Adventureland (video game)

Adventureland is the first text adventure game for microcomputers,[*][1] released by Scott Adams in 1978. It was very successful and led Adams to form Adventure International,[*][2] which went on to publish twelve similar games in different settings.

The game involves the search for thirteen lost artifacts in a fantasy setting.

4.3.1 Gameplay

Gameplay involves moving between the various locations found within the game, collecting found objects (and often subsequently using them, generally in another location), and the solving of puzzles.

The game commands take the form of either simple, two-word, verb/noun phrases, such as "climb tree," or one-word commands, such as those used for player character movement, including north, south, east, west, up, and down. Although the game has a vocabulary of about 120 words,[*][3] the parser only recognised the first three letters.[*][3] This meant that the parser occasionally identified a word incorrectly, but also that commands could be truncated, for example "lig lam" would be interpreted as "light lamp."

In order to complete the game, the player has to collect the thirteen lost artifacts: A statue of Paul Bunyan's blue ox, *Babe*, the jeweled fruit, the golden fish, a dragon's egg, a golden net, a magic carpet, a diamond necklace, a diamond bracelet, a pot of rubies, the "royal honey", a crown, a magic mirror, and a "firestone." Unlike succeeding adventure games, *Adventureland* has no story or plot, it is simply a treasure hunt.

The game was available on a number of platforms, including the Apple II series of computers, and various computers released by Atari, Commodore International, and Texas Instruments. A cut-down, three treasure version entitled 'Adventure 0: Special Sampler' was also made available at a special low price.[*][4]

In 1982, *Adventureland* was re-released with graphics, thus enabling the player to view visible representations of the scenery and objects to be found within the game.[*][5]

4.3.2 Development

Adventureland, Adams' first program, is a slightly scaled-down, machine-language game similar to the "original" *Adventure* program.[*][6] The source code for *Adventureland* was published in *SoftSide* magazine in 1980[*][7] and the database format was subsequently used in other interpreters such as Brian Howarth's *Mysterious Adventures* series.[*][8]

4.3.3 References

[1] Griffin, Brad (March–April 1983). "Scott Adams Adventures 1–12". *A.N.A.L.O.G. Computing* (10).

[2] "Game Set Interview: Adventure International's Scott Adams", Game Set Watch, July 19th, 2006, retrieved on April 20th, 2009

[3] "Great Scott". *GamesTM* (88). 2009. pp. 152–157.

[4] 0:Adventureland Demo "Scott Adams Classic Adventures," (retrieved on May 4th, 2009).

[5] "Scott Adams Classic Adventures", Adventureland. retrieved April 20th, 2009

[6] Herro, Mark (October 1980). "The Electric Eye" (PDF). *The Dragon* (42): 42–43 hardcopy (aka electronic pagenumbers 44–45 in archived PDF version). Retrieved 2015-07-11. ... an extremely complex, challenging game .. in an imaginary world. The object of ADVENTURE is to gather treasures, which is tough enough, but you sometimes must also take the treasures to a specific location ... You tell the computer what to do with simple, two-word commands, like GO NORTH, EXAMINE BOOK, or ENTER CAVE. The computer has a fairly large vocabulary... The roots of AD-VENTURE go back (I'm told) to a computerized version

of D&D developed by a consulting firm in Massachusetts. The game went through several modifications and additions, and the present form of the 'original' ADVENTURE game is generally credited to Will Crowther and Don Woods. This 'original' version is still found, on many time-sharing computer systems. (I once ran a 'system status' check on a large computer and found ADVENTURE was the second most popular game in the system, just behind a particularly good version of STAR TREK.) Then along came Scott Adams, who converted ADVENTURE for use with home computers. His first program, ADVENTURELAND, is a slightly scaled-down, machine-language version of the 'original' ADVENTURE program. Then he came out with PIRATE ADVENTURE, which has a completely different plot. With the success of these two programs, Scott wrote even more, and he has now become the acknowledged 'king' of the ADVENTURE game, with ten different versions being marketed. And there is talk of more on the way! ...During a game, the computer's video screen in divided into two parts. The upper half of the screen always displays the description of the location the player is in at the moment. It also lists the obvious directions the player may go (there may be other exits, such as climbing a tree if in a forest, or entering a specific location). The bottom half of the screen is reserved for the player's input, such as giving commands. ...there is very little bloodshed of any kind in the Adams ADVENTURE series. It's brain instead of brawn that counts here. Indeed, there are many funny occurrences ...I can't recommend ANY version of Scott Adams' ADVENTURE series highly enough. Beg, borrow, or steal a chance to play ADVENTURE!!!!!

[7] Adams, Scott (July 1980). "Adventureland". *SoftSide*. p. 36. Retrieved 13 April 2015.

[8] Graham, Nelson (2001). *The Inform Designer's Manual* (PDF). Dan Sanderson. p. 358. ISBN 0-9713119-0-0.

4.3.4 External links

- Playable Adventureland in Java

- *Adventureland* at MobyGames

- *Scott Adams' Graphic Adventure #1: Adventureland* at MobyGames

4.4 Alpha Blaster

Alpha Blaster is a fixed shooter video game developed by Sumlock for the Commodore VIC-20 home computer, and published by LiveWire Software in 1983.[*][2][*][3] Aackosoft International published their MSX adaptation of the game in the following year.[*][4][*][5]

4.4.1 Gameplay

In *Alpha Blaster*, a derivative of *Galaxian* (1979), the player defends against an alien invasion fleet from the planet Alpha. Piloting a lone Federation battle cruiser, the player scores points by shooting flying saucers and other spacecraft with a laser cannon.[*][6] The craft's laser energy and fuel are limited. After surviving two waves of attacks, the player must then dodge and blast through debris as the cruiser flies through an asteroid belt. If the player passes through the asteroid belt, the cruiser docks with a supply ship to rearm and refuel. After each such cycle, the game's difficulty level increases.

4.4.2 See also

- List of Commodore VIC-20 games

- List of MSX games

- *Megamania* (1982)

- *Space Invaders* (1978)

4.4.3 References

[1] Starr, Michael; Chapple, Craig (2008). *Vintropedia Collector Handbook: Vintage Computer & Retro Console Price Guide 2009*. Raleigh: Lulu Press. p. 241. ISBN 978-1409212775. Retrieved 31 May 2013.

[2] "*Alpha Blaster*". *Universal Videogame List*. Retrieved 31 May 2013.

[3] *Alpha Blaster* at MobyGames

[4] http://www.gamefaqs.com/msx/934729-alpha-blaster

[5] *Alpha Blaster* at GameSpot

[6] *Alpha Blaster*. Manchester: LiveWire Software. 1983. p. 1. Retrieved 31 May 2013.

4.5 Apple Panic

Apple Panic is a 1981 platform game for the Apple II programmed by Ben Serki and published by Brøderbund Software. *Apple Panic* is an unauthorized version of the arcade game *Space Panic*.

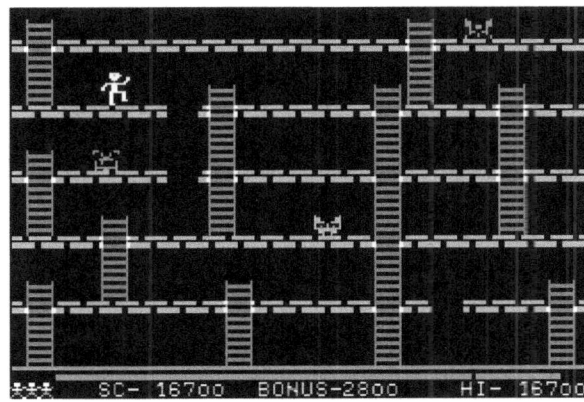

A screenshot showing the three enemies.

4.5.1 Description

Electronic Games described *Apple Panic* in 1983 as "delicious true to" the gameplay of Universal's arcade game *Space Panic*.[1] As its introduction succinctly puts it, "The object is to dig holes and pound the apples through the holes." Using the keyboard, the player controls a character that walks left and right along platforms made of green brick, and climbs up and down ladders between them. The player can use a shovel to dig holes through the platforms, into which enemies will fall and become trapped. Once an enemy is stuck in a hole, the player must strike it repeatedly with the shovel until it falls through and hits the level below. This must be done quickly, because after about 17 seconds an enemy will be able to free itself, filling in the hole in the process. The player can also refill holes they've dug, or drop through them. The game is an example of the "trap-em-up" genre, which also includes games like *Heiankyo Alien*, *Lode Runner*, and *Boomer's Adventure in ASMIK World*.

There are three types of enemy in the game, the first and most numerous being the "apples". An apple will die if it falls a single level. As the player advances, green and blue enemies will start to appear, which must be dropped through at least two or three levels, respectively. This is accomplished by digging a series of holes, one directly below another, and trapping the enemy in the uppermost hole. The player earns extra points if they drop one monster on top of another (killing them both).

On each level the player has only a limited time to dispatch all the enemies, tracked by a bar at the bottom of the screen. There are four distinct configurations of platforms and ladders through which the game cycles, but in every one there will always be five platforms in which the player can dig.

4.5.2 Other versions

Apple Panic was rewritten in the Seed7 programming language in 2004-5.[2]

A version for the TRS-80 family exists that was written in 1982 by Yves Lempereur of Funsoft, Inc.

Versions for the Atari 800 and IBM PC were written by Olaf Lubeck.

4.5.3 Reception

Unlike *Space Panic*,[1] *Apple Panic* was very successful. *Softline* reported in 1983 that it was among the top 30 best-selling Apple software for almost two years, in contrast to the "two to four month life span" of the typical arcade game.[3]

BYTE in 1982 called *Apple Panic* "one of the most creative and novel games to be invented for a microcomputer".[4] *PC Magazine* in 1983 stated "Yes, *Apple Panic* is a pretty dumb game. It's also fun to play and pretty to watch ... a welcome change from the endless stream of shoot-em-ups in space".[5]

4.5.4 References

[1] Pearl, Rick (June 1983). "Closet Classics". *Electronic Games*. p. 82. Retrieved 6 January 2015.

[2] *Seed7 Panic, with source code*

[3] Tommervik, Margot Comstock (March 1983). "By Golly, That's a Good Game! / Masters of the Mousetrap Maxim Tell Why". *Softline*. pp. 30–32. Retrieved 28 July 2014.

[4] "The Coinless Arcade". *BYTE*. December 1981 pp. 38–41. Retrieved 19 October 2013.

[5] Sandler, Corey (March 1983). "At Ease With PC". *PC Magazine*. p. 213. Retrieved 21 October 2013.

4.5.5 External links

- *Apple Panic* at MobyGames

4.6 Arcadia (video game)

Arcadia was a two-dimensional shoot 'em up released in time for Christmas 1982 on the Sinclair Spectrum, Vic 20, and later on the Commodore 64.

ZX Spectrum screenshot of level 1

4.6.1   Description

Published by Imagine Software, *Arcadia* was a *Space Invaders* style fixed shooter that also took elements of *Gorf* and *Galaxian* to create a simple - yet for its time - fast action game. The player controlled a space ship as the aliens scrolled and moved freely down the screen. The game consisted of 12 different levels of descending aliens. After level 12 the game looped back to level 1 with no extra difficulty. An extra life was rewarded after every four levels. Advancing to the next level involved staying alive until the timer in the top left corner ticked from 99 to 0, once zero was reached the surviving aliens descended rapidly down the screen.

Points awarded per alien destroyed were in line with the current level: Shoot down an alien on level 1 and you were awarded 1 point, roll around the levels and the same alien killed on level 13 was now worth 13 points.

4.6.2   Reviews

Arcadia was well received by the review magazine of the time - ZX Computing said it was *"highly addictive and well presented"* ,[2] Computer & Video Games said *"it lives up to the advertisement blurb and gives you a good addictive game of Space Attack"* - rating it 8 out of 10.[3] Popular Computing Weekly were particularly impressed with the graphics, stating that they *"have no equal in the Spectrum field,"* and *"lift this game into a class of its own"* , rating it 86%.[4][5]

4.6.3   Bugs

The ZX version was written to be compatible with the Fuller Sound Box - which included a joystick port. If the box

was not connected the nonexistent port was read incorrectly making the space ship occasionally move and fire of its own free will. This was potentially hazardous on some screens - such as level 4 *"The Pins"* - where it was tactically sound to leave a single alien falling rather than shoot it and have an entire squadron descend on the player again.

4.6.4   References

[1] Spectrum & Vic20 bestseller now on Commodore 64

[2] ZX Computing review of Arcadia

[3] Computer & Video Games review of Arcadia

[4] Popular Computing review of Arcadia page 1

[5] Popular Computing review of Arcadia page 2

4.6.5   External links

- *Arcadia* at World of Spectrum

4.7   Artillery Duel

Artillery Duel is a strategy game[2][3] and artillery clone for home console and computer systems developed by Xonox. *Artillery Duel* was featured in a few double-ender configurations as well as in a single cartridge.

4.7.1   Gameplay

Artillery Duel takes gameplay common to many games of the time and adapts it to the limitations of the Atari 2600. The game consists of dueling cannons on either side of a hill or mountain of varying height and shape. Each player has control of the incline and force behind the shell launched, the objective being to score a direct hit on the opposing target. Where many versions gave the player a few tries on the same course, *Artillery Duel* switches to a new mountain after each turn. When the player does manage to hit the opposing cannon, the reward is a brief animation of comically marching soldiers at the bottom of the screen.

4.7.2   See also

- Chuck Norris Superkicks

- Ghost Manor

- Xonox

4.7.3 References

[1] "Release date information". GameFAQs. Retrieved 2010-07-06.

[2] "MobyGames Description of Artillery Duel". MobyGames. Retrieved 2007-12-30.

[3] "http://www.consoleclassix.com/colecovision/artillery_duel.html". Console Classix. Archived from the original on 1 January 2008. Retrieved 2007-12-31.

4.7.4 External links

- *Artillery Duel* at MobyGames

4.8 Atlantis (video game)

This article is about the 1982 shoot 'em up game. For the 1997 adventure game, see Atlantis: The Lost Tales.

Atlantis is a fixed shooter video game produced by Imagic in July 1982, for the Atari 2600 video game console. The game was subsequently ported to the Atari 8-bit family of home computers, the Commodore VIC-20 home computer, the Intellivision, and the Magnavox Odyssey^2.[2]

4.8.1 Gameplay

Atlantis is a variation on the artillery and shooting game genre popular in the early 1980s. The player controls the last defenses of the City of Atlantis against the Gorgon invaders. The city has seven bases, which are vulnerable to attack. Three of these have firepower capabilities to destroy the Gorgon ships before they manage to drop bombs on one of the settlements. The gun bases have fixed cannons; the center base fires straight up, while the far left and far right bases fire diagonally upwards across the screen. The enemy ships pass back and forth from left to right four times before they enter bombing range, giving an ample opportunity to blow them away. Lost bases can be regained by destroying enough Gorgon ships. However, regardless of the player's efforts to avert the tragedy, Atlantis is doomed. The only way the game can end is when all bases are destroyed. However, a tiny ship then rises from the rubble and speeds away, foreshadowing the events of the sequel *Cosmic Ark* [3]

The Intellivision version features two gun turrets with a movable cursor that can be aimed onto enemy ships. There is also a deploy-able ship to take on enemies one-on-one. The game features day, dusk, and night settings, with the night setting limiting visibility to two moving searchlights.

Imagic created Destination Atlantis, a high score competition in which players were invited to send in pictures of their high score screens. Those with the highest scores were rewarded with a copy of Atlantis II.

It should be noted that Atlantis II was a competition cart for the top ten Defend Atlantis contestants. There were originally only supposed to be four people in this competition, but because Imagic needed the top ten players who scored above five million points to be only four players, they came up with this tie breaker cartridge to weed out the weaker players. Only ten of these cartridges were specially made with the same rom set, costing Imagic only a few thousand dollars to produce by making slight changes to the original mask, as opposed to making a brand new Rom from scratch. Those who turned the game score over to 1,000,000 points five times or more would receive the competition cartridge. The score would reflect this with the last digit turning over each time one million points was scored, whereas the ones place on the screen would advance by one place. So a person with a score of 273,007 would have turned the game over seven times for a total score of 7,273,007; therefore, this player would earn a place among the top ten finalists. This method was how Imagic knew who would be in the top ten finalists when the contestants submitted the photos to Imagic via postal mail. The top four players were never flown to Bermuda to compete in a contest to win a chest of gold worth $10,000, due to Imagic's acquisition by Activision in early 1983. Imagic continued doing business under their own name until 1986 when Activision absorbed their assets.

Atlantis II's value has been rated at $6,000.00 by a few websites, but has sold for as much as $18,000.00 and may well be worth more because there are only ten of them in existence. Until recently it was never officially known how many of these competition carts were produced.

4.8.2 Reception

Atlantis was well received. *Video Games* favorably reviewed the Intellivision version of *Atlantis*, calling it "a great shoot-'em-up for" the console,[4] and the Atari 2600 version received a Certificate of Merit in the "Video Game of the Year" category at the 4th annual Arkie Awards.[5]:30

4.8.3 Similar games

The most obvious comparison to this game is with *Colony 7* which was released by Taito in a way a combination of *Space Invaders* and *Missile Command*, which also features the defense of a number of bases, only some armed, against an irresistible invading force. The principal differences lie in the controls. While *Missile Command* features a complex targeting and positioning system that used a trackball in the arcades, *Atlantis* returns to an earlier era where the only control is in the timing of the gun fire that must be lined up with the movement of the enemy ships in order to successfully destroy the invaders. Still, like many of the early titles produced by Imagic, the game was popular and considered a hit for the company.

4.8.4 References

[1] Weiss, Brett Alan.   "Atlantis - Overview - allgame" . *allgame*. Retrieved 2014-03-02.

[2] "Atlantis". Moby Games. Retrieved on March 8th, 2009.

[3] "Game Trivia for Atlantis". Moby Games. Retrieved on March 8th, 2009.

[4] Wiswell, Phil (March 1983).   "New Games From Well-Known Names" . *Video Games*. p. 69. Retrieved 26 May 2014.

[5] Kunkel, Bill; Katz, Arnie (February 1983).   "Arcade Alley: The Fourth Annual Arcade Awards" . *Video* (Reese Communications) **6** (11): 30, 108. ISSN 0147-8907.

4.8.5 External links

- *Atlantis* at AtariAge

- LiveVideo.com: Making of Atlantis Video Game

- An unofficial 2009 Commodore 64 port

- An unofficial iOS Port

4.9 Avenger (1981 video game)

For the Capcom arcade game, see Avengers (arcade game).

VIC Avenger is a Space Invaders clone published by Commodore International for the VIC-20. A year later it was released for the Commodore 64.

Cartridge Number VIC-1901.

4.9.1 External links

- *Avenger* at MobyGames

4.10 Battlezone (1980 video game)

For articles with similar titles, see Battle zone (disambiguation).

Battlezone is an arcade game from Atari released in November 1980.[*][1] It displays a wireframe view (using vector graphics rather than raster graphics) on a horizontal black and white (with green and red sectioned color overlay) vector monitor. Due to its novel gameplay and look, this game was very popular for many years.

4.10.1 Development

The vector technique is similar to the visuals of games such as *Asteroids*. The game was designed by Ed Rotberg, who designed many games for Atari Inc., Atari Games, and Sente.

A version called *The Bradley Trainer* (also known as *Army Battlezone* or *Military Battlezone*) was also designed for use by the U.S. Army as targeting training for gunners on the Bradley Fighting Vehicle.[*][2] Approaching Atari in December 1980, some developers within Atari refused to work on the project because of its association with the Army,[*][3] most notably original Battlezone programmer Ed Rotberg.[*][4] Rotberg only came on board after he was promised by management that he would never be asked to do anything with the military in the future.[*][5] Only two were produced; one was delivered to the Army and is presumed lost, and the other is in the private collection of Scott Evans,[*][6][*][7] who found it by a dumpster in the rear parking lot at Midway Games. The gunner yoke was based on the Bradley Fighting Vehicle control and was later re-used in the popular *Star Wars* game.[*][5] The Bradley Trainer differs dramatically from the original Battlezone as it features helicopters, missiles, and machine guns; furthermore, the actual tank does not move—the guns simply rotate.

Because of its use of first-person pseudo 3D graphics combined with a "viewing goggle" that the player puts his face into, *Battlezone* is widely considered the first virtual reality arcade game.[*][8] Likewise, The Bradley Trainer is considered the first VR training device used by the U.S. Army.

4.10.2 Gameplay

Gameplay is on a plane with a mountainous horizon featuring an erupting volcano, distant crescent moon, and various geometric solids (in vector outline) like pyramids and blocks. The player views the screen, which includes an overhead radar view to find and destroy the rather slow tanks, or the faster moving supertanks. Saucer-shaped

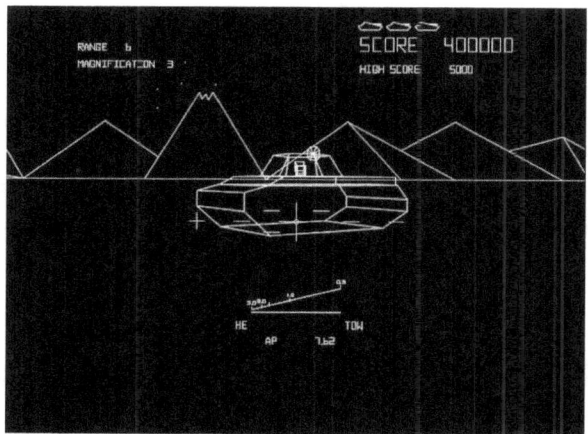

A standard enemy tank in the player's sights in the military training version The Bradley Trainer.

The cabinet of this arcade version of Battlezone *has a "periscope" and two joysticks, for controlling the movement of the player's tank.*

UFOs and guided missiles occasionally appear for a bonus opportunity. The saucers differ from the tanks in that they do not fire upon the player, and do not appear on radar. The player can hide behind the solids or maneuver in rapid turns once fired on to buy time with which to fire himself. Common play in the US could run from 25 cents to a dollar per game, depending on machine setting. The typical setting is for 25 cent play, with three tanks.

No additional tanks are awarded until the score counter rolls over at ten million, and additional bonus tanks are again awarded at indicated scores of 15,000 and 100,000. The game only includes one hostile enemy on the game board at all times; the player never has to battle two enemy tanks at once, or a tank and guided missile. The UFO can appear on the screen at the same time as an enemy tank, and it can occasionally be destroyed by enemy fire.

The geometric solid obstacles are indestructible, and can block the movement of a player's tank. However, they are also useful as shields as they block enemy fire as well.

The music heard in the high score initials prompt is from Tchaikovsky's *1812 Overture.*[9]

4.10.3 Cabinet

Battlezone was housed in a standard upright arcade cabinet with a novel "periscope" viewfinder which the player used to view the game. The game action could also be viewed from the sides of the viewfinder for spectators to watch. A later, less common version of the cabinet removed the periscope to improve visibility to non-players and improve the ergonomics for players who could not reach the periscope. This modification also was welcomed by some operators, who felt that the small windows present in the "periscoped" version did not attract enough attention to the

game when played.

A smaller version of the cabinet (known as a "cabaret cabinet") also existed with the screen angled upwards, and no periscope. A cocktail table version was tested as a prototype but not produced; it lacked the color overlays as the display would have to flip for opposing players.

The controls consisted of left and right joysticks, which could only be moved in the Y (vertical) axis, each controlling the treads on that side of the player's tank. One joystick contained a button used to fire projectiles at enemy targets.

4.10.4 Ports and clones

Ports

Throughout the 1980s, *Battlezone* was ported to several home computer systems (usually on the Atarisoft label), including DOS, the Apple II, the Commodore 64, the Sinclair ZX Spectrum, and the Atari XEGS. The Atari ST port contains large parts of the original 6502 code which is emulated in real time.*[10]

An Atari 2600 port was also released, but has colored raster graphics due to limitations and the view is behind the tank rather than inside it.

A Game Boy port was made which included a port of *Breakout*.

An Atari 5200 port was in the works, scheduled for release in November 1983, but was cancelled.*[11]

The Atari Lynx had the deluxe port *Battlezone 2000* (within that version is a hidden game with scaled sprites instead of vector graphics).

Battlezone was included in *Microsoft Arcade*.

An upright cabinet

On April 16, 2008 an updated port of Battlezone was released on Xbox Live Arcade. The game was developed by Stainless Games and published by Atari Inc.. It features 1080i graphics, Dolby 5.1 audio and an online mode to play against 2 - 4 friends in Deathmatch and Capture the Flag modes, and incorporates Xbox Live Vision support.[*][12] This version received an ESRB: E (Everyone) rating.

Main article: Battlezone (2008 video game)

Battlezone was also re-released to Microsoft's *Game Room* download service for the Xbox 360 and Windows-based PCs in May 2010.

In July, 2013, Rebellion bought the Battlezone franchise from the Atari bankruptcy proceedings.[*][13]

Clones

A *Battlezone*-inspired game named *Stellar 7* was released on several home computer platforms in the 1980s. Its sequel, *Nova 9*, was later released on the Amiga and DOS by Sierra Online. *Stellar 7* has a number of features which *were* to be in the never-released *Battlezone II* or *Battlezone Deluxe* by Atari, including a variety of enemies and multiple enemies on the field at once.

In the mid-1980s, Electronic Arts released a similar game named *Arcticfox* for several platforms, with multiple enemies on the field at once like *Stellar 7*, but with a varied landscape of mountains and valleys and crevasses to traverse, and other features not found in *Battlezone*.

A game by the name of *Robot Tank* was released by Activision in 1983 for the Atari 2600, and was very similar to the Atari 2600 version of Battlezone.

A *Battlezone* clone named *3D Tank Zone* was released on the Acorn Electron and BBC Micro in 1983 by Dynabyte.

A *Battlezone* clone named *3D Tank Duel* was released on the Sinclair Spectrum in 1984 by Realtime Games Software, shortly before the Atari-authorized version was released by Quicksilva.

A clone of *Battlezone* for DOS called "bzone.exe" circulated through the BBS community in the 1980s.

The TRS-80 Color Computer clone is called *Rommel 3D* and was released in 1985.

A *Battlezone* clone for Apollo Domain/OS called *bzone* was written by Justin S. Revenaugh in 1986 and re-written for the X Window System by Todd Mummert in 1990. The X Window System version, *cbzone,* differed from the original arcade version in that the player could be attacked by more than one enemy tank at the same time. This version of the game was also ported to the Macintosh in the 1990s and was included in the UMich software archive.[*][14]

Another clone from Design Design software called Tank Busters[*][15] was released in the mid-80s for the Amstrad CPC.

The 1991 Macintosh tank game *Spectre* and its sequels such as *Spectre VR* owed much to *Battlezone* for their gameplay and appearance.

The SGI workstations had a *Battlezone* derived game in the early 1990s called *BZ* which added network play.[*][16] *BZ* also had guided missiles, where the player would fly the missile after launch, returning to the tank on impact.

Activision, the video game publishing giant, released a game for Microsoft Windows inspired by and named *Battlezone* in 1998. Aside from the name, however, the game bears little resemblance to the original arcade game.

A "re-imagining" of *Battlezone* was developed by Paradigm Entertainment and released for the PlayStation Portable.[*][17]

4.10.5   Reception

Battlezone was well received, earning an Honorable Mention for "Best Commercial Arcade Game" in 1982 at the Third Annual Arkie Awards.[*][18][*]:76

4.10.6 See also

- *Battlezone*, a 3D remake from 1998 which changed the game from an arcade game to a more complicated tank piloting strategy game.

- *Battlezone II: Combat Commander*, another sequel to *Battlezone* released by Pandemic Studios in 1999.

4.10.7 References

[1] "Production Numbers" (PDF). Atari. 1999. Retrieved 19 March 2012.

[2] "www.safestuff.com/bradley.htm". Retrieved 2007-09-17.

[3] Jung, Robert. "The Army Battlezone Q & A". Archived from the original on 31 October 2007. Retrieved 2007-09-17.

[4] Hague, James. "Halcyon Days: Ed Rotberg". Archived from the original on 27 September 2007. Retrieved 2007-09-17.

[5] Kent, Steven L. (2001). *The Ultimate History of Video Games*. Prima Publishing. pp. 153–155. ISBN 0-7515-3643-4.

[6] Evans, Scott. "Bradley Trainer". Retrieved 2007-09-17.

[7] "MAWS Bradley Trainer ROM set info". Archived from the original on 16 October 2007. Retrieved 2007-10-09.

[8] Dan Harries (2002). *The New Media Book*. British Film Institute.

[9] International Arcade Museum

[10] http://www.klapauzius.net/Old_Games.html#Battlezone

[11] Reichert, Matt. "Battlezone". *AtariProtos.com*. Retrieved 2007-07-05.

[12] "Xbox – Battlezone Game Detail Page". Archived from the original on April 14, 2008.

[13] http://www.develop-online.net/news/44381/Wargaming-and-Rebellion-claim-Atari-IPs

[14] "/mac/game/war/00index.txt".

[15] "home-computing-gaming-heroes-design.html"

[16] BZ(6D)

[17] Dobson, Jason (May 4, 2006). "Pre-E3: Battlezone Re-imagined, Charlotte's Web, Codemasters Finds Bliss". Gamasutra. Retrieved 2010-01-29.

[18] Kunkel, Bill; Laney, Jr., Frank (January 1982). "Arcade Alley: The Third Annual Arcade Awards". *Video* (Reese Communications) **5** (10): 28, 76–77. ISSN 0147-8907.

4.10.8 External links

- *Battlezone* at the Killer List of Videogames

- *Battlezone* at the Arcade History database

- *Battlezone* guide at StrategyWiki

- *Battlezone* at World of Spectrum

- Arcade Games–this article on arcade games names *Battlezone* as "the first truly interactive 3-D environment"

- Battlezone Series at DMOZ

- *Battlezone* at Coinop.org

4.11 Blitz (video game)

Blitz is an arcade-style game for the VIC-20 personal computer.

The first game of this genre was written for the Commodore PET by Peter Calver and published under the name *Air Attack* by his company, Supersoft, in 1979. At that time it was common for computer magazines to publish games listings which readers could type into their own computer, and *Air Attack* was also published as a listing in the December 1979 issue of Personal Computer World.

For a short period a 4-coloured transparent overlay was available which simulated colour on the black-and-white screen of the PET computer (a similar technique had previously been used in some arcade machines, notably Breakout). However, when Commodore switched from black-and-white to green screens the effect was no longer as convincing, and the overlay was discontinued.

Although the game was prompted by a verbal description of the arcade game *Canyon Bomber* (Atari, 1977), it was not until many years later that Peter Calver saw the original game. The inspired change from a canyon filled with rock pillars to a city of skyscrapers was copied by all later clones including *Blitz*, *City Bomber* and *City Lander* (for ZX Spectrum, C64 and ZX-81). 'Blitz' was published by Commodore themselves for the Vic-20, then Blitz-64 for the Commodore 64 and Blitz-16 for the Commodore-16. Blitz was taken by Commodore from 'Vic New York' written by Simon Taylor/TaySoft. Taylor later produced versions for the CBM-64 and CBM-16.

4.11.1 Game description

A plane moves across the screen at a steady speed. When the plane reaches the end of the screen it moves to the other side and drops down one line, with the speed increasing each time the plane drops a line. Below is a cityscape composed of blocks. The player has to drop bombs from a plane, and each bomb which hits a building removes one or more blocks. As the plane descends it risks hitting any remaining blocks so priority has to be given to bombing the tallest "buildings". The level is completed when all blocks are removed and the plane has descended safely to the bottom of the screen.

4.11.2 Newer Versions

A more modern version of this game has been developed called [SuperBlitz] for Apple IOS devices. This was developed by [HungryOrange]. This version features various variations as well as the ability to play the original classic game.

4.12 Cannonball Blitz

Cannonball Blitz is a game by Olaf Lubeck and released in 1982 by Sierra On-Line (then known as "On-Line Systems") for Apple II, VIC-20, and TI-99/4A computers. The game is a *Donkey Kong* clone, although cannonballs and cannons replace barrels and a soldier replaces the large ape. On the first level, the player character catches a flag instead of rescuing a girl.

There were three different levels, the third of which is particularly challenging. After completing the third level, the player views a small celebration scene and then restarts at the first level. Repeated levels only differ from those of the first round in the harsher timing patterns of the game.

Cannonball Blitz achieved some notoriety in the Apple hacking community as being rather difficult to crack. Track 17, sector D of the game contained the message "YOU'LL NEVER CRACK IT".*[1]

4.12.1 Reception

Ahoy! wrote that "*Cannonball Blitz* [for the VIC-20], make no mistake about it, is *Donkey Kong* in dress blues. Not a bloody thing new here. However, you're going to find it a barrel of fun". The magazine favorably reviewed the animation and the "*unbelievable*" sound effects, and concluded that it was "a very good version of a fine game".*[2]

4.12.2 References

[1] Kracowicz' Kracking Korner: The Basics of Kracking, Part II

[2] Meade, E. C. (January 1984). "Cannonball Blitz". *Ahoy!*. pp. 56–57. Retrieved 27 June 2014.

4.13 Chariot Race

For chariot racing, see chariot race.

Chariot Race is a racing game released in 1983 by Micro Antics. A two dimensional game which involved the player racing to the end of the track while trying to take out his opponent.

4.13.1 Description

Chariot Race ran on the unexpanded Commodore VIC-20 which had 5k of memory. This game also featured the ability to play with another player on the same computer (which was a quite unusual for a 2 player game for the time and machine). The object of the game was for the player to race his Chariot along the track avoiding side walls and other oncoming chariots. Competing chariots could take him out of the race by either pushing the player into another chariot or making him crash into the arena walls.

Unlike other games where the player just levels up Chariot Race did have a finish line which allowed one of the two players to be the winner.

4.13.2 External links

- Personal Computer News Review of *Chariot Race*

4.14 Choplifter

Choplifter (stylized as ***Choplifter!***) is a 1982 Apple II game developed by Dan Gorlin and published by Brøderbund. It was ported to other home computers and, in 1985, Sega released a coin-operated arcade game remake, which in turn received several home ports of its own. While many arcade games have been ported to home computers and consumer consoles, *Choplifter* was one of the few games to take the reverse route: first appearing on a home system and being ported to the arcade.

4.14.1 Overview

In *Choplifter*, the player assumes the role of a combat helicopter pilot. The player attempts to save hostages being held in prisoner of war camps in territory ruled by the evil

The title screen of the Apple II game Choplifter

Bungeling Empire. The player must collect the hostages and transport them safely to the nearby friendly base, all the while fighting off hostile tanks and other enemy combatants. According to the backstory, the helicopter parts were smuggled into the country described as "mail sorting equipment."

Although the Iran hostage crisis ended the year before the game was released, Gorlin has stated "the tie-in with current events was something that never really crossed my mind until we published." *[1]

4.14.2   Development

Choplifter was developed in six months. After Gorlin began experimenting with animating a helicopter on the Apple II, he added scenery, tanks, and planes, with the hostages last. He stated that as "A story developed ... movie camera techniques seemed appropriate", including the final message be "The End" instead of "Game Over". Gorlin's first demonstration to Broderbund was "too realistic. too much a helicopter simulation", and the company helped him make it easier to fly.*[2] The original Choplifter art (not shown) for the Broderbund Commodore 64 release was produced by Marc Ericksen, who created the art for Broderbund's original first five covers.

4.14.3   Description

The helicopter (named "Hawk-Z" in the Master System version manual) can face three directions: left, right, or forward (facing the player). It may shoot at enemies in any of these directions and need not fly in the same direction it is facing. The forward-facing mode is used primarily to shoot tanks. Care must also be taken to both protect the hostages from enemy fire and not accidentally shoot them oneself.

The player rescues the prisoners by first shooting one of the

hostage buildings to release them, landing to allow the prisoners to board the sortie, and returning them to the player's starting point. Each building holds 16 hostages, and 16 passengers can be carried at a time, so several trips must be made. When the chopper is full, no more hostages will attempt to board; they will wave the helicopter off and wait (hopefully) for its return. Usually, each trip back is more risky than the previous one since the enemy is alerted and has deployed a counter-attack.

If the player lands directly on top of a hostage, or completely blocks the building exit, the hostage(s) will be killed. In the Apple II and Atari 7800 versions, hostages will also die if the vehicle is not landed correctly (it is slightly tilted), being crushed as they attempt to board the chopper. While grounded, the helicopter may be attacked by enemy tanks, which it can shoot at only by returning to the air. Also, the enemy scrambles jet fighters which can attack the vehicle in the air with air-to-air missiles or on the ground with bombs.

4.14.4   Platforms

Tanks and a jet target the helicopter while hostages flee a burning building in the original Apple II game.

Choplifter was ported to many other home systems of the era. These versions were ports of the original Apple II game, not the later arcade version. These systems include the Atari 5200, Atari 7800, Atari 8-bit family (and a graphically updated version for the Atari XEGS), ColecoVision, Commodore 64, Commodore VIC-20 and MSX. German publisher Ariolasoft published the European Commodore 64 version.

In 1985, Sega, looking for properties marketable in the west, produced remakes of *Choplifter* and *Pitfall II* on their System 8 hardware. Sega's version added scoring elements, music, a fuel gauge, and several new environments including a naval battle, a cave, and futuristic city. This version was also notable at the time for its heavy use of parallax scrolling.*[3]

In 1986, ports of the arcade version back to home versions were developed for the Nintendo Entertainment System and Sega Master System. These versions include some gameplay and scoring changes of their own, but use the environments, music, and approximate scoreplay of the arcade remake. The arcade version is listed in the Killer List of Videogames Top 100 and one of the four best games in 1985.

On 17 June 2009, inXile Entertainment announced it was working on a remake of the classic game for Xbox Live Arcade, PlayStation Network, and Windows.[4] This game is entitled *Choplifter HD* and was released on January 11, 2012.

4.14.5 Version differences

In the original Apple II game, play continues until all three helicopters are destroyed or all prisoners are either rescued or killed. There is no scoring system other than the counters at the top of the screen, which indicate how many of the 64 total hostages have been killed (red), how many are on board the helicopter (blue), and how many have been rescued (green). The best possible result is to rescue all the hostages, for which the game will award you a triple crown, Brøderbund's emblem. The Commodore 64 version is the same.

In the arcade version, a point system is used, giving points for enemies killed and hostages rescued. Furthermore, the arcade version has only eight hostages per building rather than 16. In order to move from one level to the next, the player must rescue at least 20 hostages (40 in the Sega Master System version). The arcade version also forces the player to restart a level if too many hostages are killed, but does not restore any helicopters lost. (In the Sega Master System version, this automatically ends the current game in progress.) Another difference in the arcade version is the addition of a fuel meter. This was essentially a time limit because there was only one way to replenish the meter— saving hostages.

The original game provides a safe zone around the player's launch area where the player was largely free from attack. A fence indicates the border between friendly and enemy territory. While the fence is still present in the arcade version, enemy jets will pursue the player's helicopter all the way to his landing pad.

In the original game, a new enemy is added with each trip the player makes. First, the player faces only tanks which are limited to attacking only when the helicopter has landed or is extremely close to the ground. The next trip introduces jet fighters that shoot missiles at the helicopter in the air and bomb it when it's on the ground. The last enemies are "air mines" which attempt to collide with the player's helicopter, and which on the fourth trip gain the ability to shoot. The arcade game has a larger variety of enemies which vary more according to each level's landscape rather than the number of trips the player has made. The most significant of these are anti-aircraft guns which make the arcade version much harder than the original. It retains the tanks and jet fighters, but does not include air mines which follow the player's helicopter. The arcade version also gives the ability, once the helicopter is shot down and while it is falling in flames, to make the hostages jump with parachutes by repeatedly pressing the 'turn' button, these hostages can be rescued again in subsequent sorties and will not count as dead hostages.

The original Apple version (and perhaps other platforms) allows the player to use the helicopter to lead hostages back to base on foot, but the game does not count such rescues. Without all hostages accounted for as either dead or rescued, the game will never end short of the destruction of all of the player's helicopters.

4.14.6 Reception and legacy

Softline in 1982 called the game "what may well be the first Interactive Computer-Assisted Animated Movie. A fusion of arcade gaming, simulation, and filmic visual aesthetics, *Choplifter* is destined to occupy a place in the software Hall of Fame". The magazine praised the animation and the helicopter's "subtle flight control", and concluded that seeing the hostages' "hope and excitement, their faith in you" made the game "hard to play. It hurts to see one of those lively people killed".[2] In 1983 its readers named *Choplifter* fourth on the magazine's Top Thirty list of Atari 8-bit programs by popularity.[5] *BYTE* called *Choplifter* "great fun",[6] and *Computer Gaming World* highly praised the graphics and animation.[7] The Apple II version of the game received a Certificate of Merit in the category of "Best Computer Audiovisual Effects" at the 4th annual Arkie Awards,[8]:33 and shortly afterward *Billboard* named it Computer Game of the Year.[9] *The Addison-Wesley Book of Atari Software 1984* gave the game an overall A+ rating, calling it "a masterpiece". The book concluded that "the concept, graphics, and animation make this a delightful game".[10]

II Computing listed *Choplifter* seventh on the magazine's list of top Apple II games as of late 1985, based on sales and market-share data.[11] It had two sequels:

- **Choplifter II** for the Game Boy (1991), and remade on Game Boy and Game Gear as Choplifter III in 1994.

- **Choplifter III** for Super NES, a different game than

the handheld games of the same name.

Sega also released a pair of spiritual successors without the *Choplifter* brand:

- **Air Rescue** (1991) for System 32 hardware was a first-person, pseudo-3D take on the concept.

- **Air Rescue** (1992) for Sega Master System more closely resembled classic 2D Choplifter, but had stages that scrolled in all directions.

On the Commodore 64 and the MSX, games related to *Choplifter* were *Lode Runner* and *Raid on Bungeling Bay*, all three games featuring the fictional Bungeling Empire.

The world record holder, as listed by the *1987 Guinness Book of World Records*, is Charles Collins of Madison, WI., with a score of 1,781,000.[*][12]

4.14.7 In popular culture

- *Choplifter* is played by U.S. submarine crew members in Tom Clancy's book *The Hunt for Red October*, and the Sonar man aboard the USS Dallas, Jones, has the high score.

- In *Mario's Picross*, "Easy Picross Level 6A" is a helicopter similar in appearance to the one in *Choplifter*. When the puzzle is solved, the caption describes the image as "Choplifter".

- Web comic *Up Up Down Down* created a comic based on *Choplifter* in 2011.[*][13]

4.14.8 See also

- *Choplifter HD*, a 2012 remake
- *Rescue Raiders*

4.14.9 References

[1] Interview with Dan Gorlin in *Halcyon Days*, by James Hague.

[2] Salmons, Jim (July 1982). "The Choppers of Mercy". *Softline*. p. 18. Retrieved 17 July 2014.

[3] http://retro.ign.com/articles/932/932659p1.html

[4] "Choplifter will fly again?". *Gamespot*. 17 June 2009. Retrieved 2011-03-29.

[5] "The Most Popular Atari Program Ever". *Softline*. March 1983. p. 44. Retrieved 28 July 2014.

[6] Clark, Pamela; Williams, Gregg (December 1982). "The Coinless Arcade - Rediscovered". *BYTE*. p. 84. Retrieved 19 October 2013.

[7] Greenlaw, Stanley (July–August 1982), "Choplifter! Rescue the Hostages", *Computer Gaming World*: 30, 38

[8] Kunkel, Bill; Katz, Arnie (March 1983). "Arcade Alley: The Best Computer Games". *Video* (Reese Communications) **6** (12): 32–33. ISSN 0147-8907.

[9] Kleiner, Karen (Jul–Aug 1983). "Billboard Conference". *Softline*. pp. 44–45. Retrieved 28 July 2014.

[10] Stanton, Jeffrey; Wells, Robert P. Ph.D.; Rochowansky, Sandra; Mellid, Michael Ph.D., ed. (1984). *The Addison-Wesley Book of Atari Software*. Addison-Wesley. p. 74. ISBN 0-201-16454-X.

[11] Ciraolo, Michael (Oct–Nov 1985). "Top Software / A List of Favorites". *II Computing*. p. 51. Retrieved 23 January 2015.

[12] Russel, Alan (1987), *Guinness Book of World Records*: 393, ISBN 0-553-26408-7 Missing or empty |title= (help)

[13] *Up Up Down Down* comic based on *Choplifter*

4.14.10 External links

- *Choplifter* at the Killer List of Videogames
- *Choplifter* at MobyGames
- *Choplifter* guide at StrategyWiki

4.15 Chuck Norris Superkicks

Chuck Norris Superkicks, is a 1983 video game produced by Xonox where the player takes control of Chuck Norris himself. It was also sold as "Kung Fu Superkicks" when the license for the use of the name 'Chuck Norris' expired.[*][1] The game was produced for the Commodore 64, Commodore VIC-20, Atari 2600, and Colecovision as part of Xonox's double-ender cartridge line (cartridges with two games and two connectors that were flipped over depending on which one the user wanted to play).

4.15.1 Gameplay

The player is a martial arts expert that has to liberate a hostage. The game combines two gameplay: moving through a map, and fighting against enemies.

4.15.2 Reception

The game received mostly negative reviews.

4.15.3 See also

- Artillery Duel

- Ghost Manor

- Xonox

- Chuck Norris: Bring on the Pain

4.15.4 References

[1] Chuck Norris Superkicks entry at allgame.com

4.15.5 External links

- Screenshots at uk.cheats.ign.com

4.16 Clowns (video game)

Clowns is a 1978 multiplayer game (consisting of 1 - 2 players) similar to *Circus Atari* in which the player controls a seesaw to propel two clowns into the air, catching balloons situated in three rows at the top of the screen. "Clowns" has no definite ending - instead players can compete against previously-set high scores in beaten levels. On multi-player mode, two players can go against each other, seeing who can get the highest score or complete levels together. The game was released in the arcade in 1978,[*][1] on cartridge for VIC-20 home computers in 1982, and then released a year later for the Commodore 64.

4.16.1 Gameplay

Players start with two clowns and they get to control where they go with a seesaw. The goal is to prevent them from falling to the ground, or else the level is lost. Players get three lives and can earn more after getting a certain amount of points. Getting to the center ring at the top of the sky, while gaining points from popping balloons, will help players complete levels. As the player advances to the next level, hazards and objects will start appearing in the air trying to hurt the clowns.

4.16.2 References

[1] http://www.ggdb.com/GameByManufacturer.aspx?c= Coin-Op&s=Arcade&m=Bally%20Midway&vid=558

4.17 Cops 'n' Robbers

This article is about the Atlantis computer game. For other uses, see Cops and Robbers (disambiguation).

Cops 'n' Robbers is a game for home computers published by Atlantis Software originally in 1985 for the Commodore VIC-20 and in virtually identical form on the Commodore 64. It was ported to the Commodore 16/Commodore Plus/4 (1986), Acorn Electron and BBC Micro (1987) and the Atari 8-bit family of computers (1988). The game was controversial when released as the player is the 'robber' and must shoot the 'cops'.

4.17.1 Gameplay

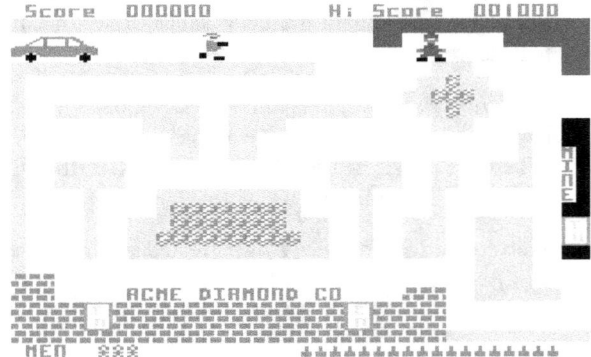

In game shot of the opening screen (Electron)

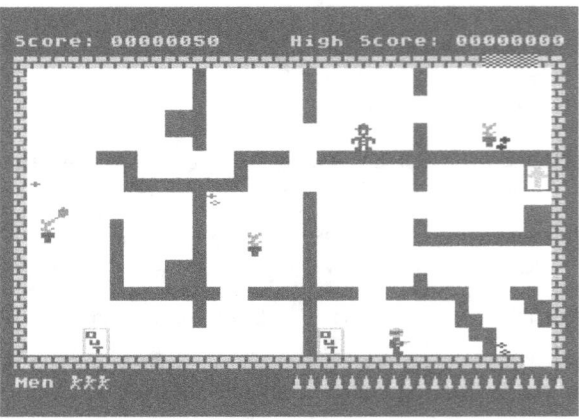

In game shot inside the Acme Diamond Company (Atari)

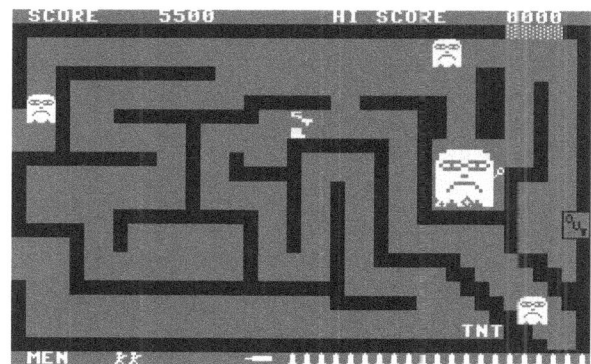

In game shot inside the haunted mine (C64)

The player takes the role of a diamond thief named Fingers Lonegan. The game starts at Lonegan's car in the top left hand corner of the opening screen. The player must make their way to the Acme Diamond Company building at the bottom of the screen by negotiating the garden maze. Police appear from the Police station (at the top right of the screen) and will home in on and arrest Lonegan. To avoid being arrested, the player must shoot the police. Bullets are limited but can be replenished by going back to the car. The player may also enter the mine or the police station (not all versions) from this opening screen.

There are many diamonds to collect which are on many levels of the Acme Diamond Company building as well as the mine. Police enter the building but not the mine where the only enemies are ghosts. The player may also enter the police station (on some versions) and free fellow robbers.

There are basic puzzles such as collecting keys, TNT, torches (some rooms are dark so the walls of the maze cannot be seen without the torch) and the code for the safe to advance in the game.

4.17.2 Critical reaction

The game was criticised for the fact that you shoot and kill many policemen. At the time, this was highly controversial as video games were very much seen as being for children. Rog Frost in Electron User wrote *"I find this game wholly inappropriate. It strikes me as abhorrent that success is measured by the ability to shoot policemen or steal diamonds. It should have been given a fantasy setting where the nasties which must be shot are not recognisable as creatures from the Earth"* .[1] This view was shared by Ray Sharp of Atari User as *"shooting policemen is not my idea of a good thing to teach children to do. Aliens from a distant planet OK but not your neighbourhood bobby"* .[2]

The game also gained almost universally negative reviews from critics at the time. Despite being a budget game, due

to its origins on the VIC-20, the graphics were very basic for most the other machines it was ported to. Also, the original C64 and VIC-20 releases were on the Atlantis Gold label and released at the higher budget price of £2.99 (later versions and most other Atlantis games at this time were released at £1.99).

Electron User was noted for its usually over enthusiastic reviews but even so only awarded an overall mark of 4/10, complaining *"The graphics aren't really up to par, even for software at this price. The sprites are simple and undergo a sort of jerky animation... The scenery that supports the action can probably best be described as plain or perhaps boring"* .[1] Atari User awarded only 2/10 overall (with a score of 0/10 for value for money despite only costing £1.99) concluding *"The graphics are pathetic and the sound effects dismal. The game is playable but not for long... it's a disaster"* .[2] Zzap!64 gave an even lower score of 9% claiming it to be *"The worst program we've seen on the 64"* .[3]

Despite the terrible reviews, the game sold well and was ported to many systems over a number of years. After getting past the limited graphics and sound, buyers seemed happier with the gameplay and it is fondly remembered among retro gamers. The user ratings on current websites are much higher with the current user rating on Commodore 64 website *Lemon 64* being a respectable 6.6/10.[4]

4.17.3 References

[1] "Dodgy Scenario" , Electron User, Vol. 5, No. 3, January 1988

[2] "Cops'n Robbers" , Atari User, Vol. 4, No. 3, July 1988

[3] "Cops 'N' Robbers" , Zzap!64, Issue 7, November 1985

[4] "Cops and Robbers" at Lemon 64, retrieved 28 March 2009

4.17.4 External links

- *Cops 'n' Robbers* at *Lemon 64*

4.18 The Count (video game)

The Count is a text-based adventure program written by Scott Adams and published by Adventure International.

4.18.1 Gameplay

The player character has been sent to defeat the vampire Count Dracula by the local Transylvanian villagers, and must obtain and use items from around the vampire's castle in order to defeat him.[*][1] Players move from location to location, picking up any objects found and using them somewhere else to solve puzzles. The interface is text-based; commands took the form of verb and noun, e.g. "Climb Tree". Movement from location to location was limited to North, South, East, West, Up and Down.

The game differs from earlier Scott Adams adventures due to the use of time. Set over three days, certain problems needed to be solved on particular days, and events would happen at particular times on certain days. The protagonist also had to avoid being attacked on the first two nights to finish the game.

4.18.2 References

[1]

4.18.3 External links

- The Adventure International Memorial page

4.19 Deadly Duck

Deadly Duck is an Atari 2600 game released on January 20, 1982[*][1] in North America. The game was programmed by Sirius Software and released by 20th Century Fox Games.[*][2]

4.19.1 Description

In the game Deadly Duck, there are cranky crabs that are attempting to get the ducks out of their ponds. The crabs have the ability to fly into the air while throwing bricks and bombs aimed at the ducks. To fight back against the crabs, the ducks are armed with a bill that is also a gun barrel that shoots bullets at the crabs.[*][3] The player has four lives and a bonus life is awarded when all eight crabs in a level have been shot. If the player is hit by a brick they lose a life. When bricks land at the bottom of the play area they impede player movement for a temporary period.[*][4] The game is designed for the Atari 2600 (VCS). It was later ported to the VIC-20 and Commodore 64 as a game cartridge.[*][5]

4.19.2 Information

The game Deadly Duck is played with a joystick. The genre of this game is recorded to be shoot 'em up. It was released in the year of 1983 in the United States of America. It is a single player game. The programmer of the game was Ed Hodapp.[*][6][*][7]

4.19.3 References

[1] "Deadly Duck for Atari 2600 from". 1UP. Retrieved 2013-10-05.

[2] "Atari 2600 - Deadly Duck (20th Century Fox)". AtariAge. Retrieved 2013-10-05.

[3] "Deadly Duck". GameFAQs. Retrieved 5 October 2013.

[4] https://atariage.com/manual_page.html?SystemID=2600&SoftwareLabelID=124&maxPages=6&currentPage=2

[5] "AGH Atari 2600 Review: DEADLY DUCK". Atari HQ. Retrieved 5 October 2013.

[6] "Deadly Duck". Atari Mania. Retrieved 5 October 2013.

[7] Halfhill, Tom R (July 1983). "Deadly Duck Cartridge Game for Unexpanded VIC-20". *Compute! Gazette*. pp. 63–64. Retrieved 4 December 2013.

4.20 Demon Attack

Demon Attack is a video game written by Rob Fulop and published by Imagic. It was originally for the Atari 2600, then ported to the Intellivision, Odyssey², Atari 8-bit, Commodore VIC-20, Commodore 64, PC (booter), TRS-80 and TRS-80 Color Computer. There was also a port to the TI-99/4A titled **Super Demon Attack**.

Demon Attack is supposedly based on the 1979 arcade shooter *Galaxian*, though it closely resembles several waves from the 1980 arcade game *Phoenix*.[*][2] The similarities prompted a lawsuit from Atari, who had purchased the latter's home video game rights.[*][3] Imagic settled out of court, and *Demon Attack* became Imagic's best-selling game as of 1983.[*][4]

4.20.1 Gameplay

Marooned on the ice planet Krybor, the player uses a laser cannon to destroy legions of demons that attack from above. Visually, the demons appear in waves similar to other space-themed shooters, but individually combine from the sides of the screen to the area above the player's cannon.

Each wave introduces new weapons with which the demons attack, such as long streaming lasers and laser clusters. Starting in Wave 5, demons also divide into two smaller, bird-like creatures that eventually attempt descent onto the

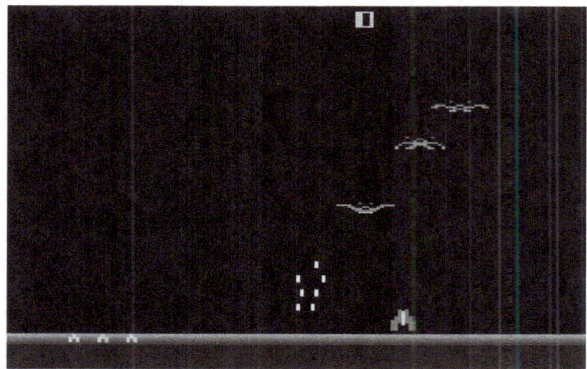

Demon Attack *player under attack.*

player's cannon. Starting in Wave 9, the demons' shots follow directly beneath the monsters, making it difficult for the player to slip underneath to get in a direct shot.

4.20.2 Development

The game was originally programmed to end after the 84th wave, as Fulop did not expect anyone to "wrap" the game. Two days after its initial release, a kid was able to beat the game. After this initial run of cartridges, Fulop went back and changed one line of code so that the game never ends, but never gets harder after the 84th wave.[2]

The Odyssey² version was the first third-party game for the console.[5]

4.20.3 Reception

Commenting on *Demon Attack* 's critical reception, AllGame described the game as "one of the most critically praised Atari 2600 games of all time."[6] The VCS version is considered to be a classic by many Atari fans.[7]

Video magazine reviewed the VCS version of *Demon Attack* in 1982, describing it as "quite simply excellent", and characterizing it as a "true coin-op-level program".[8] Covering the game again in its 1982 Guide to Electronic Games, *Video* editors called the cartridge "a state-of-the-art invasion game" and suggested that its "slick graphics" represented "a quantum leap for the VCS",[9]:52 however *Video* reserved higher praise for the Intellivision version of the game which was described as "even more thrilling graphically than the original VCS edition".[9]:53 *Video Games* praised the Intellivision version of the game, stating that "while the VCS version is a very good TV-game, this one is even better".[4] *Ahoy!* called the VIC-20 version "excellent ... it's a super-grabber type of twitch game, and good for a few long nights".[10] *Demon Attack*

won the 1983 Arcade Award for "Best Videogame of the Year",[5] with the judges commenting that the game had "turned out to be yardstick against which gamers measured the quality of each new cartridge during 1982".[11]:30

AllGame gave the game a four and a half star rating out of five, referring to it as "an excellent game in the ubiquitous "slide-and-shoot" genre" with graphics that were "beautifully drawn and animated".[6] The review also noted that "As with so many shooters of its type, the action can get repetitive, but no matter -- the game is fun, the enemies are incredibly varied, and the sound effects are solid."[6]

4.20.4 References

[1] *Demon Attack* at MobyGames

[2] Stilphen, Scott. "DP Interviews... Rob Fulop", *Digital Press.*

[3] "Player 3 Stage 1: Pixel Boxes". The Dot Eaters. Retrieved 2007-12-18.

[4] Wiswell, Phil (March 1983). "New Games From Well-Known Names". *Video Games.* p. 69. Retrieved 26 May 2014.

[5] Katz, Arnie; Kunkel, Bill (June 1983). "Programmable Arcade". *Electronic Games.* pp. 38–42. Retrieved 6 January 2015.

[6] Weiss, Brett Alan. "Demon Attack". AllGame. Archived from the original on November 14, 2014. Retrieved January 6, 2015.

[7] Barton, Matt and Bill Loguidice. "A History of Gaming Platforms: Atari 2600 Video Computer System/VCS". *Gamasutra.* 28 February 2008.

[8] Kunkel, Bill; Katz, Arnie (August 1982). "Arcade Alley: The Imagic Show". *Video* (Reese Communications) **6** (5): 14. ISSN 0147-8907.

[9] Kunkel, Bill; Katz, Arnie (November 1982). "Video's Guide to Electronic Games". *Video* (Reese Communications) **6** (8): 47–56, 108. ISSN 0147-8907.

[10] Salm, Walter (March 1984). "VIC Game Buyer's Guide". *Ahoy!.* p. 49. Retrieved 27 June 2014.

[11] Kunkel, Bill; Katz, Arnie (February 1983). "Arcade Alley: The Fourth Annual Arcade Awards". *Video* (Reese Communications) **6** (11): 30, 108. ISSN 0147-8907

4.20.5 External links

- *Demon Attack* at the Internet Archive Console Living Room

- *Demon Attack* (Atari 2600) at AtariAge

- *Demon Attack* at MobyGames

4.21 Dig Dug

Dig Dug (ディグダグ *Digu Dagu*) is an arcade game developed and published by Namco in Japan in 1982. It runs on Namco Galaga hardware, and was later published outside of Japan by Atari, Inc.. A popular game based on a simple concept, it was also released as a video game on many consoles.

4.21.1 Objective

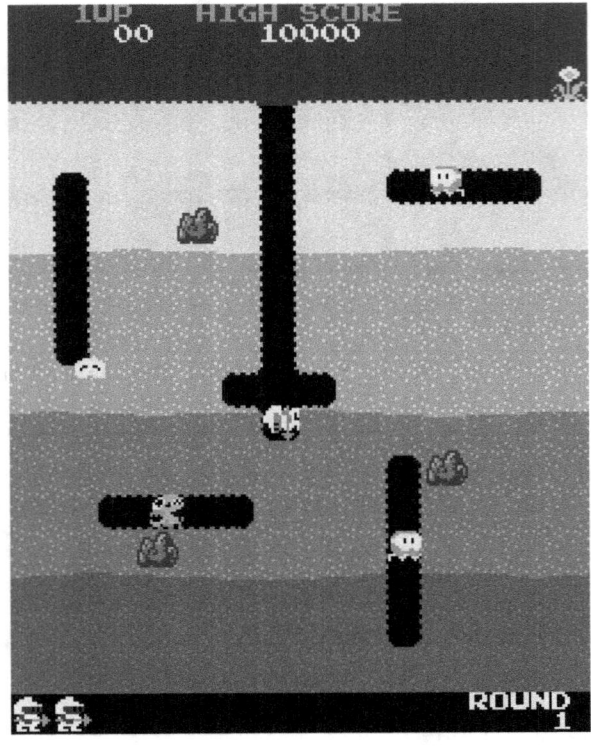

Screenshot of Round 1

The objective of *Dig Dug* is to eliminate underground-dwelling monsters by either inflating them with an air pump until they explode, or by dropping rocks on them. There are two kinds of enemies in the game: "Pookas" (a race of round red monsters, said to be modeled after tomatoes, that wear yellow goggles) and "Fygars" (a race of green dragons that can breathe fire while their wings flash).

The player's character is the eponymous Dig Dug, dressed in white and blue and able to dig tunnels through destructible environments. Dig Dug will be killed if he is caught by either a Pooka or a Fygar, burned by a Fygar's fire, or crushed by a rock he has loosened.

A partially inflated monster will gradually deflate and recover after a few seconds, during which time Dig Dug can

pass safely through it. The monsters normally crawl through the tunnels in the dirt but can turn into ghostly eyes and travel slowly through the dirt. The last enemy in a round will try to escape off the top left of the screen - and if he succeeds, the potential points are lost.

More points will be awarded for exploding an enemy further down in the dirt (the levels are color-coded). Additionally, Fygars are worth double points if exploded horizontally, since they can only breathe fire horizontally in the direction they are facing. Extra points are also awarded for dropping rocks on enemies in order to eliminate them rather than inflating them. If one enemy is killed by the rock, it is worth 1000 points. The next two add 1500 points each, and any after that, add 2000. The act of digging is itself worth points - giving 10 points for each block dug, so some players will do as much of it as possible while the threat from the remaining monsters is minimal.

After the player drops two rocks, a bonus item appears at the center of the screen, awarding points if the player can collect it before it disappears. These items consists of various fruits and vegetables, as well as the flagship from the Namco game Galaxian, and appear even if either of the dropped rocks fails to crush any enemies. In the original arcade version, the most points attainable from a single bonus item is 8000 from the pineapple, which appears in round 18 and every round thereafter.

If the player should drop a rock on an enemy at the same time it explodes, a glitch will occur whereupon all enemies will promptly disappear, but the game will not progress and the player will be free to dig through all dirt. Attaining the next level of play will then remain impossible, but the glitch can be resolved by forcing a rock to drop (unless, of course, there are no rocks remaining).

The round numbers are represented by flowers in the top right of the screen, and each new round is noted at the beginning of each round. After every fourth round, the color of the dirt will alternate (as seen in this article's screenshot graphic). In successive rounds more monsters appear on each screen, and they move quicker. A round is completed successfully when the last monster is dispatched or succeeds in fleeing. In the original Namco version, the game will end on round 256 (round zero), since the board is essentially an unplayable kill screen; at the start of the round, a Pooka will be placed directly on top of Dig Dug with no way to kill him. Therefore, the game will basically be over at this point, regardless of how many lives a player may have remaining - but the Atari version corrects this problem.

4.21.2 *Dig Dug Arrangement*

In 1996, Namco packaged both this game and an updated variant and re-released it in arcades with the title *Namco Classic Collection Vol. 2*. The updated variant was named *Dig Dug Arrangement*, and allowed two players to play simultaneously, unlike the original. Out of the six created *Arrangement* games, this version has the least amount of changes. The graphics are updated and the rounds are different. There are also new features such as giant rocks which can fall down to the bottom of the screen, and special power-up items.

Dig Dug Arrangement was re-released alongside the original *Dig Dug* and ten other Namco games on the PS2, Xbox and GameCube versions of Namco Museum.

In this version of Dig Dug, there are balls (inspired by *Cosmo Gang the Puzzle*) that can kill a lot of enemies in a row.

In 2005, Namco released another game with the title *Dig Dug Arrangement*, as part of *Namco Museum Battle Collection*. It is an entirely different game from *Namco Classic Collection Vol. 2's Dig Dug Arrangement* but still has the concept of being an updated variant of *Dig Dug* by having new graphics, obstacles, enemies, boss battles, power-ups, and so on. The *Battle Collection* edition of *Dig Dug Arrangement* was also released as part of *Namco Museum Virtual Arcade* for the Xbox 360, but with the multiplayer features removed.

Dig Dug Remix

The *Dig Dug Arrangement* from *Namco Museum Battle Collection* was also ported to iOS renamed 'Dig Dug Remix" with the original *Dig Dug* included, but all multiplayer features were removed from that port.*[3]

4.21.3 Mobile game

In 2005, Namco Networks released a version of *Dig Dug* for cell phones and Palm OS/Windows Mobile devices that is authentic to the arcade original in terms of graphics and controls, even though the levels are as they are in the NES version of *Dig Dug*. Unlike the arcade version there is no kill screen at level 256, but rather the levels go on past 500.

4.21.4 Protagonist

Although Namco has officially given the character of the original *Dig Dug* the name Dig Dug, in other games where he makes an appearance, the protagonist goes by the name

Taizo Hori (in Japanese order, HORI Taizo), and is the father of Susumu Hori, the main character in the *Mr. Driller* series. He is also the ex-husband of Toby "Kissy" Masuyo, the heroine of *Baraduke*. His name is a pun on the Japanese phrase "Horitai zo" (掘りたいぞ) or "I want to dig!" (掘り = dig, たい = want, ぞ = !) – a similar pun might be rendered in English as "Will Dig" or "Wanda (Want To) Dig". Many American gamers learned of his real name via the Nintendo DS game *Mr. Driller Drill Spirits*, where he is also a playable character. He is additionally featured in an unlockable gallery of Mr. Driller items in *Mr. Driller 2*. In the *Mr. Driller* series, Hori is known as the "Hero of the Dig Dug Incident". In Japan, he is also the Hero of the South Island incident and is the honorary chairman of the Driller Council to whom most of the characters answer. This contrasts greatly with the PC remake *Dig Dug Deeper*, where the hero is simply named *Dig Dug*.

4.21.5 Versions and ports

Cartdridge of the 1985 Famicom version of Dig Dug.

As well as the arcade version, Atari obtained the license for home versions of *Dig Dug*, and then released it for the Atari 2600, Atari 5200, Atari 7800, Intellivision, Apple II, Atari 400/800, Commodore VIC-20, Commodore 64, IBM PC, and Texas Instruments TI-99/4A. Namco ported *Dig Dug* to the Nintendo Family Computer in 1985. A Game Boy version was released in 1992 with the choices of the original and new versions was added. Another version was released on a Plug 'N Play System, along with *Galaxian*, *Pac-Man*, *Rally-X*, and *Bosconian*. In October 2006, a version of *Dig Dug* was released on the Xbox Live Arcade. Namco Networks ported *Dig Dug* to Windows (bought online) in 2009 which also includes an "Enhanced" mode which replaces all of the original sprites with the sprites from *Dig Dug: Digging Strike*, Namco Networks also made a bundle (also bought online) which includes their Windows version of *Dig Dug* as well as their port of the original *Pac-Man* called *Namco All-Stars: Pac-Man and Dig Dug*. The arcade version has also been released on the Wii Virtual Console in Japan on October 20, 2009, along with its sequel, *Dig Dug II* and the original *Dig Dug* was released as part of the *Pac-Man's Arcade Party* 30th Anniversary arcade machine in 2010. The NES version for the Virtual Console was released in 2008 for the Wii, 2013 for the Nintendo 3DS and 2015 for the Wii U, but as an import for Western regions when ported to the former.

Gakken made a table top handheld game of *Dig Dug* in 1982. It was one of a series of 3 flip-top games with VFD screen and magnifying Fresnel lens. The other two similar style Gakken handheld games were Konami's *Jungler* and *Amidar*. *Dig Dug* has been included on most Namco Museum compilations, and it is one of the three bonus games in the Wii and Nintendo 3DS versions of *Pac-Man Party*.

4.21.6 Reception

Dig Dug was rated the sixth most popular coin-operated video game of all time by the Killer List of Video Games website.[*][4]

In 1984 *Softline* readers named computer versions of *Dig Dug* the tenth-worst Apple and fourth-worst Atari program of 1983.[*][5]

4.21.7 Legacy

A 1985 sequel to this game, the overhead-view oriented *Dig Dug II*, was much less common and met with less success in the arcades. *Mr. Driller* (1999) was originally conceived as a sequel, with the working title *Dig Dug 3*, but it developed into a distinct but related series. Another sequel, *Dig Dug: Digging Strike*, was released in 2005 for the Nintendo DS. This combined the side-view play of the original with the overhead play of the sequel and added a narrative link to the *Mr. Driller* series. A 3D remake of the original, entitled *Dig Dug Deeper*, was released for PC in 2001 by Infogrames. The original *Dig Dug* was released for the Xbox 360 console via Xbox Live Arcade on October 11, 2006. The original *Dig Dug* is also available for play via the GameTap subscription gaming service, and was shown in one of the television commercials for the Gametap website in 2005. It was re-released for the Wii's Virtual Console in North America on June 9, 2008[*][6] and in Europe on August 29, 2008, at a cost of 600 Wii Points.

It has been said that the music for the game show *Starcade* was inspired from the music for *Dig Dug*.[*][7]

4.21.8 Cameos

Some bootleg arcade versions of *Dig Dug* were made, under the name *Zig Zag*. One version looked exactly like the original,[*][8] and the other changed both the sounds and colors, as well as adding a pickaxe power-up that made the player move faster.[*][9]

The character Pooka has many cameos in Namco games, most often as an enemy in Namco games such as the *Pac-Man World* series. Pooka was playable for the first time in the game *Ms. Pac-Man Maze Madness* as an unlockable character for the multiplayer modes. He is also available to play as in *Pac-Man World Rally*, as well as Fygar. Pooka also appeared in the Atari licensed version of the Namco arcade game *Pole Position* on one of the roadside billboards. In *Pac-Man World*, he appears as one of the friends of Pac-Man who was kidnapped by Toc-Man.

In *R4: Ridge Racer Type 4*, there is an American racing team with *Dig Dug* artwork on its hauler and is named the "Dig Racing Team", run by manager Robert Chrisman. It is the "expert" team of the game. Also in R4, the track "Phantomile" has a giant statue of Pooka alongside Pac-Man on the left hand side of the finishing straight. The "Pooka Line" track, which is the first in the game, has a giant screen with a Pooka and Fygar chasing *Dig Dug's* protagonist in arcade-style graphics, which changes to him inflating a Pooka when the player takes the lead. In *Ridge Racer 64*, however, "Dig Racing Team" has to be unlocked by winning all the tracks in Stage 3 in first place – and waiting for the credits to roll, then defeating the car on the "Renegade Expert" track in Car Attack mode. This same game also features Pooka as a selectable car, which is unlocked by breaking the time record in any of the Ridge Racer Extreme tracks on Time Attack mode.

The *Dig Dug* universe and some of its characters appear in the *Mr. Driller* games, starring Taizo Hori's son, Susumu. Additionally, he is also a playable character in the Japan only RPG, *Namco × Capcom*.

Dig Dug was parodied in the *Robot Chicken* episode "President Evil" and the *Drawn Together* episode "The One Wherin There Is a Big Twist, Part II".

In the Disney movie *Wreck-It Ralph*, Dig Dug, a Pooka, and a Fygar are three of the characters in Game Central Station. The Pooka and the Fygar are seen when the camera is moving from one of the tunnels, and Dig Dug starts trying to dig away from Ralph when he comes towards him.

Taizo Hori appeared in an episode of *Death Battle*, where he fought Bomberman. Although he had trouble early on from Bomberman's powerful arsenal and superior movement above ground with his Rooey mount, Taizo was ultimately able to win due to his superior movement when underground, his better control over the terrain, and ability to immobilize opponents with his harpoon- which enabled him to move the fight underground (Where he had the advantage) and eventually stun Bomberman with the harpoon long enough for him to trap him next to one of his own Super Bombs.

4.21.9 References

[1] "retrodiary: 1 April – 28 April". *Retro Gamer* (Bournemouth: Imagine Publishing) (88): 17. April 2011. ISSN 1742-3155. OCLC 489477015.

[2] http://books.google.com/books?id=BzxTtml8Jq4C& pg=PA289&dq=dig+dug+maze+game&hl=en& sa=X&ei=TvuOUZC7OsHXyAGn44CQCw&ved= 0CDYQ6AEwAA#v=onepage&q=dig%20dug%20maze% 20game&f=false

[3] Bailey, Kat (2009-05-08). "Dig Dug Remix Arrives On iPhone". 1UP.com. Retrieved 2009-07-24.

[4] McLemore, Greg. "The Top Coin-Operated Videogames of All Time". Killer List of Videogames. Archived from the original on July 3, 2007. Retrieved 2007-07-17.

[5] "The Best and the Rest". *St.Game*. Mar–Apr 1984. p. 49. Retrieved 28 July 2014.

[6] "Wii-kly Update: One WiiWare game and two Virtual Console games added to Wii Shop Channel". *MCV*. Games Press. 2008-06-09. Retrieved 2011-10-09. Nintendo adds new and classic games to the Wii Shop Channel at 9 am Pacific time every Monday. [⋯] This week's new games are: [⋯] DIG DUG (NES, 1 player, Rated E for Everyone, 600 Wii Points))

[7] "Starcade". illustriousgameshowpage.com.

[8] "KLOV Game Info". Killer List of Video Games.

[9] "Zig Zag". Video Game Museum.

4.21.10 External links

- *Dig Dug* at the Killer List of Videogames

- *Dig Dug* at the Arcade History database

- *Dig Dug* at MobyGames

- *Dig Dug* guide at StrategyWiki

- Dig Dug on mobile at NamcoGames.com

- Dig Dug: Tips and History

- Dig Dug Series at DMOZ

- Video from the C64 Version on archive.org

- Dig Dug for PC at Intel AppUp

4.22 Donkey Kong (video game)

This article is about the arcade game. For the 1994 Game Boy game, see Donkey Kong (Game Boy).

Donkey Kong (Japanese: ドンキーコング Hepburn: *Donkī Kongu*) is an arcade game released by Nintendo in 1981. It is an early example of the platform game genre, as the gameplay focuses on maneuvering the main character across a series of platforms while dodging and jumping over obstacles. In the game, Mario (originally named Mr. Video but then changed to "Jumpman") must rescue a damsel in distress named Pauline (originally named Lady), from a giant ape named Donkey Kong. The hero and ape later became two of Nintendo's most popular and recognizable characters. *Donkey Kong* is one of the most important titles from the Golden Age of Video Arcade Games, and is one of the most popular arcade games of all time.

The game was the latest in a series of efforts by Nintendo to break into the North American market. Hiroshi Yamauchi, Nintendo's president at the time, assigned the project to a first-time video game designer named Shigeru Miyamoto. Drawing from a wide range of inspirations, including *Popeye*, *Beauty and the Beast* and *King Kong*, Miyamoto developed the scenario and designed the game alongside Nintendo's chief engineer, Gunpei Yokoi. The two men broke new ground by using graphics as a means of characterization, including cutscenes to advance the game's plot, and integrating multiple stages into the gameplay.

Regardless of initial doubts by Nintendo's American staff, *Donkey Kong* succeeded commercially and critically in North America and Japan. Nintendo licensed the game to Coleco, who developed home console versions for numerous platforms. Other companies cloned Nintendo's hit and avoided royalties altogether. Miyamoto's characters appeared on cereal boxes, television cartoons, and dozens of other places. A lawsuit brought on by Universal City Studios, alleging *Donkey Kong* violated their trademark of *King Kong*, ultimately failed. The success of *Donkey Kong* and Nintendo's victory in the courtroom helped to position the company for video game market dominance from its release in 1981 until the late 1990s (1996–1999).

4.22.1 Gameplay

Following 1980's *Space Panic*, *Donkey Kong* is one of the earliest examples of the platform game genre[10]:94[11] even prior to the term behind coined; the US gaming press used *climbing game* for titles with platforms and ladders.[12] As the first platform game to feature jumping, *Donkey Kong* requires the player to jump between gaps and over obstacles or approaching enemies, setting the tem-

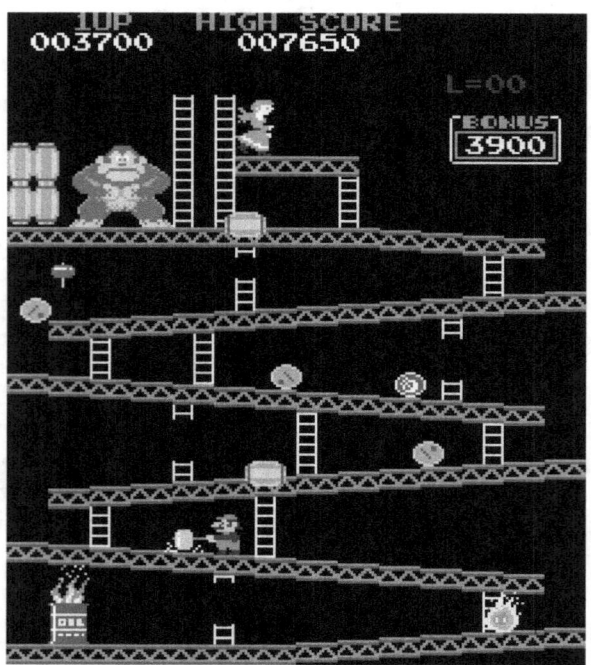

Gameplay of Donkey Kong *in the first level.*

plate for the future of the platform genre.[13] With its four unique stages, *Donkey Kong* was the most complex arcade game at the time of its release, and one of the first arcade games to feature multiple stages, following 1980's *Phoenix* and 1981's *Gorf* and *Scramble*[14]

Competitive video gamers and referees stress the game's high level of difficulty compared to other classic arcade games. Winning the game requires patience and the ability to accurately time Mario's ascent.[15]:82 In addition to presenting the goal of saving Pauline, the game also gives the player a score. Points are awarded for the following: finishing each stage; leaping over obstacles; destroying objects with a hammer power-up; collecting items such as hats, parasols, and purses (apparently belonging to Pauline); and completing other tasks. The player typically receives three lives with a bonus awarded for the first 7,000 points, although this can be modified via the game's built in DIP switches.

The game is divided into four different single-screen stages. Each represents 25 meters of the structure Donkey Kong has climbed, one stage being 25 meters higher than the previous. The final stage occurs at 100 meters. Stage one involves Mario scaling a construction site made of crooked girders and ladders while jumping over or hammering barrels and oil barrels tossed by Donkey Kong. Stage two involves climbing a five-story structure of conveyor belts, each of which transports cement pans. The third stage involves the player riding elevators while avoiding bouncing springs. The final stage involves Mario removing eight riv-

ets which support Donkey Kong. Removing the final rivet causes Donkey Kong to fall and the hero to be reunited with Pauline.[16] These four stages combine to form a level.

Upon completion of the fourth stage, the level then increments, and the game repeats the stages with progressive difficulty. For example, Donkey Kong begins to hurl barrels faster and sometimes diagonally, and fireballs get speedier. The victory music alternates between levels 1 and 2. The 22nd level is colloquially known as the kill screen, due to an error in the game's programming that kills Mario after a few seconds, effectively ending the game.[16]

4.22.2 Plot

Donkey Kong is considered to be the earliest video game with a storyline that visually unfolds on screen.[13] The eponymous Donkey Kong character is the game's *de facto* villain. The hero is a carpenter originally named Jumpman, later renamed Mario.[17] The ape kidnaps Mario's girlfriend, originally known as Lady, but later renamed Pauline. The player must take the role of Mario and rescue her. This is the first occurrence of the damsel in distress scenario that would provide the template for countless video games to come.[15]:82

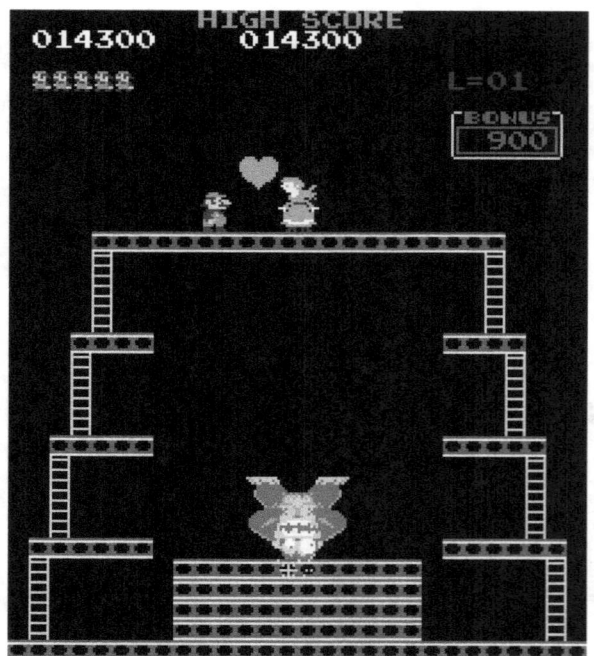

At the game's end, Jumpman and Pauline are reunited.

The game uses graphics and animation as vehicles of characterization. Donkey Kong smirks upon Mario's demise. Pauline has a pink dress and long hair,[18]:19–20 and a speech balloon crying "HELP!" appears frequently beside her. Mario, depicted in red overalls and cap, is an everyman character, a type common in Japan. Graphical limitations and the low pixel resolution of the small sprites prompted his design: drawing a mouth is infeasible, so the character was given a mustache;[19]:37 the programmers could not animate hair, so he got a cap; and to make his arm movements visible, he needed colored overalls.[15]:238 The artwork used for the cabinets and promotional materials make these cartoon-like character designs even more explicit. Pauline, for example, is depicted to be disheveled (like *King Kong* 's Fay Wray) in a torn dress and stiletto heels.[18]:19–20

Donkey Kong is the first example of a complete narrative told in video game form, and like 1980's Pac-Man, it employs cutscenes to advance its plot. The game opens with the gorilla climbing a pair of ladders to the top of a construction site. He sets Pauline down and stomps his feet, causing the steel beams to change shape. He then moves to his final perch and sneers. This brief animation sets the scene and adds background to the gameplay, a first for video games. Upon reaching the end of the stage, another cutscene begins. A heart appears between Mario and Pauline, but Donkey Kong grabs the woman and climbs higher, causing the heart to break. The narrative concludes when Mario reaches the end of the rivet stage. He and Pauline are reunited, and a short intermission plays.[19]:40–42 The gameplay then loops from the beginning at a higher level of difficulty, without any formal ending.

4.22.3 Development

As of late 1980 to early 1981, Nintendo's efforts to expand to North America had failed, culminating with the attempted export of the otherwise successful *Radar Scope*. They were left with a large amount of unsold *Radar Scope* machines, so company president Hiroshi Yamauchi thought of simply converting them into something new. He approached a young industrial designer named Shigeru Miyamoto, who had been working for Nintendo since 1977, to see if he could design such a replacement. Miyamoto said that he could.[20]:157 Yamauchi appointed Nintendo's head engineer, Gunpei Yokoi, to supervise the project.[20]:158 Nintendo's budget for the development of the game was $100,000.[21] Some sources also claim that Ikegami Tsushinki was involved in some of the development.[22][23] They played no role in the game's creation or concept, but were hired by Nintendo to provide "mechanical programming assistance to fix the software created by Nintendo".[21]

At the time, Nintendo was also pursuing a license to make a game based on the *Popeye* comic strip. When this license attempt failed, Nintendo took the opportunity to create new characters that could then be marketed and used in later games.[15]:238[24] Miyamoto came up with many characters and plot concepts, but he eventually settled on a love triangle between gorilla, carpenter, and girlfriend that mirrors the rivalry between Bluto and Popeye for Olive Oyl.[19]:39 Bluto became an ape, which Miyamoto said was "nothing too evil or repulsive".[25]:47 He would be the pet of the main character, "a funny, hang-loose kind of guy."[25]:47 Miyamoto has also named "Beauty and the Beast" and the 1933 film *King Kong* as influences.[19]:36 Although its origin as a comic strip license played a major part, *Donkey Kong* marked the first time that the storyline for a video game preceded the game's programming rather than simply being appended as an afterthought.[19]:38 Unrelated *Popeye* games would eventually be released by Nintendo for the Game & Watch the following month, and for the arcades in 1982.

Yamauchi wanted primarily to target the North American market, so he mandated that the game be given an English title, though many of their games to this point had English titles anyway. Miyamoto decided to name the game for the ape, whom he felt to be the strongest character.[19]:39 The story of how Miyamoto came up with the name "Donkey Kong" varies. A false urban myth says that the name was originally meant to be "Monkey Kong", but was misspelled or misinterpreted due to a blurred fax or bad telephone connection.[26] Another, more credible story claims Miyamoto looked in a Japanese-English dictionary for something that would mean "stubborn gorilla",[20] or that "Donkey" was meant to convey "silly" or "stubborn"; "Kong" was common Japanese slang for "gorilla".[15]:238 A rival claim is that he worked with Nintendo's export manager to come up with the title, and that "Donkey" was meant to represent "stupid and goofy".[25]:48–49 In the end, Miyamoto stated that he thought the name would convey the thought of a "stupid ape".[27]

Miyamoto himself had high hopes for his new project. He lacked the technical skills to program it alone, so instead came up with concepts and consulted technicians to see if they were possible. He wanted to make the characters different sizes, move in different manners and react in various ways. Yokoi thought Miyamoto's original design was too complex,[25]:47–48 though he had some difficult suggestions himself, such as using see-saws to catapult the hero across the screen (eventually found too hard to program, though a similar concept would appear in the aforementioned *Popeye* arcade game). Miyamoto then thought of using sloped platforms, barrels and ladders. When he specified that the game would have multiple stages, the four-man programming team complained that he was essentially asking them to make the game repeatedly.[19]:38–39 Nevertheless, they followed Miyamoto's design, creating a total of approximately 20 kilobytes of content.[20]:530 Yukio Kaneoka composed a simplistic soundtrack to serve as background music for the levels and story events.[1][28]

Hiroshi Yamauchi thought the game was going to sell well and called Minoru Arakawa, head of Nintendo's operations in the US, to tell him. Nintendo's American distributors, Ron Judy and Al Stone, brought Arakawa to a lawyer named Howard Lincoln to secure a trademark.[20]:159

North American Donkey Kong *promotional flier from 1981, showing Mario, Donkey Kong, and Pauline*

The game was sent to Nintendo of America for testing. The sales manager disliked it for being too different from the maze and shooter games common at the time,[25]:49 and Judy and Lincoln expressed reservations over the strange title. Still, Arakawa adamantly believed that it would be big.[20]:159 American staffers asked Yamauchi to change the name, but he refused. Arakawa and the American staff began translating the storyline for the cabinet art and naming the other characters. They chose "Pauline" for the Lady, after Polly James, wife of Nintendo's Redmond, Washington, warehouse manager, Don James. The name of "Jumpman", a name originally chosen for its similarity to the popular brands *Walkman* and *Pac-Man*,[19]:39 was eventually changed to "Mario" in likeness of Mario Segale, the landlord of the original office space of Nintendo of America.[25]:109 These character names were printed on the American cabinet art and used in promotional materials. *Donkey Kong* was ready for release.[19]:212

Stone and Judy convinced the managers of two bars in Seattle, Washington, to set up *Donkey Kong* machines. The managers initially showed reluctance, but when they saw sales of $30 a day—or 120 plays—for a week straight, they requested more units.[14]:68 In their Redmond headquarters, a skeleton crew composed of Arakawa, his wife Yoko, James, Judy, Phillips and Stone set about gutting 2,000 surplus *Radar Scope* machines and converting them with *Donkey Kong* motherboards and power supplies from Japan.[25]:110 The game officially went on sale in July 1981.[20]:211

4.22.4 Reception

In his 1982 book *Video Invaders*, Steve Bloom described *Donkey Kong* as "another bizarre cartoon game, courtesy of Japan".[19]:5 *Donkey Kong* was, however, extremely popular in the United States and Canada. The game's initial 2,000 units sold, and more orders were made. Arakawa began manufacturing the electronic components in Redmond because waiting for shipments from Japan was taking too long.[20]:160 By October, *Donkey Kong* was selling 4,000 units a month, and by late June 1982, Nintendo had sold 60,000 *Donkey Kong* machines overall and earned $180 million.[20]:211 Judy and Stone, who worked on straight commission, became millionaires.[20]:160 Arakawa used Nintendo's profits to buy 27 acres (11 ha) of land in Redmond in July 1982.[25]:113 Nintendo earned another $100 million on the game in its second year of release,[25]:111 totaling $280 million[29] (equivalent to $726 million in 2015).[30] It remained Nintendo's top seller into summer 1983.[20]:284 *Donkey Kong* also sold steadily in Japan.[19]:46 *Electronic Games* speculated in June 1983 that the game's home versions contributed to the arcade version's extended popularity, compared to the four to six months that the average game lasted.[31]

In January 1983, the 1982 Arcade Awards gave it the Best Solitaire Videogame award and the Certificate of Merit as runner-up for Coin-Op Game of the Year.[32] In September 1982, *Arcade Express* reviewed the ColecoVision port and scored it 9 out of 10.[33] *Computer and Video Games* reviewed the ColecoVision port in its September 1984 issue and scored it 4 out of 4 in all four categories of Action, Graphics, Addiction and Theme.[34]

4.22.5 Licensing and ports

By late June 1982, *Donkey Kong* 's success had prompted more than 50 parties in the U.S. and Japan to license the game's characters.[20]:215 Mario and his simian nemesis appeared on cereal boxes, board games, pajamas, and manga. In 1983, the animation studio Ruby-Spears produced a *Donkey Kong* cartoon (as well as *Donkey Kong Jr*) for the *Saturday Supercade* program on CBS. In the show, mystery crime-solving plots in the mode of *Scooby-Doo* are framed around the premise of Mario and Pauline chasing Donkey Kong (voiced by Soupy Sales), who has escaped from the circus. The show lasted two seasons.

Makers of video game consoles were also interested. Taito offered a considerable sum to buy all rights to *Donkey Kong*, but Nintendo turned them down after three days of discussion within the company.[25] Rivals Coleco and Atari approached Nintendo in Japan and the United States respectively. In the end, Yamauchi granted Coleco exclusive console and tabletop rights to *Donkey Kong* because he felt that "It [was] the hungriest company".*[25]*:111 In addition, Arakawa felt that as a more established company in the US, Coleco could better handle marketing. In return, Nintendo would receive an undisclosed lump sum plus $1.40 per game cartridge sold and $1 per tabletop unit. On December 24, 1981, Howard Lincoln drafted the contract. He included language that Coleco would be held liable for anything on the game cartridge, an unusual clause for a licensing agreement.*[20]*:208–209 Arakawa signed the document the next day, and, on February 1, 1982, Yamauchi persuaded the Coleco representative in Japan to sign without running the document by the company's lawyers.*[25]*:112

Coleco did not offer the game cartridge stand-alone; instead, they bundled it with their ColecoVision, which went on sale in August 1982. Six months later, Coleco offered Atari 2600 and Intellivision versions, too. Notably, they did not port it to the Atari 5200, a system comparable to their own (as opposed to the less powerful 2600 and Intellivision). Coleco's sales doubled to $500 million and their earnings quadrupled to $40 million.*[20]*:210 Coleco's console versions of *Donkey Kong* sold six million cartridges in total, grossing over $153 million,*[25]*:121*[note 1]<sup class="noprint Inline-Template "noprint Inline-Template"" style="white-space:nowrap;">[*check quotation syntax*] and earning Nintendo over $5 million in royalties.*[35] Coleco also released stand-alone Mini-Arcade tabletop versions of *Donkey Kong*, which, along with *Pac-Man*, *Galaxian*, and *Frogger*, sold three million units combined.*[36] Meanwhile, Atari got the license for computer versions of *Donkey Kong* and released it for the Atari 400/800. When Coleco unveiled the Adam Computer, running a port of *Donkey Kong* at the 1983 Consumer Electronics Show in Chicago, Illinois, Atari protested that it was in violation of the licensing agreement. Yamauchi demanded that Arnold Greenberg, Coleco's president, shelve his Adam port. This version of the game was cartridge-based, and thus not a violation of Nintendo's license with Atari; still, Greenberg complied. Ray Kassar of Atari was fired the next month, and the home PC version of *Donkey Kong* fell through.*[20]*:283–285

In 1983, Atari released several computer versions under the Atarisoft label. All of the computer ports had the cement factory level, while most of the console versions did not. None of the home versions of Donkey Kong had all of the intermissions or animations from the arcade game. Some have Donkey Kong on the left side of the screen in the barrel level (like he is in the arcade game) and others have him on the right side.

Miyamoto created a greatly simplified version for the Game & Watch multiscreen. Other ports include the Amiga, Apple II, Atari 7800, Intellivision, Commodore 64, Commodore VIC-20, Famicom Disk System, IBM PC booter, ZX Spectrum, Amstrad CPC, MSX, Atari 8-bit

Game & Watch Donkey Kong

family and Mini-Arcade versions. The game was ported to Nintendo's Family Computer (Famicom) console in 1983 as one of the system's three launch titles; the same version was an early title for the Famicom's North American version, the Nintendo Entertainment System (NES). However, the cement factory level is not included, nor are most of the cutscenes since Nintendo did not have large enough cartridge ROMs available in the beginning. This port includes a new song composed by Yukio Kaneoka for the title screen;*[1] an arrangement of the tune appears in *Donkey Kong Country* for the Super Nintendo Entertainment System. Both *Donkey Kong* and its sequel, *Donkey Kong Jr.*, are included in the 1988 NES compilation *Donkey Kong Classics*. The NES version was re-released as an unlockable game in *Animal Crossing* for the GameCube and as an item for purchase on the Virtual Console for the Wii,*[37] Wii U and Nintendo 3DS. The Wii U version is also the last game was be released to celebrate the 30-year anniversary of the Japanese version of the NES, the Famicom. The original arcade version of the game appears in the Nintendo 64 game *Donkey Kong 64*. Nintendo released the NES version on the e-Reader and for the Game Boy Advance *Classic NES* series in 2002 and 2004, respectively.*[38] The Famicom version of the game sold 840,000 units in Japan.*[39]

Donkey Kong: Original Edition is a port based on the NES version that reinstates the cement factory stage and includes some intermission animations absent from the original NES version, which has only ever been released on the Virtual Console. It was preinstalled on 25th Anniversary PAL region red Wii systems,*[40] which were first released in Europe on October 29, 2010.*[41] In Japan, a download code for the game for Nintendo 3DS Virtual Console was sent to users who purchased *New Super Mario Bros. 2* or *Brain Age:*

Concentration Training from the Nintendo eShop from July 28 to September 2, 2012.[*][42] In North America, a download code for the game for Nintendo 3DS Virtual Console was sent to users who purchased one of five select 3DS titles on the Nintendo eShop and registered it on Club Nintendo from October 1, 2012 to January 6, 2013.[*][43][*][44] In Europe and Australia, it was released for purchase on the Nintendo 3DS eShop, being released on September 18, 2014 in Europe[*][45] and on September 19, 2014 in Australia.[*][46]

4.22.6 Clones

Other companies bypassed Nintendo's licensing completely. In 1981, O. R. Rissman, president of Tiger Electronics, obtained a license to use the name *King Kong* from Universal City Studios. Under this title, Tiger created a handheld game with a scenario and gameplay based directly on Nintendo's creation.[*][20][*]:210–211 The 1982 *Logger* arcade game from Century Electronics is a direct clone of *Donkey Kong*, with a large bird standing in for the ape and rolling logs instead of barrels.[*][47]

Crazy Kong is a clone manufactured by Falcon and licensed for some non-American markets. Nevertheless, *Crazy Kong* machines found their way into some American arcades during the early 1980s, often installed in cabinets marked as *Congorilla*. Nintendo was quick to take legal action against those distributing the game in the US.[*][48][*]:119 Bootleg copies of *Donkey Kong* also appeared in both North America and France under the *Crazy Kong*, *Konkey Kong* or *Donkey King* names.

Home computer clones include *Cannonball Blitz* (1982), with a soldier and cannonballs replacing the ape and barrels, *Canyon Climber* (1982),[*][49] and *Killer Gorilla* (1983). A magazine ad for Epyx's *Jumpman* (1983) had the tagline "If you liked Donkey Kong, you'll love JUMPMAN!"[*][50] *Jumpman*, along with *Miner 2049er* (1982) and *Mr. Robot and His Robot Factory* (1984), focused on traversing all of the platforms in the level, or collecting scattered objects, instead of climbing to the top. One of the first releases from Electronic Arts was *Hard Hat Mack* (1983) for the Apple II, a three-stage game that even borrowed the construction site theme from *Donkey Kong*.

There were so many games with multiple ladder and platforms stages by 1983 that *Electronic Games* described Nintendo's own *Popeye* as "yet another variation of a theme that's become all too familiar since the success of *Donkey Kong*".[*][51] That year Sega created a *Donkey Kong* clone called *Congo Bongo*. Despite using isometric perspective, the structure and gameplay are similar.

4.22.7 *Universal City Studios, Inc. v. Nintendo Co., Ltd.*

Main article: Universal City Studios, Inc. v. Nintendo Co., Ltd.

In April 1982, Sid Sheinberg, a seasoned lawyer and president of MCA and Universal City Studios, learned of the game's success and suspected it might be a trademark infringement of Universal's own *King Kong*.[*][20][*]:211 On April 27, 1982, he met with Arnold Greenberg of Coleco and threatened to sue over Coleco's home version of *Donkey Kong*. Coleco agreed on May 3, 1982 to pay royalties to Universal of 3% of their *Donkey Kong*'s net sale price, worth about $4.6 million.[*][25][*]:121 Meanwhile, Sheinberg revoked Tiger's license to make its *King Kong* game, but O. R. Rissman refused to acknowledge Universal's claim to the trademark.[*][20][*]:214 When Universal threatened Nintendo, Howard Lincoln and Nintendo refused to cave. In preparation for the court battle ahead, Universal agreed to allow Tiger to continue producing its *King Kong* game as long as they distinguished it from *Donkey Kong*.[*][20][*]:215

Universal sued Nintendo on June 29, 1982 and announced its license with Coleco. The company sent cease and desist letters to Nintendo's licensees, all of which agreed to pay royalties to Universal except Milton Bradley and Ralston Purina.[*][52][*]:74–75 *Universal City Studios, Inc. v. Nintendo, Co., Ltd.* was heard in the United States District Court for the Southern District of New York by Judge Robert W. Sweet. Over seven days, Universal's counsel, the New York firm Townley & Updike, argued that the names *King Kong* and *Donkey Kong* were easily confused and that the plot of the game was an infringement on that of the films.[*][52][*]:74 Nintendo's counsel, John Kirby, countered that Universal had themselves argued in a previous case that *King Kong*'s scenario and characters were in the public domain. Judge Sweet ruled in Nintendo's favor, awarding the company Universal's profits from Tiger's game ($56,689.41), damages and attorney's fees.[*][20][*]:217

Universal appealed, trying to prove consumer confusion by presenting the results of a telephone survey and examples from print media where people had allegedly assumed a connection between the two Kongs.[*][48][*]:118 On October 4, 1984, however, the court upheld the previous verdict.[*][48][*]:112

Nintendo and its licensees filed counterclaims against Universal. On May 20, 1985, Judge Sweet awarded Nintendo $1.8 million for legal fees, lost revenues, and other expenses.[*][20][*]:218 However, he denied Nintendo's claim of damages from those licensees who had paid royalties to both Nintendo and Universal.[*][52][*]:72 Both parties appealed this judgment, but the verdict was upheld on July 15,

1986.*[52]*:77–78

Nintendo thanked John Kirby with the gift of a $30,000 sailboat named *Donkey Kong* and "exclusive worldwide rights to use the name for sailboats" .*[25]*:126 The court battle also taught Nintendo they could compete with larger entertainment industry companies.*[25]*:127

4.22.8 Legacy

Donkey Kong spawned the sequels *Donkey Kong Jr.* and *Donkey Kong 3*, as well as the spin-off *Mario Bros.* A complete remake of the original arcade game on the Game Boy, named *Donkey Kong* or *Donkey Kong '94* contains levels from both the original Donkey Kong and Donkey Kong Jr arcades. It starts with the same damsel-in-distress premise and four basic locations as the arcade game and then progresses to 97 additional puzzle-based levels. It is the first game to have built-in enhancement for the Super Game Boy accessory. The arcade version makes an appearance in *Donkey Kong 64* in the Frantic Factory level.

Nintendo revived the *Donkey Kong* franchise in the 1990s for a series of platform games and spin-offs developed by Rare, beginning with *Donkey Kong Country* in 1994. In 2004, Nintendo released *Mario vs. Donkey Kong*, a sequel to the Game Boy title. In it, Mario must chase Donkey Kong to get back the stolen Mini-Mario toys. In the follow-up *Mario vs. Donkey Kong 2: March of the Minis*, Donkey Kong once again falls in love with Pauline and kidnaps her, and Mario uses the Mini-Mario toys to help him rescue her. *Donkey Kong Racing* for the GameCube was in development by Rare, but was canceled when Microsoft purchased the company. In 2004, Nintendo released the first of the *Donkey Konga* games, a rhythm-based game series that uses a special bongo controller. *Donkey Kong Jungle Beat* (2005) is a unique platform action game that uses the same bongo controller accessory. In 2007, *Donkey Kong Barrel Blast* was released for the Nintendo Wii. It was originally developed as a GameCube game and would have used the bongo controller, but it was delayed and released exclusively as a Wii title with no support for the bongo accessory. *Super Smash Bros. Brawl* features music from the game arranged by Hirokazu "Hip" Tanaka*[28] and a stage called "75m", an almost exact replica of its *Donkey Kong* namesake.*[53] While the stage contains her items, Pauline is missing from her perch at the top of the stage.*[53]

Donkey Kong was said to be an inspiration for the 1983 platform game for home computers *Jumpman*, according to the game's creator.*[54]

In 2013, video game developer Mike Mika hacked the game to create a version where Pauline is the main character and rescues Mario. He created this remixed version for his three-year-old daughter who wanted to play as a heroine.*[55] *Donkey Kong* appears as a game in the Wii U game *NES Remix*, which features multiple NES games and sometimes "remixes" them by presenting significantly modified versions of the games as challenges. One such challenge features Link from *The Legend of Zelda* traveling through the first screen to save Pauline. The difficulty is increased compared to the original *Donkey Kong* because of Link's inability to jump, as seen in *Zelda*.

In popular culture

Since its original release, *Donkey Kong* 's success has entrenched the game in American popular culture. In 1982, Buckner & Garcia and R. Cade and the Video Victims both recorded songs ("Do the Donkey Kong" and "Donkey Kong", respectively) based on the game. Artists like DJ Jazzy Jeff & the Fresh Prince and Trace Adkins referenced the game in songs. Episodes of television series such as *The Simpsons*, *Futurama*, *Crank Yankers* and *The Fairly Odd-Parents* have also contained references to the game. Even today, sound effects from the Atari 2600 version often serve as generic video game sounds in films and television series. The Killer List of Videogames ranks *Donkey Kong* the third most popular arcade game of all time and places it at No. 25 on the "Top 100 Videogames" list. in February 2006, *Nintendo Power* rated it the 148th best game made on a Nintendo system.*[56] Today, *Donkey Kong* is the fifth most popular arcade game among collectors.*[57] The phrase "It's on like Donkey Kong" has been used in various works of popular culture. In November 2010, Nintendo applied for a trademark on the phrase with the United States Patent and Trademark Office.*[58]

Atari computer Easter egg

The *Atari computer* port of *Donkey Kong* contains one of the longest-undiscovered *Easter eggs* in a video game.*[59] Landon Dyer, the programmer assigned to create the port, added a secret where his initials would appear if the player died under certain conditions and then waited for the game to cycle to the title screen. This secret remained undiscovered for 26 years until Dyer revealed it on his blog, stating "there's an easter egg, but it's totally not worth it, and I don't remember how to bring it up anyway." *[60] After this announcement, the steps required to trigger the Easter egg were discovered by Don Hodges, who used an emulator and a debugger to trace through the 25,000 lines of the game's code.*[61]

Competition

The 2007 motion picture documentary *The King of Kong: A Fistful of Quarters* explores the world of competitive classic arcade gaming and tells the story of Steve Wiebe's quest to break Billy Mitchell's record.[*][62]

Among celebrity players, actor Will Forte is currently ranked on the Twin Galaxies *Donkey Kong* scoreboard,[*][63] and rapper Eminem has tweeted his *Donkey Kong* scores on two different occasions. The latter score, if it were eligible for verification by Twin Galaxies, would place him within the top 25 on the scoreboard.[*][64][*][65]

4.22.9 See also

- 1981 in video gaming

4.22.10 Notes

[1] "And we received from Coleco an agreement that they would pay us three percent of the net sales price [of all the "Donkey Kong" cartridges Coleco sold]." It turned out to be an impressive number of cartridges, 6 million, which translated into $4.6 million.

4.22.11 References

[1] *Famicom 20th Anniversary Original Sound Tracks Vol. 1* (Media notes). Scitron Digital Contents Inc. 2004.

[2] "Retro Diary". *Retro Gamer* (Bournemouth: Imagine Publishing) (104): 13. July 2012. ISSN 1742-3155. OCLC 489477015.

[3] "ファミコンミニ／ドンキーコング". Nintendo. Retrieved 2015-05-24.

[4] "Nintendo - Donkey Kong". Nintendo. Retrieved 2015-05-24.

[5] "VC ドンキーコング". Nintendo. Retrieved 2015-05-24.

[6] "バーチャルコンソールドンキーコングニンテンドー3DS Nintendo". Nintendo. Retrieved 2015-05-24.

[7] "Nintendo - Donkey Kong". Nintendo. Retrieved 2015-05-24.

[8] "ドンキーコング Wii U Nintendo". Nintendo. Retrieved 2015-05-24.

[9] "Nintendo - Donkey Kong". Nintendo. Retrieved 2015-05-24.

[10] Crawford, Chris (2003). *Chris Crawford on Game Design.* New Riders Publishing.

[11] Space Panic at Allgame

[12] "The Player's Guide to Climbing Games". *Electronic Games* **1** (11): 49. January 1983.

[13] "Gaming's most important evolutions". GamesRadar. Oct 8, 2010. p. 3. Retrieved April 11, 2011.

[14] Sellers, John (2001). *Arcade Fever: The Fan's Guide to the Golden Age of Video Games.* Philadelphia: Running Book Publishers.

[15] De Maria, Rusel, and Wilson, Johnny L. (2004). *High Score!: The Illustrated History of Electronic Games.* 2nd ed. New York: McGraw-Hill/Osborne.

[16] Seth Gordon (director) (2007). *The King of Kong: A Fistful of Quarters* (DVD). Picturehouse.

[17] McLaughlin, Rus (September 14, 2010). "IGN Presents The History of Super Mario Bros.". IGN. Retrieved April 9, 2014.

[18] Ray, Sheri Graner (2004). *Gender Inclusive Game Design: Expanding the Market.* Hingham, Massachusetts: Charles Rivers Media, Inc.

[19] Kohler, Chris (2005). *Power-up: How Japanese Video Games Gave the World an Extra Life.* Indianapolis, Indiana: BradyGAMES.

[20] Kent, Steven L. (2002). *The Ultimate History of Video Games: The Story Behind the Craze that Touched our Lives and Changed the World.* New York: Random House International. ISBN 978-0-7615-3643-7. OCLC 59416169.

[21] *Copyright law decisions.* Commerce Clearing House. 1985. Retrieved February 26, 2012. An English translation of the Japanese term Donkey Kong is "crazy gorilla." Nintendo Co., Ltd. expended over $100,000.00 in direct development of the game, and Nintendo Co., Ltd. hired Ikegami Tsushinki Co., Ltd. to provide mechanical programming assistance to fix the software created by Nintendo Co., Ltd. in the storage component of the game. The name "Ikegami Co. Lim." appears in the computer program for the Donkey Kong game. Individuals within the research and development department of Nintendo Co., Ltd., however, created the Donkey Kong concept and game.

[22] Company:Ikegami Tsushinki. *Game Developer Research Institute.* Retrieved on May 17, 2009.

[23] It started from Pong (それは『ポン』から始まった: アーケード TV ゲームの成り立ち *sore wa pon kara hajimatta: ākēdo terebi gēmu no naritachi*), Masumi Akagi (赤木真澄 *Akagi Masumi*), Amusement Tsūshinsha (アミューズメント通信社 *Amyūzumento Tsūshinsha*), 2005, ISBN 4-9902512-0-2.

[24] East, Tom (25 November 2009). "Donkey Kong Was Originally A Popeye Game". *Official Nintendo Magazine.* Official Nintendo Magazine. Retrieved 28 February 2013. Miyamoto says Nintendo's main monkey might not have existed.

[25] Sheff, David (1999). *Game Over: Press Start to Continue: The Maturing of Mario*. Wilton, Connecticut: GamePress.

[26] Mikkelson, Barbara; Mikkelson, David (February 25, 2001). "Donkey Wrong". Snopes. Retrieved August 29, 2014.

[27] "Miyamoto Shrine: Shigeru Miyamoto's Home on The Web:". *Interview with Miyamoto (May 16th 2001, E3 Expo)*. Retrieved May 31, 2007.

[28] Smash Bros. DOJO!!, Donkey Kong

[29] Jörg Ziesak (2009), *Wii Innovate – How Nintendo Created a New Market Through Strategic Innovation*, GRIN Verlag, p. 2029, ISBN 3-640-49774-0, retrieved April 9, 2011, Donkey Kong was Nintendo's first international smash hit and the main reason behind the company's breakthrough in the Northern American market. In the first year of its publication, it earned Nintendo 180 million US dollars, continuing with a return of 100 million dollars in the second year.

[30] "CPI Inflation Calculator". Bureau of Labor Statistics. Retrieved March 22, 2011.

[31] Pearl, Rick (June 1983). "Closet Classics". *Electronic Games*. p. 82. Retrieved 6 January 2015.

[32] "Electronic Games Magazine". Internet Archive. Retrieved February 1, 2012.

[33] "Arcade Express" (PDF). Reese Publishing Co. September 26, 1982. Retrieved 26 August 2015.

[34] http://retrocdn.net/index.php?title=File:CVG_UK_035.pdf&page=44

[35] Harmetz, Aljean (January 15, 1983). "New Faces, More Profits For Video Games". *Times-Union*. p. 18. Retrieved February 28, 2012.

[36] "More Mini-Arcades A Comin'". *Electronic Games* **4** (16): 10. June 1983. Retrieved February 1, 2012.

[37] Parish, Jeremy (October 31, 2006). "Wii Virtual Console Lineup Unveiled". 1UP.com. Retrieved November 1, 2006.

[38] Parish.

[39] "Japan Sales". Nintendojo. September 26, 2006. Archived from the original on July 30, 2008. Retrieved October 9, 2008. (Translation)

[40] Kemps, Heidi (November 16, 2010). "Europe gets exclusive 'perfect version' of NES Donkey Kong in its Mario 25th Anniversary Wiis". GamesRadar.

[41] Axon, Samuel (October 12, 2010). "Nintendo Announces 25th Anniversary Mario Consoles for Europe". *Mashable*. Retrieved 27 January 2015.

[42] Gantayat, Anoop (July 20, 2012). "Nintendo Kicks off Download Game Sales With Campaign". *Andriasang*. Retrieved 27 January 2015.

[43] "Donkey Kong Free Game Giveaway". *Club Nintendo*. Nintendo. Archived from the original on October 3, 2012. Retrieved 27 January 2015.

[44] Schreier, Jason (October 1, 2012). "Buy One Of Five 3DS Games Online And You Get A Free Copy Of Donkey Kong: Original Edition". *Kotaku*. Retrieved 27 January 2015.

[45] Warmuth, Christopher. "Europe: Original Donkey Kong Edition & Japan: Mario Pinball Land". *Mario Party Legacy*. Retrieved 27 January 2015.

[46] Vuckovic, Daniel (September 18, 2014). "Nintendo Download Updates (19/9) Beats, Rhythm and Warriors". *Vooks*. Retrieved 27 January 2015.

[47] "Logger". *Killer List of Video Games*.

[48] *Universal City Studios, Inc. v. Nintendo, Co., Ltd.*. United States Second Circuit Court of Appeals, October 4, 1984

[49] "Canyon Climber". *YouTube*.

[50] "Epyx Jumpman ad". *Electronic Games Magazine*: 81. June 1983.

[51] Sharpe, Roger C. (June 1983). "Insert Coin Here". *Electronic Games* p. 92. Retrieved 6 January 2015.

[52] *Universal City Studios, Inc. v. Nintendo, Co., Ltd.*. United States Second Circuit Court of Appeals, July 15, 1986

[53] "Smash Bros. DOJO!! – 75m". Smash Bros Dojo. Retrieved March 8, 2008.

[54] "I talked to Randy Glover about Jumpman.". Archived from the original on 15 January 2008. Retrieved 3 June 2007.

[55] "Dad Hacks 'Donkey Kong' to Make Pauline the Hero for Gamer Daughter - ABC News". Abcnews.go.com. 2013-03-12. Retrieved 2013-05-29.

[56] Michaud, Pete (February 2006). "NP Top 200". *Nintendo Power* **197**: 58.

[57] McLemore, Greg, et al. (2005). "The Top Coin-operated Videogames of All Time". Retrieved October 11, 2011.

[58] Gross, Doug (November 10, 2010). "Nintendo seeks to trademark 'On like Donkey Kong'". CNN. Retrieved November 10, 2010.

[59] "Donkey Kong Easter Egg Discovered 26 Years Later". Kotaku.com Retrieved 2013-05-29.

[60] "Donkey Kong and Me". Dadhacker.com. 2008-03-04. Retrieved 2013-05-29.

[61] Hodges, Don (July 1, 2009). "Donkey Kong Lays an Easter Egg".

[62] "The King of Kong: A Fistful of Quarters > Overview". *Allmovie*. Retrieved May 4, 2009.

[63] Life After MacGruber and SNL: Catching up with Will Forte Movieline

[64] "Donkey Kong Ambitions". Donkey Blog. Retrieved August 15, 2012.

[65] Day, Walter (2007). "Donkey Kong: Points [Hammers Allowed] [Default, TGTS]". Twin Galaxies, LLC. Archived from the original on August 22, 2008. Retrieved August 29, 2014.

- Consalvo, Mia (2003). "Hot Dates and Fairy-tale Romances". *The Video Game Theory Reader*. New York: Routledge.

- Fox, Matt (2006). *The Video Games Guide*. Boxtree Ltd.

- Mingo, Jack. (1994) *How the Cadillac Got its Fins* New York: HarperBusiness. ISBN 0-88730-677-2

- Schodt, Frederick L. (1996). *Dreamland Japan: Writings on Modern Manga*. Berkeley, California: Stone Bridge Press.

4.22.12   External links

- *Donkey Kong* at the Killer List of Videogames

- Arcade-History.com entry for *Donkey Kong*

- *Donkey Kong* at MobyGames

- *Donkey Kong* guide at StrategyWiki

4.23   Dragonfire (video game)

Dragonfire is a 1982 video game developed and published by Imagic in which the player grabs treasure guarded by a dragon while avoiding fireballs. It was originally released for the Atari 2600 then ported to other systems.

The game's source code was made available by developer Bob Smith on May 24, 2003.[1]

4.23.1   Gameplay

Each level of *Dragonfire* consists of two stages. The first stage is a side view of the character trying to cross a drawbridge to reach a castle. To traverse the bridge, the player must duck under high fireballs and jump over low fireballs (or perform the improbable-looking kneeling jump). Upon success, the second stage begins, which has a more top-down point of view and wherein the player must guide the character around the room collecting treasure and dodging

more fireballs spewed by a dragon that patrols the bottom of the screen. Every piece of treasure in the room must be collected in order for a door to appear, which takes the player to the next level. A single hit from a fireball in either stage will deplete one of the player's seven initial lives. Gameplay in each level is identical, except that the character and fireballs get progressively faster.

4.23.2   Reception

Electronic Games in 1983 described *Dragonfire* as "especially useful as an introduction to fantasy gaming for younger players —while still having enough thrills to please the rest".[2] The game would go on to receive a Certificate of Merit in the category of "1984 Best Videogame Audio-Visual Effects (Less than 16K ROM)" at the 5th annual Arkie Awards.[3]:42

4.23.3   See also

- List of dragon video games

4.23.4   References

[1] Dragonfire_source

[2] "The Players Guide to Fantasy Games". *Electronic Games*. June 1983. p. 47. Retrieved 6 January 2015.

[3] Kunkel, Bill; Katz, Arnie (January 1984). "Arcade Alley: The Arcade Awards, Part 1". *Video* (Reese Communications) **7** (10): 40–42. ISSN 0147-8907.

4.23.5   External links

- The Atari 2600 version of *Dragonfire* can be played for free in the browser at the Internet Archive

4.24   Frantic (video game)

Frantic is a Commodore VIC-20 space shoot-em-up video game published by Imagine Software in 1982. The game involves the player piloting a space ship whilst trying to keep an X and Y axis centered on the enemy, which enters the field of play at varying speeds and directions. Slower enemies appear horizontally and quicker enemies diagonally. The game's title alludes to the fact that the game is timed, as fuel levels deplete during play. Centering the X and Y axis to target enemies involves engaging thrusters which in turn burns fuel. The game can be played with either a joystick or keyboard.

The game was released in cassette form only and was initially priced at £5.50.

4.24.1 Other formats

According to the World of Spectrum, *Frantic* was due for a Sinclair Spectrum 48K release, but this never happened.[1]

4.24.2 References

[1] Frantic. World of Spectrum.

4.25 Frogger

This article is about the 1981 video game. For other uses, see Frogger (disambiguation).

Frogger (フロッガー) is a 1981 arcade game developed by Konami and licensed for North American distribution by Sega-Gremlin. It is regarded as a classic from the golden age of video arcade games, noted for its novel gameplay and theme. The object of the game is to direct frogs to their homes one by one by crossing a busy road and navigating a river full of hazards. The Frogger coin-op is an early example of a game with more than one CPU, as it used two Z80 processors.[4]

By 2005, *Frogger* in its various home video game incarnations had sold 20 million copies worldwide, including 5 million in the United States.[5]

4.25.1 Gameplay (Arcade Version)

The game starts with three, five, or seven frogs (lives), depending on the settings used by the operator. The player guides a frog which starts at the bottom of the screen, to his home in one of 5 slots at the top of the screen. The lower half of the screen contains a road with motor vehicles, which in various versions include cars, trucks, buses, dune buggies, bulldozers, vans, taxis, bicyclists, and/or motorcycles, speeding along it horizontally. The upper half of the screen consists of a river with logs, alligators, and turtles, all moving horizontally across the screen. The very top of the screen contains five "frog homes" which are the destinations for each frog. Every level is timed; the player must act quickly to finish each level before the time expires.

The only player control is the 4 direction joystick used to navigate the frog; each push in a direction causes the frog to hop once in that direction. On the bottom half of the

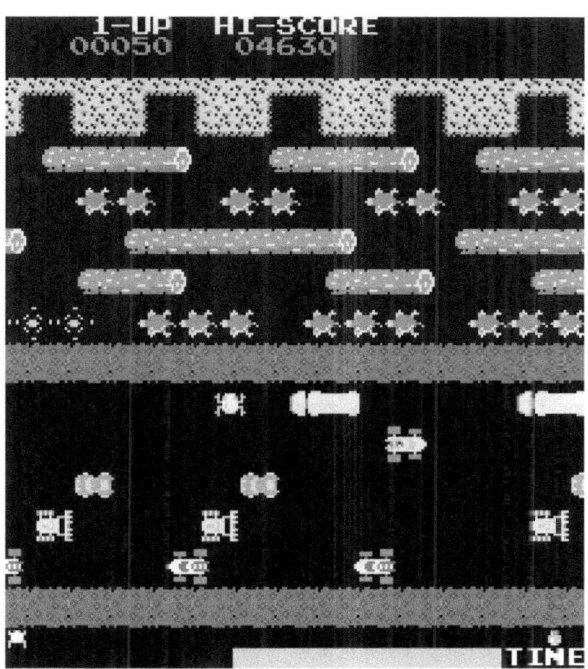

Screenshot of Frogger

screen, the player must successfully guide the frog between opposing lanes of trucks, cars, and other vehicles, to avoid becoming roadkill.

The middle of the screen, after the road, contains a median where the player must prepare to navigate the river.

By jumping on swiftly moving logs and the backs of turtles and alligators except the alligator jaws, the player can guide his or her frog safely to one of the empty lily pads. The player must avoid alligators sticking out at one of the five "frog homes", snakes, and otters in the river, but may catch bugs or escort a lady frog for bonuses. When all five frogs are directed home, the game progresses to the next level with an increased difficulty. After five levels, the game gets briefly easier yet again gets progressively harder to the next fifth level.

Softline in 1982 stated that "*Frogger* has earned the ominous distinction of being 'the arcade game with the most ways to die'".[6] There are many different ways to lose a life (illustrated by a "skull and crossbones" symbol where the frog was), including:

1. Being hit by or running into a road vehicle

2. Jumping into the river's water

3. Running into snakes, otters or into an alligator's jaws in the river

4. Jumping into a home invaded by an alligator

5. Staying on top of a diving turtle until it has completely submerged

6. Riding a log, alligator, or turtle off the side of the screen

7. Jumping into a home already occupied by a frog

8. Jumping into the side of a home or the bush

9. Running out of time

Frogger is available as a standard upright or cocktail cabinet. The controls consist solely of a 4-direction joystick used to guide the frog's jump direction. The number of simultaneous players is one, and the game has a maximum of two players.

The game's opening tune is the first verse of a Japanese children's song called *Inu No Omawarisan* (The Dog Policeman). The song remained intact in the US release. Other Japanese tunes that are played during gameplay include the themes to the anime Hana no Ko Lunlun and Araiguma Rascal.

4.25.2   Legacy

In addition to inspiring numerous clones, this game inspired an unofficial sequel by Sega in 1991 called *Ribbit* which featured improved graphics and simultaneous two-player action.

4.25.3   Ports

Frogger was ported to many contemporary home systems. Parker Brothers received the license from Sega for cartridge versions, while Sierra gained the magnetic media rights. Several platforms were capable of accepting both ROM cartridges and magnetic media, thus these systems, such as the Commodore 64, received multiple versions of the game.[7] Sierra also sublicensed their magnetic-media rights to developers who published for systems not normally supported by Sierra (e.g. Cornsoft published the official TRS-80, Timex Sinclair 1000 and Timex Sinclair 2068 ports); because of this, even the Atari 2600 received multiple releases: a cartridge from Parker Bros. and a cassette for the Supercharger from Starpath. The Tomy Tutor version was directly licensed from Konami themselves, although it is not clear if they developed it.

Parker Bros. produced cartridge ports of *Frogger* for the Atari 2600, Intellivision, Atari 5200, ColecoVision, Atari 8-bit computers, Commodore VIC-20 and 64. Sierra released disk and/or tape ports for the C64 (which as a result ended up with two versions of the game), Apple II,

the original 128k Macintosh, IBM PC, Atari 2600 Supercharger, and the above-mentioned versions for the TRS-80 Color Computer and Sinclair developed by UK-based Cornsoft. Parker Bros. spent $10 million on advertising *Frogger*, along with *The Empire Strikes Back*, larger than the $6 million marketing budget for a movie at the time.[8] Parker Brothers sold 3 million cartridges of both *Frogger* and *The Empire Strikes Back*, with *Frogger* alone being the company's most successful first-year product, beating the sales and revenues of Merlin, their previous best-seller.[9] Coleco also released stand-alone Mini-Arcade tabletop versions of *Frogger*, which, along with *Pac-Man*, *Galaxian*, and *Donkey Kong*, sold three million units combined.[10]

A prototype game based on gameplay elements of Frogger was developed for Sega Game Gear, but never released—presumably due to legal issues between Sega and Konami. The prototype wasn't a direct port of the arcade game, as it had additional features and redesigned levels.

Frogger *disk by Sierra for PC.*

Hasbro Interactive released a vastly expanded remake of the original for Microsoft Windows and the PlayStation in 1997 (in this game, Frogger is green with an orange stripe). Unlike the original, the game consisted of multiple levels, each different than the preceding one. It was a commercial success, with the PC version alone selling nearly one million units in less than four months.[11]

In 1998, Hasbro released a series of ports of the original game for the Sega Genesis, Super NES, Game com, Game Boy, and Game Boy Color. Each port featured the game with different graphics, with the Sega Genesis port in particular featuring the same graphics of the original arcade game. The Sega Genesis and SNES versions are notable for both being the last games released for those consoles in North America. Despite using the same box art of the 1997 remake, the ports are otherwise unrelated to that game.

In 2005, InfoSpace teamed up with Konami Digital Entertainment to create the mobile game *Frogger for Prizes*,*[12] in which players across the U.S. compete in multiplayer tournaments to win daily and weekly prizes. In 2006, the mobile game version of *Frogger* grossed over $10 million in the United States.*[13] A Java port of the game is available for compatible mobile phones.

A port of *Frogger* was released on the Xbox Live Arcade for the Xbox 360 on July 12, 2006. It was developed by Digital Eclipse and published by Konami. It has two new gameplay modes: Versus speed mode and Co-op play. Some of the music, including the familiar Frogger theme, was removed from this version and replaced with other music. This version was included in the compilation Konami Classics Vol. 1.

4.25.4 Clones

In addition to these official releases, there are numerous unofficial clones including *Ribbit* for the Apple II in 1981, *Froggy* for the ZX Spectrum released by DJL Software in 1984, Acornsoft's *Hopper* (1983) for the BBC Micro and Acorn Electron, A&F Software's *Frogger* (1983) for BBC Micro and ZX Spectrum, Solo Software's *Frogger* for the Sharp MZ-700 in the UK in 1984, and a version for the NewBrain under the name *Leap Frog*.

The 1981 Atari 2600 game *Freeway* is often considered a clone of *Frogger*, but each game was developed independently of the other, and both were released in 1981.

Pacific Coast Highway (1982), for the Atari 8-bit family, splits the gameplay into two alternating screens: one for the highway, one for the water.*[14]

Preppie! (1982), for the Atari 8-bit family, changes the frog to a preppy retrieving golf balls at a country club.

Frostbite (1984), for the Atari 2600, uses the *Frogger* river gameplay with an arctic theme.

Crossy Road (2014), for iOS and Android, has a randomly generated series of road and river sections. The game is one endless level, with only one life and a single point given for each forward hop

4.25.5 Sequels

Unlike the arcade version, the home versions had numerous sequels, including:

- *Frogger II: ThreeeDeep!* (1984)

- *Frogger* (1997)

- *Frogger 2: Swampy's Revenge* (2000)

- *Frogger: The Great Quest* (2001)

- *Frogger's Adventures: Temple of the Frog* (2001)

- *Frogger Advance: The Great Quest* (2002)

- *Frogger Beyond* (2002)

- *Frogger's Adventures 2: The Lost Wand* (2002)

- *Frogger's Journey: The Forgotten Relic* (2003)

- *Frogger's Adventures: The Rescue* (2003)

- *Frogger: Ancient Shadow* (2005)

- *Frogger: Helmet Chaos* (2005)

- *Frogger Puzzle* (2005)

- *Frogger's 25 Anniversary* (Xbox 360) (2006)

- *Frogger 25th, Frogger Evolution* (mobile game) (2006)*[15]

- *My Frogger Toy Trials* (Nintendo DS) (2006)

- *Frogger Launch* (2007)

- *Frogger Hop Trivia* (arcade) (2007)

- *Frogger 2* (Xbox 360) (2008), the third game to call itself "*Frogger 2*", for Xbox Live Arcade

- *Frogger Returns* (Wii/PlayStation 3) (2009)*[16]

- *Frogger Beats 'n' Bounces* (2008)

- *Frogger Inferno* (iOS) (2010)

- *Frogger* (Windows Phone) (2010)

- *Frogger 3D* (Nintendo 3DS) (2011)

- *Frogger Decades* (iOS) (2011)

- *Frogger Free* (iOS) (2011)

- *Frogger: Hyper Arcade Edition* (Wii) / (PlayStation 3) / (Xbox 360) (2012)

- "Frogger Free" (iPhone/ Android) (2012)

- *Frogger's Crackout* (Windows Store) (2013)

In many of the recent games (starting with *Frogger: The Great Quest*), Frogger is shown as bipedal, wearing a shirt with a crossed-out truck.

4.25.6 In popular culture

In film and television

- In 1983, *Frogger* made its animated television debut as a segment on CBS' *Saturday Supercade* cartoon lineup. On the series, Frogger was voiced by Bob Sarlatte, and worked as an investigative reporter. Frogger was joined by two frog characters created for the series, his gruff boss Tex and his female colleague Shelly. After only one season, Frogger and the Pitfall Harry segment were replaced by Kangaroo and Space Ace. *Saturday Supercade* has yet to be released on DVD or streaming.

- In 1995, a *VR Troopers* episode featured a usurping frog monster called "Amphibidor" that hated being called "Frogger" by Ryan, because he does not "hop across roads or rivers to avoid cars and alligators just to get home".

- In 1998, the game was featured in the *Seinfeld* episode "The Frogger".[17] Jerry and George visit a soon-to-be-closed restaurant they frequented as teenagers and discover the *Frogger* machine still in place, with George's decades-old high score still recorded. He buys the machine and tries to get it home without letting it lose power, which would erase the score with his initials "GLC" (in reality, *Frogger* does not actually let players enter their initials). After rigging the machine up with a battery, his attempt to navigate it across a busy New York street is a direct parody of the game (which uses the same sound effects and is shown from a top down view) and ends with the machine being smashed. George's score was 860,630 points, a score once thought to be unachievable on an actual Frogger arcade machine, not to mention that the real game has only a 5 digit score counter.

- In the MTV Movie Awards 2003 sketch, "The MTV Movie Awards Reloaded" has the Architect (Will Ferrell) saying that, while having created *Q*bert* and *Dig Dug*, he did not create *Frogger* but he came up with the name for it because it was going to be called "Highway Crossing Frog". The last half of the joke is actually true - "Highway Crossing Frog" was the working title for *Frogger*.[18]

- Frogger appears in the Disney animated film *Wreck-It Ralph*. In the second trailer for the film, he can be seen in the Game Central Station hopping away from Ralph upon seeing him.[19]

- Frogger appeared in the film *Pixels*.[20]

- In the *Burn Notice* episode Bloodlines, season 5 episode 2 (air date 30/06/2011), the character Sam Axe (Bruce Campbell) likens the Japanese Yakuza people trafficker Takeda to having played a game of Frogger, as he ran across busy car traffic in the road to escape his pursuers.

In music

- In 1982, Buckner & Garcia recorded a song called "Froggy's Lament", using sound effects from the game, and released it on the album *Pac-Man Fever*. The song begins:

  Froggy takes one step at a time
  The way that he moves has no reason or rhyme
  He hops and jumps, dodges and ducks

  Cars and buses, vans and trucks.

- Bad Religion has also recorded a song called "Frogger" about the traffic in Los Angeles, in which the singer claims to be "playing *Frogger* with my life".

- Lagwagon's compilation titled "Let's Talk About Leftovers" featured the song from the game played on bass on loop at the end of the CD.

- Paul and Storm wrote and performed a comedic song called Frogger! The Frogger Musical.

- The line "Frogger bass" appears in Deee-lite's song "'Say Ahhh...'".

- Frogger is also named in the song *Abiura di me* of Italian rapper *Caparezza*.

- In the 2010 music video of "My Feelings For You" by Avicii and Sebastien Drums, the main character of the video is seen chasing another character through the gameplay of Frogger.[21]

- The line "When you see me better cross the street; Frogger" appears in the song *Semicolon* in the 2013 album *The Wack Album* by The Lonely Island.

Other

- In 2006, a group in Austin, Texas used a modified Roomba dressed as *Frogger* to play a real-life version of the game. Although the group expected the Bluetooth controlled machine to be crushed on its first time across, the modified Roomba was able to get across the street 10 times (40 lanes) and survive for 15 minutes before it was "killed" by an SUV.[22]

- Frogger is also the name given to a transposon ("jumping gene") family in the fruitfly Drosophila melanogaster.*[23]

- On November 5, 2011, a live-action tribute to Frogger called Field Frogger was played at the Come Out and Play Festival in San Francisco.*[24]

4.25.7 Highest score

On July 15, 2012, Michael Smith of Springfield, Virginia, USA, scored a Frogger world record high score of 970,440 points.*[25] This beat Pat Laffaye's score of 896,980 from December 22, 2009. These are the only two scores that have been verified as having beaten the fictional George Costanza *Seinfeld* score of 860,630 points.*[26]*[27]

4.25.8 References

[1] *Frogger (Konami)* at the Arcade History database

[2] *Frogger (Sega)* at the Arcade History database

[3] Frogger

[4] *Frogger* at the Killer List of Videogames

[5] "Konami's Frogger and Castlevania Nominated for Walk of Game Star" (Press release). Konami. 2005-10-11. Archived from the original on 2013-02-02. Retrieved 2009-03-21.

[6] Rose, Gary and Marcia (November 1982). "Frogger". *Softline*. p. 19. Retrieved 27 July 2014.

[7] Moriarty, Tim (May 1984). "Frogger". *Ahoy!*. pp 52–53. Retrieved 27 June 2014.

[8] Harmetz, Aljean (January 15, 1983). "New Faces, More Profits For Video Games". *Times-Union*. p. 18. Retrieved 28 February 2012.

[9] Rosenberg, Ron (December 11, 1982). "Competitors Claim Role in Warner Setback". *The Boston Globe*. p. 1. Retrieved 6 March 2012.

[10] "More Mini-Arcades A Comin'". *Electronic Games* **4** (16): 10. June 1983. Retrieved 1 February 2012.

[11] Reidy, Chris (March 17, 1998). "Hasbro Unit Pays $5m for Atari Arcade Game Rights Plans Include New Versions for Users of PCs, Playstation". *The Boston Globe*. Retrieved 6 March 2012. Just before the holidays, Hasbro Interactive introduced a PC version of Frogger; in less than four months, it has sold nearly one million units

[12] Video Game News - Konami Digital Entertainment and InfoSpace Partner to Create Mobile Game Frogger for Prizes

[13] "Frogger Mobile Games Exceed $10 Million In The US". GameZone. September 12, 2006. Retrieved 20 April 2012.

[14] "Atari 8-bit - Pacific Coast Highway [Datasoft] 1982". *YouTube*.

[15] Konami Mobile: Frogger Archived September 29, 2007 at the Wayback Machine

[16] "Konami reveals new screenshots for *Frogger Returns*" (PDF). Konami Digital Entertainment. 4 November 2009. Retrieved 9 November 2009.

[17] ""Seinfeld" The Frogger (1998)". Imdb.com. Retrieved 2011-02-09.

[18] "Frogger Timeline and Biography". Twoop.com. Retrieved 2011-02-09.

[19] disneyanimation (2012-09-13). "Wreck-It Ralph Trailer". *YouTube*. Retrieved 2012-10-13.

[20] "Classic video game characters unite via film 'Pixels'". *Philstar*. July 23, 2014. Retrieved July 23, 2014.

[21] "Dance music gives nod to 8-bit era". Megabitsofgaming.com. Retrieved 2012-06-20.

[22] Terdiman, Daniel. "Roomba takes Frogger to the asphalt jungle - CNET News.com". News.com.com. Retrieved 2011-02-09.

[23] "FlyBase Transposon Report: Dmel\Frogger".

[24] "Come Out and Play San Francisco 2011".

[25] "Classic Frogger arcade world record squashed once again". Retrieved 17 November 2012.

[26] "Twin Galaxies' Frogger High Score Rankings". Retrieved 1 January 2010.

[27] Moriarty, Tim (May 1984). "Frogger". *Ahoy! Magazine*. pp. 52–53. Retrieved 25 April 2015.

4.25.9 External links

- *Frogger* at the Killer List of Videogames

- *Frogger* at the Arcade History database

- *Frogger* at MobyGames

- *Frogger* guide at StrategyWiki

- *Frogger Series* at DMOZ

- Comparison of Sierra Frogger home computer releases

- Randall, Neil (August 1983). "Frogger for the 64". *Compute! Gazette*. p. 58. Retrieved 4 December 2013.

4.26   Galaxian

For other uses, see Galaxian (disambiguation).

Galaxian (ギャラクシアン *Gyarakushian*) is an arcade game that was developed by Namco and released in October 1979. It was published by Namco in Japan and imported to North America by Midway that December. A fixed shooter game in which the player controls a spaceship at the bottom of the screen, and shoots enemies descending in various directions, it was designed to compete with Taito Corporation's successful earlier game *Space Invaders* (which was released in the previous year, and also imported to the US by Midway Games).

The game was highly popular for Namco upon its release, and has been a focus of competitive gaming ever since. It spawned a successful sequel, *Galaga*, in 1981, and the lesser known *Gaplus* and *Galaga '88* in 1984 and 1987 respectively, as well as many later ports and adaptations. Along with its immediate sequel, it was one of the most popular games during the golden age of arcade video games.

4.26.1   Description

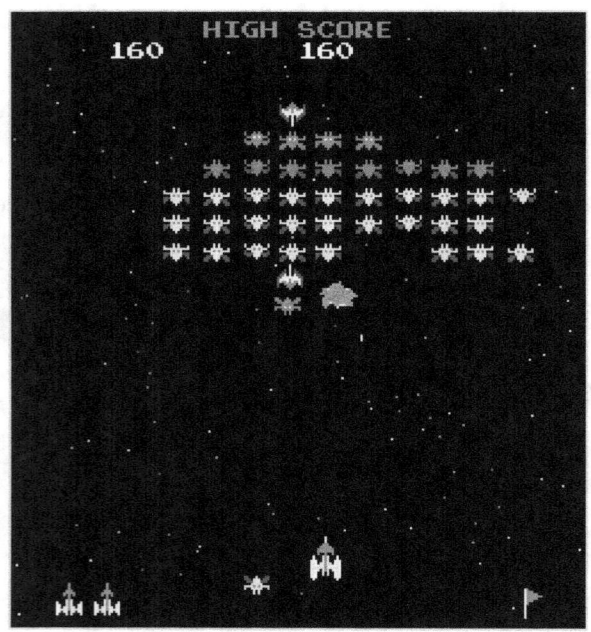

Gameplay screenshot

Galaxian expanded on the formula pioneered by *Space Invaders*. As in the earlier game, *Galaxian* features a horde of attacking aliens that exchanged shots with the player. In contrast to *Space Invaders*, *Galaxian* added an element of drama by having the aliens periodically make kamikaze-like dives at the player's ship, the *Galaxip*.[1] This made it the first game to feature enemies with individual personalities.[2] The game's plot consists of a title screen that displayed the message "WE ARE THE GALAXIANS / MISSION: DESTROY ALIENS".[3]

Galaxian was very successful for Namco and introduced several "firsts". Although not the first color video game, Galaxian took RGB color graphics a step further with multicolored animated sprites and explosions, different colored fonts for the score and high score, the scrolling starfield, and graphic icons that show the number of lives left and how many stages the player had completed. It also features a crude theme song and more prominent background "music." These elements combine to create a look and feel that would set the standard for arcade games in the 1980s such as *Pac-Man*.

4.26.2   Gameplay

The gameplay is relatively simple. Swarm after swarm of alien armies attack the player's ship that moves left and right at the bottom of the wraparound screen. The ship can only have one shot on screen at a time. The player defeats one swarm, only to have it replaced by another more aggressive and challenging swarm in the next stage. A plain and repetitive starfield scrolls in the background.

4.26.3   Development and release

The game was developed by Namco in 1979, and released in Japan that year. It was designed to build and improve upon the formula of Taito's game *Space Invaders*, which revolutionized the gaming industry upon its release a year earlier. *Galaxian* incorporated new technology into its dedicated arcade system board, the Namco Galaxian. Unlike *Space Invaders*, which was black and white and featured enemies that could only move vertically and horizontally as they descended, *Galaxian* had a color screen and enemies that descended in patterns and came from various directions. The result was more complex and difficult game play.[4]

Soon after the Japanese release Namco partnered with the American company Midway to release the game in North America. Midway had previously published *Space Invaders* in the market, but had to seek new foreign partners when Taito decided to market their games themselves.

Standard arcade games

- *Galaga* (1981)

- *Gaplus* (1984)

A Galaxian arcade game displayed at the mNACTEC museum in Terrassa, Catalonia

- *Galaga '88* (1987)

- *Galaga Arrangement* (1995) - released as part of *Namco Classic Collection Vol. 1*

- *Galaxian* was one of the most widely pirated motherboards during the early '80s. Numerous hacks were made of the game and featured slightly redesigned enemy characters and special bonus stages. The scrolling starfield and death explosion were still familiar as those from *Galaxian*, however. These hacks include: *Galaxian Part 4*, *Galaxian Part X*, *Galaxian Turbo*, and *Super Galaxians*.

Arcade laserdisc

- *Galaxian^3* (1990) - Galaxian^3: Project Dragoon (Theatre 6) for six players on two 110-inch RGB projectors - 18-foot-wide (5.5 m) screen

- *Attack of the Zolgear* (1994) - a ROM and laserdisc upgrade for *Galaxian^3*

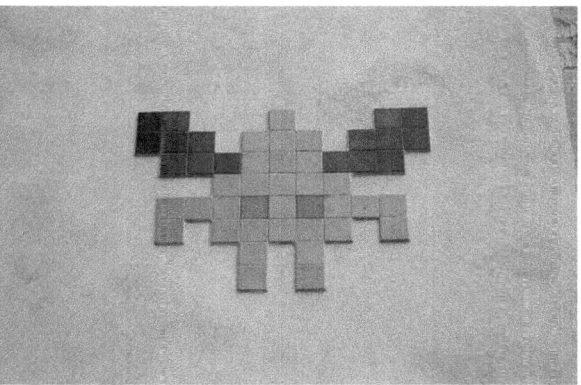

Street art mosaic of one of the characters (upside-down while in formation, this way up while attacking) made by Invader

4.26.4 Reception

Video magazine in 1982 reviewed the Astrocade version of *Galaxian* (named *Galactic Invasion*), noting that the graphics were inferior to the coin-op and PC versions, but praising the play-action as "magnificent" compared to other console versions.[5]:43 The Astrocade version would later be awarded a Certificate of Merit for "Best Arcade-to-Home Video Game Translation" at the 4th annual Arkie Awards.[6]:108 *Arcade Express* reviewed the Atari 5200 version in November 1982 and scored it 7 out of 10.[7] *Home Computing Weekly* in 1983 gave the Spectrum version of *Galaxian* 3/5 stars describing it as a well-written version and praising the graphics as fast although flickery [8] *Softline* in 1983 criticized the Atari 8-bit version of the game for being shipped on cartridge, which raised its cost, and stated that "this game becomes tedious very quickly".[9] *Famicom Tsūshin* in 1995 scored the Game Boy version of the game a 24 out of 40.[10]

4.26.5 Legacy

Galaxian has spawned several follow-up games. The most popular of these was its immediate successor, *Galaga*, which largely eclipsed its predecessor in popularity, introducing aliens attacking in intricate formations, multiple shots, and bonus stages. A third official sequel, *Gaplus*, was released in 1984. As with *Galaga*, this was a fixed shooter, with limited vertical movement (like *Centipede*). However, by 1984 the novelty of the *Space Invaders* formula had faded, and it was no longer successful. A fourth sequel, *Galaga '88*, was released in 1987, and imported to North America by Atari Games; and a fifth and final sequel, *Galaga Arrangement*, was released as part of the *Namco Classic Collection Vol. 1*, in 1995.

Ports

The original arcade version of *Galaxian* has been ported to many different systems. These include:

- Apple II

- Atari 400/800

- Atari 2600

- Atari 5200

- Bally Astrocade (*Galactic Invasion*)

- Coleco VFD table top

- ColecoVision

- Commodore VIC-20

- Commodore 64

- Dreamcast

- Game Boy (bundled with its direct successor, *Galaga*)

- IBM PC

- Mobile Java

- MSX (Europe and Japan only)

- NEC PC-8801

- Nintendo Famicom (Japan only)

- Sharp X1

- Virtual Console (Japan only)

- ZX Spectrum

Coleco also released stand-alone Mini-Arcade tabletop versions of *Galaxian*, which, along with *Pac-Man*, *Donkey Kong*, and *Frogger*, sold three million units combined.*[11] A port for the Game Boy Color was planned, but never released.

Galaxian has also been released as part of the *Namco Museum* series of collections across several platforms:

- Dreamcast (*Namco Museum*)

- Game Boy Advance (*Namco Museum Advance*)

- Nintendo 64 (*Namco Museum 64*)

- Nintendo DS (*Namco Museum DS*)

- PlayStation (as part of *Namco Museum Volume 3*)

- PlayStation 2, Xbox, Nintendo GameCube and Microsoft Windows (*Namco Museum: 50th Anniversary Arcade Collection*)

- PlayStation Portable (*Namco Museum Battle Collection*)

- Wii (*Namco Museum Remix*)

Galaxian was also released on Microsoft Windows in 1995 as part of *Microsoft Return of Arcade*. The game was also released as part of the *Pac-Man's Arcade Party* 30th Anniversary arcade machine.

The game has also been seen in Jakks Pacific's "Plug It In & Play" TV game controllers.

Galaxian, along with *Galaga*, *Gaplus*, and *Galaga '88*, was "redesigned and modernized"*[12] for an iPhone app compilation called the *Galaga 30th Anniversary Collection*, released in commemoration of the event by Namco Bandai.*[13]

Games featuring elements of Galaxian

- Entex Industries released a handheld electronic game called *Galaxian 2* in 1981. The game is called *Galaxian 2* because it has a two-player mode. It is not a sequel, as there is no *Entex Galaxian*.*[14]

- The video game *Gorf*, by Bally Midway, has a *Galaxian* stage.

- A version of the game can be unlocked in Midway's *Mortal Kombat 3* on the Sega Genesis.

- The game can be seen on the home stretch of various *Ridge Racer* circuits. On the PlayStation version, while the player is waiting for *Ridge Racer* to load, they can play a quick game of *Galaxian*. Also, on *Ridge Racer 64* and *Ridge Racer DS*, a car is available called the "Galaxian Paradise" (in Ridge Racer 64, the car is named "White Angel" like *Ridge Racer* and *Ridge Racer Revolution*).

- The boss of the Space Zone in the game *Pac-Man World* for PlayStation is inspired by the game *Galaxian*. However, the stage itself is similar to *Galaga*.

- Japanese RTS game *New Space Order* by Namco Bandai Games is set in the same U.G.S.F. universe as the setting of Galaxian.

Games featuring the Galaxian flagship The *Galaxian* flagship (also known as the "Galboss") has made numerous cameo appearances in other Namco games (like

Namco's signature character *Pac-Man* and the Special Flag from *Rally-X*, which also went on to become recurring items in other Namco games).

- *Pac-Man* (1980): The flagship makes an appearance as a bonus item on Rounds 9 and 10, and is worth 2000 points when it is eaten.

- *Galaga* (1981): The flagship makes an appearance as one of the "transform" ships, and attacks by splitting into two, then three clones of itself. If all three are killed, they are worth 3000 points, and this was the first time they reappeared as an evil character.

- *Dig Dug* (1982): The flagship makes an appearance as a bonus item on rounds 16 and 17, and is worth 7000 points when collected.

- *Super Pac-Man* (1982): All regular edible items on Rounds 15, 31, 47, and 63 are flagships, and they are worth 150 points each. Starting from their second appearance on Round 31, they are 160 points instead (given that every regular item from Round 16 onwards is).

- *Pac-Man Plus* (1982): The flagship's role is exactly the same as the role it was given in the original *Pac-Man* arcade game.

- *Pac & Pal* (1983): The flagship makes an appearance as one of the "special items" (that make Pac-Man turn blue when eaten), and allows him to stun the ghosts for a short while by spitting a *Galaga*-style tractor beam. It is worth 1000 points if it is eaten.

- *Pac-Land* (1984): When the ghosts fly past in airplanes, they sometimes drop flagships instead of miniaturized ghosts, and they are worth 7650 points (765 being Namco's goroawase number in Japanese) if Pac-Man jumps up and eats them before they hit the ground.

- *Super Xevious* (1984): The flagship makes its first appearance as an enemy since *Galaga* and in a silver form, and sometimes several of them attack the Solvalou at once by flying towards it from the top of the screen. They are worth 300 points each when killed.

- *Genpei Tōma Den* (1986): The flagship makes an appearance in an alternate colour palette as one of the four special items which are left behind by the flame-spitting stone lions in the Small Mode stages when they are killed. It is worth 1000 points when collected.

- *Quester* (1987): The blocks on the fifth round are arranged to look like a flagship - but this is not an official reappearance.

- *Pac-Mania* (1987): The flagship makes a 3D appearance as a special item in two different forms, the second one being the silver form from *Super Xevious*. The regular versions are worth 7650 points if eaten and the silver versions are worth 9000 points if eaten.

- *Pistol Daimyo no Bōken* (1990): The flagship makes an appearance as an enemy, along with the other *Galaxian* characters, and they attack by flying towards Pistol Daimyo while firing shots at him.[15]

- *Tinkle Pit* (1993): The flagship makes an appearance with two of the other *Galaxian* characters (Red Alien and Galaxip), but this time they appear as three of the game's forty-six hidden bonus items (on Stages 6, 17 and 24). It is worth 800 points if collected.

- *Tekken* (1994 - Arcade, 1995 - PlayStation) and *Tekken 2* (1995 - Arcade, 1996 - PlayStation): Winning at least seven rounds in "Arcade Vs." mode will reveal the *Galaxian* flagship on the lower left (or right) hand corner of the screen. In order for this to be seen, the arcade operator should have set the "Number of Wins Shown By" option to "Fruit".

- *Namco Classic Collection Vol. 1* (1995): The flagship makes an appearance in *Galaga Arrangement*, as a Challenging Stage enemy in Space-Plant Zone (Stage 20) and as a regular stage enemy in Space-Flower Zone (Stage 26). They are worth 150 points when killed during their appearances as regular enemies, but are worth 300 points when killed during their appearances as Challenging Stage enemies.

- *Namco Classic Collection Vol. 2* (1996): The flagship makes its appearances in both *Pac-Man Arrangement* and *Dig Dug Arrangement*. In *Pac-Man Arrangement*, the Galaxian Flagship makes its appearance in Rounds 15 and 16, and it is worth 5000 points if Pac-Man eats it - and in *Dig Dug Arrangement*, it appears in Rounds 17 and 18, and it is worth 7000 points if Dig Dug picks it up.

- *Pac-Man World* (1999): The flagship again appears in a *Pac-Man* game, and it must be collected in order to access the mazes.

- *Pac-Man World 2* (2002): The flagship teleports Pac-Man to mazes. Its point value will be the same as the points earned for the maze it teleported him to (if he manages to complete it) - with an added bonus of 2000 points (its value from the original *Pac-Man*).

- *Pac-Man World 3* (2005): The flagship's role is exactly the same as it was given in *Pac-Man World 2* three years previously.

- *Namco Museum Battle Collection* (2005): The arrangement versions of Pac-Man and Dig Dug, later called Pac-Man Remix and Dig Dug Remix in the iOS version, feature the flagship. Pac-Man Remix features both the flagship (worth 3200), and the Red Galaxian (worth 2800) as bonus fruits, while in Dig Dug Remix, the flagship is a bonus vegetable and is worth 7000 (just like it was in the original Dig Dug).

- *Dig Dug: Digging Strike* (2005): Just like the original *Dig Dug*, the flagship appears as a vegetable on Round 13, except it is only worth 6000 points when Dig Dug picks it up this time (as opposed to the 7000 it was worth when he picked it up in the original).

- *Pac-Man Championship Edition* (2007): The flagship reappears as a bonus fruit, but this time it's joined by the Galaga Boss, King Gaplus and two drones, one each from *Galaxian* and *Galaga*. Their respective point values are all unknown.

- *Pac-Man Championship Edition DX* (2010): The flagship, the *Galaxian/Galaga* drones, the Galaga Boss, and the King Gaplus serve exactly the same purpose as they did in the original *Championship Edition* three years previously (they are only bonus fruits).

- *Tekken Tag Tournament 2* (2012): The character customization allows the players to add decals to their fighters' clothes. These decals include the flagship, along with other classic Namco sprites (like *Pac-Man*, Pooka and Fygar from *Dig Dug*, and *Mappy*).

- *Super Smash Bros.   for Nintendo 3DS and Wii U* (2014): The flagship appears as one of the sprites Pac-Man summons in his "Bonus Fruit" attack, referencing its appearance in Pac-Man's original arcade game. The flagship, along with the player-controlled ship from Galaxian, also have a chance of being summoned during Pac-Man's "Namco Roulette" taunt.

In the competitive arena

The Galaxian world record has been the focus of many competitive gamers since its release. The most famous Galaxian rivalry has been between British player Gary Whelan and American Perry Rodgers, who faced off at Apollo Amusements in Pompano Beach, Florida, USA, on April 6–9, 2006. Whelan held the world record with 1,114,550 points,[16] until beaten by newcomer Aart van Vliet, of the Netherlands, who scored 1,653,270 points on May 27, 2009 at the Funspot Family Fun Center in Weirs Beach, New Hampshire, USA.[17]

4.26.6   References

[1] *Galaxian* at the Killer List of Videogames

[2] "Arcade Games". *Joystick* **1** (1): 10. September 1982.

[3] "Galaxian Screen Grab, Killer List of Videogames". 2010-06-01.

[4] Kent, Steven (2001). "The Golden Age". *The Ultimate History of Video Games*. Random House Digital. ISBN 0-7615-3643-4.

[5] Kunkel, Bill; Katz, Arnie (May 1982). "Arcade Alley: Astrovision's Rising Star". *Video* (Reese Communications) **6** (2): 42–43. ISSN 0147-8907.

[6] Kunkel, Bill; Katz, Arnie (February 1983). "Arcade Alley: The Fourth Annual Arcade Awards". *Video* (Reese Communications) **6** (11): 30, 108. ISSN 0147-8907.

[7] http://www.digitpress.com/library/newsletters/arcadeexpress/arcade_express_v1n7.pdf#page=6

[8] Harris, Ron ed. *Spectrum Software Reviews - Testing, testing... 10 programs for the Spectrum: Galaxian Spectrum £4.95*. Home Computing Weekly. Issue 4. Pg.41. 29 March 1983.

[9] Bang, Derrick (May–Jun 1983). "Beating the Classics". *Computer Gaming World*. p. 43. Retrieved 28 July 2014.

[10] NEW GAMES CROSS REVIEW: ギャラガ ＆ ギャラクシアン. Weekly Famicom Tsūshin. No.344. Pg.32. 21 July 1995.

[11] "More Mini-Arcades A Comin'". *Electronic Games* **4** (16): 10. June 1983. Retrieved 1 February 2012.

[12] "Jesse David Hollington, "Namco releases Galaga 30th Anniversary Collection"". 2011-06-09.

[13] "Galaga 30th Anniversary Collection information from Apple iTunes". 2011-06-09.

[14] Morgan, Rik. "Entex Galaxian 2". Retrieved 2010-10-23. Entex Galaxian 2, based on Bally/Midway's Galaxian arcade game.

[15] ピストル大名の冒険

[16] "Guinness World Records 2008 - Gamer's Edition", page 243

[17] "Twin Galaxies' Galaxian High Score Rankings". 2009-12-27.

4.26.7 External links

- *Galaxian* at the Killer List of Videogames

- *Galaxian* at the Arcade History database

- *Galaxian* at MobyGames

- *Galaxian* guide at StrategyWiki

- Galaxian at DMOZ

- *Galaxian* at World of Spectrum

4.27 Ghost Manor

This article is about the video game. For the comic book, see Ghost Manor (comics).

Ghost Manor is a horror video game that was released by Xonox in 1983 for the Atari 2600 and the Vic-20. It was generally packaged in a double ended cartridge and a cassette tape along with one of three other games in an effort to appeal to budget conscious buyers who would purchase two games for the price of one cartridge and one cassette tape. There was also a more limited release of single ended cartridges and cassette tapes containing Ghost Manor by itself. The double ended cartridges and cassette tapes paired *Ghost Manor* with the platform game *Spike's Peak*, the fighting game *Chuck Norris Superkicks*, and a strategy game called *Artillery Duel*.*[1]

A release entitled *Ghost Manor* was released by TTI for the TurboGrafx-16. It has no relation to the Atari 2600 or Vic-20 release of *Ghost Manor*.

4.27.1 Characters

- Friendly characters:*[2]

  - Girl: Playable on both the 2600 and the 7800, she sets out to rescue the helpless boy from Ghost Manor. Set the 2600's TV Type switch to "Color" to play as the Girl.

  - Boy: Playable on the 2600 only, he sets out to rescue the girl from Ghost Manor. Set the 2600's TV Type Switch to "B/W" to play as the Boy. Atari 7800 users will not be able to play as the Boy, since the 7800 does not have a TV Type switch.

- Neutral Characters:*[2]

  - Bones: A invincible skeleton in the graveyard who will give you spears for use at the Gate.

  - Rainbow Ghost: A ghost who is otherwise identical to Bones.

- Enemies:*[2]

  - Chopping Mummy: Found in the game's second stage, the Chopping Mummy is the leader of the Spooks at the Gate. He will try to stop the Boy or Girl from entering Ghost Manor. The Chopping Mummy is invincible until all of the Spooks are defeated. Contact with the Chopping Mummy's blade ends the game instantly.

  - Spooks: Any of the eight beings flying around the Gate. They consist of two green scorpions, two white skulls, three black bats, and the black knight. Defeat them using your Spears

  - Dracula: The sinister being who guards the Prison (and your friend) in the game's final stage. He is vulnerable only to the Crosses you find in the game. Contact with Dracula is lethal to your character.

4.27.2 Items

Most of the items in Ghost Manor help the character out, but one will end your game.*[2]

- Hourglass: Counts down with the game timer. If you cannot rescue your friend within four minutes, you lose.

- Spears: Used to shoot down Spooks at the Gate, also used to kill the Chopping Mummy. Only one spear is needed for each kill. Spear tokens appear to the left of the hourglass to indicate how many Spears you have collected, and a counter on the lower left indicates the exact number of Spears remaining.

- Lamp: Only present in the easy and medium difficulty settings. Used to light the interior of Ghost Manor. Without it, you must rely on lightning strikes to see.

- Cross: Used to repel Dracula. Press and hold the button during the last stage to use the power of a Cross. Release the button to stop using the Cross's power. Each Cross has about three seconds' worth of total usage before it is used up and disappears from the character's inventory. One Cross can be found in a coffin on Stage 3, and another Cross is hidden in a coffin on Stage 4. One bonus Cross is awarded when the character enters the Prison. Remaining Crosses are displayed to the right of the hourglass above where the Lamp appears if the character has the Lamp. The Crosses are hidden in different coffins each time you play.

- Moving walls: Found on Stage 3 and 4, touching these from behind will award you points. Colliding with them while they're moving toward you will end your game immediately.

4.27.3 Gameplay

The game consists of five stages:*[2]

Stage 1 (the Graveyard)

The player must guide their character around and touch Bones or the Rainbow Ghost a set number of times. In order for the contact to count, the contact must be made on a gravestone and while the player is moving. Each contact that meets these criteria awards the player a single Spear. The number of Spears that can be awarded depends on the difficulty setting of the game.

Stage 2 (the Gate)

The player must use the Spears to shoot down Spooks. When all seven Spooks are shot down, the player must use one final spear to shoot the Chopping Mummy. During this time, the Chopping Mummy will chop at the player in an attempt to touch the character with his blade and end the game. The Chopping Mummy does chop to a rhythm. Learning this rhythm enables players to avoid the Mummy's blade entirely with some practice. It is possible to run out of Spears. If that happens, the player cannot defeat any more Spooks or the Chopping Mummy, and the game will run until either the Chopping Mummy's blade comes in contact with the character or the player resets their console.

Stages 3 and 4

Although the layouts and colors change, these two stages are nearly identical. The player must move their character to the staircase in Stage 3, and then from one staircase to the other in Stage 4. Touching any of the stationary walls may stun the character. The character can press up against the trailing side of the moving walls to earn extra points, however, touching the leading side of a moving wall will end the game. Players may search the coffins in these stages by touching them from the correct side. A tone is played for each successful search. One coffin in each stage contains a Cross, which is used to repel Dracula in the final stage of the game. If the character has a Lamp, the playfield will be illuminated the entire time the character is in either stage 3 or 4. If the character does not have a Lamp, the playfield will only be illuminated for brief periods when lightning strikes near Ghost Manor.

Stage 5 (the Prison)

Situated atop Ghost Manor, the Prison is where the game's final battle takes place. Players must guide their character underneath Dracula and press the red button on their Joystick to force Dracula upward into one of the cells at the top of the screen. Players cannot force Dracula to move any direction other than up, however when Dracula is not being repelled he will follow the character left or right. To successfully restrain Dracula, players must use their character to bait him into moving underneath a cell, then press the button to force him upward into that cell. Repelling Dracula requires the character to have at least one Cross. Each Cross only lasts a few seconds, so players must keep the battle as short as possible. As soon as Dracula is restrained in either cell, the character's friend is released and will follow the character to the stairwell. Once both the character and their friend reach the stairwell, the player wins the game.

Ending

When the game is over, the Boy and Girl are both displayed in the Graveyard holding hands. If the rescue was not successful, both characters will sink into the ground along with the tombstones and Ghost Manor itself. The tune "Taps" plays during this ending. If the rescue was successful, the characters will remain standing in the Graveyard while everything else sinks. A "happy melody" plays for this ending. After a few minutes the game will restart automatically unless the power is turned off.

4.27.4 Scoring

Although winning or losing Ghost Manor depends on rescuing your friend, the game does keep score.

- 10 points are awarded along with each Spear in the Graveyard.
- Spooks are worth varying amounts of points.
  - The lower Scorpion is worth 800 points, and the upper Scorpion is worth 300 points.
  - The lower Skull is worth 700 points, and the upper skull is worth 600 points.
  - The lowest Bat is worth 500 points, the middle Bat is worth 200 points, and the highest Bat is worth 0 points.
  - The Chopping Mummy is worth 1100 points.
- Touching the trailing side of the moving walls awards points while contact is maintained.

- Searching coffins awards points as follows, even if no Cross is found. The amount ranges from 100 to 10,000 points depending on how far away the moving wall is and which direction it is moving. Generally the closer the wall when it's moving toward the player, the more points are awarded.

4.27.5 Difficulty Settings

Difficulty settings in Ghost Manor affect how fast enemy characters move and whether or not the Lamp is available for use. At higher difficulties, fewer spears are awarded by Bones and/or the Rainbow Ghost. Each difficulty level is set as follows:*[2]

- Easy: Set both switches to the B position.

- Medium: Set the Left Difficulty switch to B and the right difficulty switch to A.

- Hard: Set the Left Difficulty switch to A and the right difficulty switch to B.

- Hardest: Set both switches to the A position.

Note that on Sears Tele Games consoles that the Difficulty Switches are called "Skill" switches. Expert is the same as the A position, and Novice is the same as B position. Further, on the Atari 7800, the difficulty switches are located on the front of the console between the controller ports. Slide the switch to the left for B position and to the right for A position.

4.27.6 Xonox Development

Xonox games were developed by a team of people across the United States. Credit is given to the individual programmers in the accompanying instruction manuals, but not on the game's label or the box. Xonox worked with design houses for each of their games, each house having its own expert staff. Typically the development cycle of each game proceeded from script to storyline to the programmers and finally to a target audience of kids. From there, the games were play tested by kids, then sent to Xonox and the programmers in turn for tweaking.*[3]

Ghost Manor was originally intended to be released for the Atari 2600, Colecovision, the Vic-20 and the Commodore 64, and licensed to third party software companies for release on other systems. The final releases, however, were only for the Atari 2600 and the Vic-20.

4.27.7 Marketing

Ghost Manor was typically sold alongside one of Xonox's other games in a double ended cartridge. The idea was to give the customer a "two for the price of one" purchase of new games. The games packaged with Ghost Manor were selected based on common elements shared between the Ghost Manor and the paired games.*[3]

4.27.8 See also

- Artillery Duel

- Chuck Norris Superkicks

- Xonox

4.27.9 References

[1] Atari Age rarity guide

[2] Ghost Manor HTML manual at AtariAge.com.

[3] "Double Play Combination: Will Xonox Video Games be K-tel's Greatest Hits?" From AtariHQ

4.28 Gorf

This article is about the arcade game. For the comic book creator and cartoonist nicknamed "Gorf", see Jordan B. Gorfinkel.

Gorf is an arcade game released in 1981 by Midway Mfg., whose name was advertised as an acronym for "Galactic Orbiting Robot Force". It is a multiple-mission fixed shooter with five distinct modes of play, essentially making it five games in one. It is well known for its use of synthesized speech, a new feature at the time.

4.28.1 Gameplay

The player controls a spaceship that can move left, right, up and down around the lower third of the screen. The ship can fire a single shot (called a "quark laser" in this game), which travels vertically up the screen. Unlike similar games, where the player cannot fire again until the existing shot has disappeared, the player can choose to fire another shot at any time; if the previous shot is still on screen, it disappears.

Gorf consists of five distinct "missions", each with its own patterns of enemies. The central goal of each mission is to destroy all enemies in that wave, which takes the player to the next mission. Successfully completing all five missions will increase the player's rank and loop back to the first mission, where play continues on a higher difficulty level. The game continues until the player loses all their lives. The player can advance through the ranks of *Space Cadet*, *Space Captain*, *Space Colonel*, *Space General*, *Space Warrior*, and *Space Avenger*, with a higher difficulty level at each rank. Along the way, a robotic voice heckles and threatens the player, often calling the player by their current rank (for example, "Some galactic defender you are, Space Cadet!"). Some versions also display the player's current rank via a series of lit panels in the cabinet.

The missions are:

1. Astro Battles: The first mission is almost an exact clone of *Space Invaders*. This is the only mission that is not set in space, but rather against a sky-blue background. A small force of enemies (24 in *Gorf* vs. 55 in *Space Invaders*) attacks in the classic pattern set by the original game. The player is protected by a glittering parabolic force field that is gradually worn away by enemy fire. The force field switches off temporarily while the player's shots pass through it.

2. Laser Attack: In this mission, the player must battle two formations of five enemies each. Each formation contains three yellow enemies that attempt to dive-bomb the player, a white gun that fires a single laser beam, and a red miniature version of the Gorf robot.

3. Galaxians: This mission is a clone of *Galaxian*, with the key differences being the number of enemies (24 in *Gorf* vs. 46 in *Galaxian*) and the way the enemies fire (pellets in *Gorf*, missiles in *Galaxian*). Gameplay is otherwise similar to the original game.

4. Space Warp: Mission 4 places the player in a sort of wormhole, where enemies fly outward from the center of the screen and attempt to either shoot down or collide with the player's ship. It is possible to shoot enemy shots in this level.

5. Flag Ship: The Flag Ship is protected by its own force field (similar to the one protecting the player in Mission 1), and it flies back and forth and fires at the player. To defeat it, the player must break through the force field and destroy the ship's core: if a different part of the ship is hit the player receives bonus points and the part breaks off and flies in a random direction, potentially posing a risk to the player's ship. If the player successful hits the Flag Ship's core, the Flag Ship explodes in a dramatic display, the player advances to the next rank, and play continues on Mission 1, with the difficulty increased.

4.28.2 Features

Gorf is well known for introducing or popularizing two new features to the video game market. Its most notable feature is its robotic synthesised speech, powered by the Votrax speech chip.

Gorf was one of the first games to allow the player to buy additional lives before starting the game. Most games offer a predetermined number of lives (usually 3) and allow the player to earn additional lives throughout the game. *Gorf*, which was usually set to offer two lives per coin, allows the player to insert extra coins to buy up to seven starting lives.

The underlying hardware platform for *Gorf* allowed arcade operators to easily swap the pattern, CPU and RAM boards with other similar games, such as *Wizard of Wor*. Only the game logic and ROM boards are specific to each game.

4.28.3 History

Gorf was originally intended to be a tie-in with *Star Trek: The Motion Picture*, but when the game designers read the film's script, they realized that the concept would not work as a video game.[*][1] Even so, the player's ship bears a passing resemblance to the Starship Enterprise viewed from above.

4.28.4 Sequel

The sequel, *Ms. Gorf*, was never released. It was programmed in the programming language Forth. The source code for the prototype is owned by *Gorf* programmer Jamie Fenton. The game exists only as source code stored on a set of 8-inch floppy disks, and is difficult to retrieve.[*][1][*][2]

4.28.5 Ports

Gorf was ported to the Atari 2600, Atari 5200, and ColecoVision game consoles and the Atari 8-bit, Commodore 64 and VIC-20 personal computers in 1982. Due to copyright issues, the *Galaxians* mission was removed from all ports.

Video Games stated that the Atari 2600 version of *Gorf* "truly is a dog", criticizing both gameplay and graphics,[*][3] however this version of the game received a Certificate of Merit in the category of "Best Solitaire Video Game" at the 4th annual Arkie Awards,[*][4]:30 and received the "1984 Best Computer Game Audio-Visual Effects" award at the 5th Arkies the following year. At the 5th Arkies the judges pointed out that the Atari versions had out-polled both the ColecoVision and Commodore 64 versions of the game, and they suggested that it is the game's "varied action" that "keeps players coming back again and again" .[5]*:28

Regarding the VIC-20 version, *Electronic Games* wrote that "this fast-moving colorful entry is a must ... one of the best games available for the VIC-20" ,*[6] and *Ahoy!* stated that the VIC-20 version "still has my vote for the best of the bunch ... The graphics are excellent" .*[7]

4.28.6 Competitive play

On July 17, 2011, Keith Swanson of Orlando, Fla, set a new Gorf world record score of 1,129,660 points recognized by Twin Galaxies. It took a total of 6 hours 30 minutes of game play to achieve that high score. Keith Swanson is the first person to ever score a million points on 3 ship settings. His game lasted 826 missions. The previous world record was set by John McCann in 2009 with a score of 943,530*[8]

4.28.7 References

[1] "ClassicGaming Expo 2000: Arcade Games Get A Personality" from Classic Gaming.com

[2] Secrets of Ms. Gorf

[3] Wiswell, Phil (March 1983). "New Games From Well-Known Names" . *Video Games*. p. 69 Retrieved 26 May 2014.

[4] Kunkel, Bill; Katz, Arnie (February 1983). "Arcade Alley: The Fourth Annual Arcade Awards" . *Video* (Reese Communications) **6** (11): 30, 108. ISSN 0147-8907.

[5] Kunkel, Bill; Katz, Arnie (February 1984). "Arcade Alley: The 1984 Arcade Awards, Part II" . *Video* (Reese Communications) **7** (11): 28–29. ISSN 0147-8907.

[6] Komar, Charlene (June 1983). "Gorf" . *Electronic Games*. p. 69. Retrieved 6 January 2015.

[7] Salm, Walter (March 1984). "VIC Game Buyer's Guide" . *Ahoy!*. p. 49. Retrieved 27 June 2014

[8] "Twin Galaxies' Gorf High Score Rankings" . Retrieved 17 April 2013.

4.28.8 External links

- *Gorf* at the Killer List of Videogames

- *Gorf* at the Arcade History database

- *Gorf* guide at StrategyWiki

- The home versions of *Gorf* at MobyGames

- Article at The Dot Eaters, featuring a history of *Gorf*

- Video from the C64 Version on archive.org

4.29 Hareraiser

Hareraiser is a computer game, originally released in 1984 in the UK for most home computer platforms. It was released in two parts; *Prelude* and *Finale*. A prize worth £30,000 was on offer if the game could be solved.

4.29.1 Gameplay

In game shot with the hare (ZX Spectrum)

The game takes the form of a series of graphical screens depicting grass, sky and trees with occasional text 'clues'. The only interaction is pressing the cursor keys to follow a hare which moves across the screen and disappears off one of the sides.

There are no hints as to how the puzzle can be solved but it may be the case that the text and the position of the trees, sun, clouds and other elements (which are different on each screen) can be interpreted in some way. The solvers of *Prelude* would also then have to buy and solve *Finale*. They could then enter a competition to locate the prize, the bejewelled 18 carat "Golden Hare" pendant featured on the cover.

4.29.2 Background

The golden hare had previously been the prize for solving the book *Masquerade*, by the British artist Kit Williams.*[1] It had been buried at a secret location (Ampthill Park in Bedfordshire), the object of the game being to solve the clues in the book that would lead the successful treasure-hunter to this location and the golden prize. Several sources state, erroneously, that *Hareraiser* is based on *Masquerade* - in fact, the only thing apart from the prize that the two have in common is that both feature a hare.

Haresoft was founded by Dugald Thompson, the controversial winner of *Masquerade*, and his business partner John Guard.[*][2]

4.29.3 Release and critical reception

The game was released on Acorn Electron, Amstrad CPC, BBC Micro Model B, Commodore 64, Commodore VIC-20 EX, Dragon 32, MSX, Oric Atmos 48k and Sinclair ZX Spectrum in 1984 at £8.95 for each part.[*][3] Haresoft claimed the game was released in two parts *"To make it fun and enable competitors of all ages to participate"*.[*][4] Sinclair User however suggested it was simply to make more money.[*][4]

The game was awarded 3/10 in Sinclair User with reviewer Richard Price struggling to find any reason to play the game except *"the sincere need to get rich"*.[*][5]

Haresoft continued to promote the game by sending 'prolific' press releases including stating that an additional clue had been revealed in Harrods by TV personality Anneka Rice.[*][4] The nature of the clue remains unknown, as does whether or not Ms. Rice was aware that she had revealed such a clue.

4.29.4 Legacy

The game did not sell well and Haresoft went into liquidation. *Hareraiser* was never solved, and the hare was sold at a Sotheby's auction by the creditors in 1988. Although only given a guide price of £3,000-6,000,[*][6] it did in fact exceed Haresoft's stated value, selling for £31,900.[*][2] Although it was rumoured to have been sold again in the early 90s, its whereabouts were unknown for over 20 years until July 2009 when an appeal was made on BBC Radio 4. The current owner's granddaughter got in touch and Kit Williams was reunited with the hare for a BBC TV documentary.[*][7]

It is possible to download copies of some of the versions of *Prelude* and *Finale*, and play them using emulators.

4.29.5 References

[1] *"Play to Win"* in Retro Gamer, Issue 17

[2] *"Masquerade & the Mysteries of Kit Williams FAQ"* at bunnyears.net, accessed 3 May 2009

[3] Original advertisement for the game

[4] *"Gremlin"*. *Sinclair User*. January 1985.

[5] *"Business not pleasure"*. *Sinclair User*. December 1984.

[6] *"The Masquerade Hare"* in Sotheby's auction catalogue, 5 December 1988

[7] *"Artist reunited with golden hare"* by Torin Douglas, BBC News, 20 August 2009

4.29.6 External links

- *Hareraiser* at *Lemon 64*
- Haresoft at *Acorn Electron World*

4.30 Hunchback (video game)

Hunchback is an arcade game developed by Century Electronics in 1983. The player controls Quasimodo from the Victor Hugo novel *The Hunchback of Notre Dame*.

The game is set on a castle wall. The player must cross the screen from left to right avoiding obstacles in order to ring the bell at the far right. Obstacles include pits which must be swung over on a long rope, ramparts which must be jumped (some of which contain knights with spears) and flying fireballs and arrows (to be ducked or jumped). To impose a time limit on each screen a knight climbs the wall, costing the player a life should he reach the top. Eventually, after completing a number of screens, the player must rescue Esmeralda. If this final screen is completed, the game begins again at a faster speed.

The hunchback character was originally to be Robin Hood, hence the green costume and the game stages with arrows. The artist who drew the Robin Hood character left the company before the decision to change the theme to Hunchback. By the time a new artist was taken on, the green costume had become accepted and no-one questioned it (someone commented that the Robin Hood character, as drawn, looked like a hunchback).[*][1]

Ports were made for most home computer systems of the time by Ocean Software in 1984. It was their first arcade port.[*][1] The exceptions to this are the BBC Micro version (which had already been released by Superior Software) and a later port for the MSX (1985). The Spectrum version of the game reached number one in the UK sales charts.[*][2]

There is a 1989 fan-made remake for DOS by Robert Schmidt, based on the BBC version.[*][3]

4.30.1 References

[1] Arcade History

[2] http://www.worldofspectrum.org/showmag.cgi?mag=PersonalComputerGames/Issue06/Pages/PersonalComputerGames0600027.jpg

[3] Fireball Software

4.30.2 External links

- *Hunchback* at MobyGames

- *Hunchback* at World of Spectrum

4.31 Jetpac

For the 1993 video game, see Jetpack (video game). For other meanings, see Jetpack (disambiguation).

Jetpac is a 1983 shooter video game developed and published by Ultimate Play The Game and released for the ZX Spectrum and VIC-20. It was later released for BBC Micro in 1984. The game is the first instalment in the *Jetman* series, and is the first game to be released by the company, who were later known as Rare. The game follows Jetman as he must rebuild his rocket in order to explore different planets, whilst simultaneously defending himself from aliens. *Jetpac* has since been included in other Rare games such as an unlockable in *Donkey Kong 64* and part of a compilation in *Rare Replay*. The game later spawned two sequels and a 2007 remake, *Jetpac Refuelled*, which was released for the Xbox Live Arcade service.

The game was written by Chris Stamper and graphics were designed by Tim Stamper. *Jetpac* was one of the very few Spectrum games also available in ROM format for use with the Interface 2, allowing "instantaneous" loading of the game when the normal method of cassette loading could take several minutes.*[4] The game was met with critical acclaim upon release, with reviewers praising the game's presentation and playability. It later won the "Game of the Year" title at the Golden Joystick Awards in 1983.

4.31.1 Gameplay

The game world is presented in a horizontal wraparound and consists of three platforms which Jetman can manoeuvre onto. Jetman must assemble his rocket (which spawns in instalments scattered around the map), and then fill it with fuel before taking off to the next planet, where the procedure is broadly repeated.*[4] In addition, the player has to defend themselves from the planet's aliens, and for bonus points collect valuable resources which occasionally fall from above.*[5]

After the first level, the rocket stays assembled and just requires refuelling. However, every four levels, the rocket resets (giving the player an extra life) and the replacement has

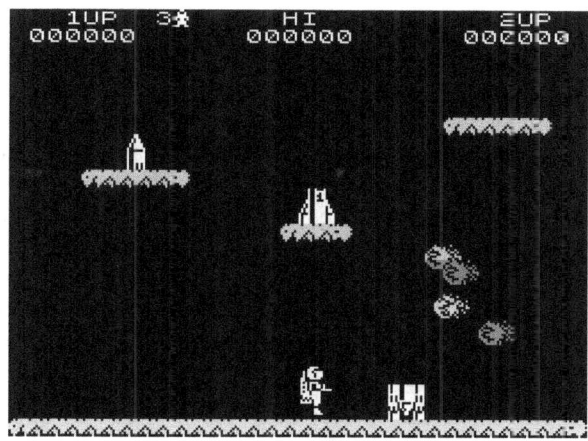

Three rocket sections need to be assembled before Jetman can leave this planet.

to be built before it can be re-fuelled for take off.*[4] Each new model has a new design with a higher number written on it, although the gameplay remains unchanged.*[5] The enemies change forms each level (cycling back to the first after eight levels) and each alien has a different pattern of movement which means they can be dealt with in a different manner.*[6]

4.31.2 Development and release

The Spectrum *Jetman* titles each offered a unique approach to fetch-for-survival gameplay.

"

„

Ste Pickford in a retrospective interview with *Retro Gamer**[7]

Ashby Computers and Graphics was founded by brothers Tim and Chris Stamper, along with Tim's wife, Carol, from their headquarters in Ashby-de-la-Zouch in 1982. Under the trading name of Ultimate Play The Game, they began producing multiple video games for the ZX Spectrum throughout the early 1980s.*[8] The company were known for their reluctance to reveal details about their operations and upcoming projects. Little was known about their development process except that they used to work in "separate teams"; one team would work on development whilst the other would concentrate on other aspects such as sound or graphics.*[8] Whilst developing *Jetpac*, the Stamper brothers closely studied the emerging Japanese video gaming market and had started to practice developing games for their upcoming Famicom console, later predicting that the ZX Spectrum had a limited lifespan.*[9]

Jetpac was one of the few Spectrum games also available in ROM format for use with the Interface 2, allowing "instantaneous" loading of the game when the normal method of cassette loading could take several minutes.[4] The game used the common technique of placing planar sprites with image sprites atop another, which often created graphical errors and overlapped colours on both ZX Spectrum and BBC Micro versions of the game.[10] The game was also able to run on the 16K version of the Spectrum.[8] *Jetpac* also inspired several clones and unofficial remakes, such as Archer MacLean's Atari 800 game *DropZone*, which was released the following year after *Jetpac*.[7]

The game sold a total of 300,000 units for the ZX Spectrum and generated £1 million in revenue for Ultimate Play The Game, which enabled the Stamper brothers to gain a foothold in the early video gaming market.[7] After the game's release, *Jetpac* was parodied in a long-running *Crash* comic strip named *Lunar Jetman*. The strip, designed by John Richardson, lasted from July 1984 to October 1991 and gained popular reception from readers. To develop the comic, photographs had to be processed manually on a photo-mechanical tone and then transferred to paper, later being fully colourised in the late 1980s.[7]

4.31.3 Reception

The game was critically acclaimed upon release. *Crash* praised the graphics and presentation, citing that they were of "the highest standard" and added that it was "difficult to find any real faults" with the game.[5] *CVG* similarly praised the graphics, stating that the presentation was "superb" and the gameplay was considered addictive.[6] In a retrospective review, Chris Wilkins of Eurogamer noted that the colourful graphics and sound effects were advanced for the time, but what truly made for a "faultless" experience was its simple gameplay.[11]

ZX Computing praised the game's playability and replay-value, stating that *Jetpac* was "a very well put together piece of software".[14] The game was number one in the first Spectrum sales chart published by *CVG*.[15] The ZX Spectrum version was voted number 73 in the *Your Sinclair Readers' Top 100 Games of All Time* in 1993[16] and was voted the 14th best game of all time by the readers of Retro Gamer for an article that was scheduled to be in a special *Your Sinclair* tribute issue.[17] The game won the title "Game of the Year" at the 1983 Golden Joystick Awards.[13]

Legacy

Jetpac 's popularity further spawned two sequels, *Lunar Jetman* (1983) and *Solar Jetman: Hunt for the Golden Warp-*

ship (1990). The latter, however, was not released for the ZX Spectrum due to disappointing sales of the original NES version, although a version for the Commodore 64 was finished but never released.[18]

Since its release, *Jetpac* has been included in other games developed by Rare. The game is playable in *Donkey Kong 64*, where it could be unlocked to play in Cranky Kong's laboratory after obtaining 15 Banana Medals. Beating Cranky Kong's high score rewards the player with the Rareware Coin, which is necessary to beat the game. The game was retained in the April 2015 Virtual Console re-release of *Donkey Kong 64* on the Wii U, despite it being technically owned by Microsoft.[19] An enhanced remake of *Jetpac*, entitled *Jetpac Refuelled*, was released on the Xbox Live Arcade in March 2007.[20] Microsoft's E3 2015 press conference unveiled the compilation title *Rare Replay*, which has a selection of thirty games from Rare's lifetime game library, including *Jetpac* and its sequels and remake.[21]

4.31.4 References

[1] "PSST is this the Ultimate?", *Personal Computer Games* (1), June 1983: 5

[2] "Coming Soon...". *Personal Computer Games* (2): 7. November 1983.

[3] "Jetpac review, BBC Micro version". *Computer and Video Games* (38): 36. December 1984.

[4] "Interface Games are Fast but not Furious", Sinclair User (EMAP) (24), March 1984: 54–55

[5] "Jetpac review - Crash Magazine". *Crash Magazine* (4): 65. April 1984. Retrieved 3 August 2015.

[6] "Jetpac - Review", *Computer and Video Games* (Future Publishing) (21), January 1983: 136

[7] "1983: A Spaceman's Odyssey - The History of Jetman" (PDF). *Retro Gamer* (Imagine) (96): 50. November 2011. Retrieved 22 August 2015.

[8] "The Best of British - Ultimate". Crash. Retrieved 13 August 2015.

[9] "The Ultimate Hero: The Complete History of Sabreman" (PDF). *Retro Gamer* (Imagine) (73): 27. February 2010. Retrieved 23 August 2015.

[10] "Game Design". *Crash*. June 1986. Retrieved 22 August 2015.

[11] Wilkins, Chris (25 October 2007). "Jetpac review". Eurogamer. Retrieved 3 August 2015.

[12] "Reaction games across the Spectrum - Jet Pac", *Home Computing Weekly* (16), June 1983: 15

[13] "C&VG's Golden Joystick Awards 1983". *Computer and Video Games* (Future Publishing) (29): 15. June 1985. Retrieved 15 January 2012.

[14] "The soft touch - Jet Pac", *ZX Computing* (8), August 1983: 106-107

[15] "Chart Toppers", *C+VG* (Future Publishing) (23), September 1983: 37

[16] "Readers' Top 100 Games of All Time", *Your Sinclair* (Future plc) (93), September 1993: 11

[17] "The 50 Best Speccy Games Ever!". ysrnry.co.uk. November 2004.

[18] "Solar Jetman - 1991 storm". Games That Weren't. Retrieved 3 August 2015.

[19] "Donkey Kong 64". *GameSpot*. CNET. Retrieved 30 May 2006.

[20] "Jetpac Refuelled". *Xbox.com*. Microsoft. Archived from the original on 22 February 2008. Retrieved 25 February 2008.

[21] "Rare Celebrates Its 30th Anniversary with a Massive 30-Game Collection". *XBox News*. 15 June 2015. Retrieved 19 August 2015.

4.31.5 External links

- *Jetpac* at MobyGames
- *Jetpac* at World of Spectrum
- *Jetpac* at Ultimate Wurlde
- *Jetpac* at Classic Web Games

4.32 Jungle Hunt

Jungle Hunt (ジャンル・ハント) is side-scrolling arcade game produced and released by Taito in 1982. It was initially released as *Jungle King*. *Jungle Hunt* is one of the first video games to use parallax scrolling.

The player controls an unnamed jungle explorer sporting a pith helmet and a safari suit. The player attempts to rescue his girl from a tribe of hungry cannibals by swinging from vine to vine, swimming a crocodile-infested river, jumping over or ducking beneath rolling rocks, then releasing the girl before she is lowered into a boiling cauldron.

Home versions were released for the Apple II, Atari 2600, Atari 5200, Atari 8-bit family, Commodore 64, ColecoVision, VIC-20, and IBM PC. The PC version was developed by Sierra On-Line and is incompatible with anything except an original IBM PC/XT with a CGA video card.

In the Atari-ported versions the hero is named Sir Dudley, and the girl, married to Sir Dudley, is Lady Penelope.

4.32.1 History

Jungle Hunt changed names several times during development. The original prototypes were called *Jungle Boy* and later became **Jungle King** for release to the arcades.*[2] In these earlier versions the playable character was a bare-chested man with a loincloth who resembled Tarzan.

Taito were sued by the Edgar Rice Burroughs estate for copyright infringement for using the character's likeness. This led to a rerelease as *Jungle Hunt*,*[3] with the following changes made to the game:

- The character was replaced with a jungle explorer wearing a pith helmet and safari outfit.

- In the first scene, ropes replace the vines as the objects that the player has to swing on.

- The Tarzan yell was removed and cannot be heard throughout the game. It was replaced by the music from the end of the second scene.

Taito Brazil (Taito do Brasil) released a version of the game in Brazil in 1983 under the title 'Jungle Hunt' which included the bare-chested character and the Tarzan yell. Yet another variant of the game was called *Pirate Pete*. Gameplay in this version was identical to Jungle Hunt but the character was replaced with a pirate (complete with eye patch) and the levels had a pirate theme.*[4]

4.32.2 Gameplay

The gameplay is split into four scenes, which have different objectives.

In Scene 1, the explorer is required to swing from vine to vine. This is accomplished by pressing the action key when two vines swing closely enough together. Timing is critical, and missing the vine causes the explorer to fall to the jungle floor, losing a life.

Scene 2 has the explorer navigating a crocodile-infested river. The explorer can attack the crocodiles from below with his knife, unless their mouths are open. The explorer must return to the surface periodically to breathe, where he cannot attack the crocodiles. Bubbles periodically rise from the bottom of the river, which can trap the explorer and carry him to the surface, potentially hitting crocodiles on the way.

Scene 3 involves the explorer dodging various-sized boulders rolling and bouncing towards him as he runs up the side of a volcano. Timing is critical as the differently sized boulders bounce at different speeds and heights, and the explorer can be trapped between them.

In the final scene, the explorer must evade cannibals while attempting to get to a woman being lowered into a flaming cauldron. After the player rescues the woman, the word "Congratulations!" appears, which is then followed by a message saying "**I Love You!!!**" followed by the woman kissing the explorer.

Further gameplay repeats the scenes with increased difficulty. On rounds other than the first, a cannibal appears in the tree of the cauldron scene and throws spears at the player.*[5]

4.32.3 Ports

Jungle Hunt was ported to the following platforms:

- Apple II

- Atari 2600

- Atari 5200

- Atari 8-bit computers

- ColecoVision via Atarisoft

- Commodore 64

- Commodore VIC-20

- IBM PC via Atarisoft, and as part of *Taito Legends*

- PlayStation 2 as part of *Taito Legends*

- Texas Instruments TI-99/4A

- Xbox as part of *Taito Legends*

4.32.4 Reception

Jungle Hunt was well received, gaining a Certificate of Merit in the category of "1984 Best Adventure Videogame" at the 5th annual Arkie Awards.*[6]*:42

4.32.5 References

[1] Fujihara, Mary (1983-07-25). "Inter Office Memo". Atari. Retrieved 18 March 2012.

[2] "Jungle King / Jungle Boy"

[3] "Jungle Hunt"

[4] "Pirate Pete"

[5] *Jungle Hunt* at the Killer List of Videogames

[6] Kunkel, Bill; Katz, Arnie (January 1984). "Arcade Alley: The Arcade Awards, Part 1". *Video* (Reese Communications) **7** (10): 40–42. ISSN 0147-8907.

4.32.6 External links

- *Jungle Hunt* guide at StrategyWiki

- Classic Game Reviews covers Jungle Hunt

4.33 Log Run

Log Run is a 1982 VIC-20 game written by Mark Brennand and published by Terminal Software of Prestwich. *Log Run* is similar to the Century Electronics arcade game *Logger*, which itself is a clone of *Donkey Kong* with the main character replaced by a bird.

4.34 Lunar Leepers

Lunar Leepers, also released as **Lunar Leeper**, is a 1982 video game written by Chuckles (Chuck Bueche) and published under Sierra On-Line's SierraVision label.

The game takes place on the planet Opthamalia, in the Valley of the Leepers, which the manual describes as omnivorous creatures having "two long rubbery legs, a single eye and a massive green beak". In the first phase of the game, the player pilots a spaceship to rescue crew members stranded in the valley among the Leepers. The Leepers must be avoided or shot, lest they leap up and consume the spaceship or the crew member it carries. Once all the crew members are either rescued or killed, the second phase begins. In this phase the player navigates the ship through a cave in search of a giant eyeball which must be destroyed. The cave is trapped with automatic lasers and guarded by creatures known as "trabants".

The Apple II version of the game was copy-protected using Sierra's Spiradisc system, though this could be circumvented with a sector editor.*[1]

4.34.1 Reception

Lunar Leeper was favourably reviewed in the inaugural issue of *Personal Computer Games*.*[2] *Softline* in 1983 called the game "very addictive, and you'll probably lose hours of sleep over it".*[3] *Ahoy!* in 1984 stated that the VIC-20 version was "original, 'cute', and hard as hell",*[4] and that the Commodore 64 version had a great "personality", but "with no graphic or gameplay breakthroughs and a lack of variation of long-range playability ... there is less here than meets the eye".[5]

4.34.2 References

[1] Etarip, Rich (1990). "Softkey for *Lunar Leepers*". *Computist* (SoftKey Publishing) (83): 10–11.

[2] Cross, Nigel (1983). "Screen Scroll: Lunar Leepers". *Personal Computer Games* (1): 97.

[3] Mankovitz, Alan (January 1983). "Lunar Leeper". *Softline*. p. 41. Retrieved 27 July 2014.

[4] Salm, Walter (March 1984). "VIC Game Buyer's Guide". *Ahoy!*. p. 49. Retrieved 27 June 2014

[5] Moriarty, Tim (April 1984). "Lunar Leeper". *Ahoy!*. p. 59. Retrieved 27 June 2014.

4.34.3 External links

- Lunar Leeper at MobyGames

4.35 Mole Attack

Mole Attack is a video game for the Commodore VIC-20.

4.35.1 Plot

Moles pop up from nine holes, and the player has 60 seconds to send them fleeing back underground by bopping them on the head with a hammer.[1]

4.35.2 Gameplay

Players have the option of using a joystick to position the hammer, but can also use a keyboard.[1]

4.35.3 Reception

Electronic Games reviewed *Mole Attack* in 1983 by Charlene Komar. The reviewer commented: "*Mole Attack* will probably be a favorite among younger arcaders. Even though the eye-catching graphics combine well with the time-limit excitement, adults will probably find the game too simple and repetitive to get too many repeat plays."[1]

4.35.4 References

[1] Komar, Charlene (June 1983). "Mole Attack". *Electronic Games*. p. 70. Retrieved 6 January 2015.

4.36 Mountain King

This article is about the video game. For the computer game company, see Mountain King Studios.

Mountain King is a scrolling platform video game released by CBS Electronics in 1983. It was available on the Atari 2600, Atari 5200, Atari 400/800, ColecoVision, Commodore-64 and VIC-20.

4.36.1 Gameplay

The player takes the role of an explorer searching a diamond mine and the temple of an ancient civilization. The object of the game is to discover and collect the Golden Crown and take it to the peak of the mountain.[1]

The mountain environment is made up of platforms and ladders, with the Perpetual Flame located on the top of the summit. This is where the Golden Crown must be taken. To retrieve the Crown, the player must first collect one thousand points. Diamonds are worth 25 points each and treasure chests are worth 260 points. Diamonds are scattered throughout the level, and treasure chests can be located by using the explorer's flashlight to see in the darkness. While it is not necessary for the player to descend to the very bottom of the mountain, there are many treasure chests to find there, but these resources are not without peril: a giant spider lurks in the lowest level, and will give chase if the player spends more than a few seconds exploring the bottom. Once the player has the required score, he must locate the Flame Spirit by following the musical cues. The Flame Spirit is the key to the Temple Chamber, where the Crown is kept. If the player attempts to enter the Temple before he has the Flame Spirit, the Skull Spirit guarding the entrance will stop him. When the Temple Chamber is unlocked, the player can collect the Crown and take it to the Perpetual Flame - however, there will now be Cave Bats stalking the player. If they catch him, they will steal the Crown and the player will have to start collecting points again.

There are three ways to lose the game. The first and most probable conclusion to the game is the expiration of the timer. When time reaches zero, the game is over. The second, more unfortunate fate involves the giant spider. If the spider catches the player on the lowest level, the player is trapped in a web cocoon. The player can escape from this web by rapidly moving the joystick left and right. If the player does not escape by the time the spider returns, the game will end, ostensibly with the player having been eaten by the spider. The third, when the player touches any regular fires, he shall die and the game will end.

4.36.2 Hidden level

Mountain King has a "hidden level" better known to players as "Glitch Heaven" which can be reached by jumping from the tip of a ledge located on the mountain peak at the top of the game screen . Timed correctly, the player will catch the very bottom rung of a higher ladder. The ladder leads into an hidden level of ladders and platforms. At the very top of this area are two strange ghost-like figures, but there is no reward or bonus for entering this level, and the level is extremely difficult to navigate (with falling from the various ladders a common occurrence).

4.36.3 Reception

Softline stated that *Mountain King* "is also something more [than a game], something like a fantasy or adventure" , and "a game worth playing" .[2] *Antic* praised *Mountain King* for its gameplay and intelligent use of music and sound. Its graphics were criticised for being plain and blocky.[3]

4.36.4 References

[1] Mountain King instruction booklet - CBS Electronics, 1983.

[2] Durkee, David (Nov–Dec 1983). "Mountain King" . *Softline*. pp. 23–24. Retrieved 29 July 2014.

[3] Antic magazine, Volume 2, Number 9 (December 1983).

4.37 Mutant Herd

Mutant Herd is a 1983 action computer game developed by and published by Thorn EMI.

4.37.1 Description

Mutant Herd involves protecting a powerhouse from an invasion of mutants. The mutants crawl towards you from multiple *burrows* located at each corner of the screen. Using laser beams, you need to prevent the mutant herds from destroying your powerhouse.

The game was even more complicated than this, however, as it was possible to enter the *burrows* and to shut them down. Once a burrow was entered, you started another phase of the game. A queen mutant and her eggs are hiding in each burrow. You must get to the eggs and plant an explosive charge, while avoiding the queen mutant, who can kill you or move your charge away from her eggs. Once you have succeeded in blowing up the eggs, the burrow is sealed. When you return to the powerhouse screen, you will have one less burrow spewing mutants at you.

The game was even more complicated yet, in that, after each burrow is sealed, your lasers get weaker and the next burrow will have more eggs and be more challenging. You will have won the game when you have sealed all the burrows and successfully protected your powerhouse. [1][2]

4.37.2 Reception

Compute! wrote that "There are enough little problems to always keep you thinking about what you'll have to do next" .[1] *Electronic Games* stated that *Mutant Herd* "may not be the ultimate VIC-20 game, but the graphics are fun, the action is challenging, and it takes at least one step off the beaten game-track" .[2] *Ahoy!* called it "a two-part arcade game with a unique twist" , favorably reviewing the graphics, sound, animation, and gameplay.[3]

4.37.3 References

[1] Roberts, Tony (September 1983). "Mutant Herd for the VIC" . *Compute!*. pp. 190–192. Retrieved 4 December 2013.

[2] Komar, Charlene (September 1983). "Mutant Herd". *Electronic Games Magazine*. pp. 73–74. Retrieved 4 December 2013.

[3] Kevelson, Morton A. (February 1984). "Mutant Herd" . *Ahoy!*. pp. 56–57. Retrieved 27 June 2014.

4.37.4 External links

- Atari Magazines Review

4.38 Mystery Fun House (video game)

Mystery Fun House is a text-based adventure program written by Scott Adams, "Adventure 7" in the series released by Adventure International.[1] The game setting was a fun house that the player had to explore in order to locate a set of secret plans, solving puzzles along the way.[2] *Mystery Fun House* was produced in only one week[3] and was among the less easy games in the series.[4]

4.38.1 Description

Published by Adventure International, this text-based adventure game was one of many from Scott Adams.*[5]

Gameplay involved moving from location to location, picking up any objects found there, and using them somewhere else to unlock puzzles. Commands took the form of verb and noun, e.g. "Take Wrench". Movement from location to location was limited to North, South, East, West, Up and Down.

The player of this game must navigate through a maze*[5] and a shooting gallery, charm a mermaid,*[6] and turn off a steam calliope that is so loud the player's instructions are misunderstood – a reference to the "Loud Room" in *Zork I*.*[7] Violent solutions to puzzles are discouraged by a gameplay feature which sees the player character ejected from the fun house by a bouncer whenever certain commands are typed.*[7]

4.38.2 Releases

Mystery Fun House was among a number of classic Scott Adams adventures made available for free download by non-profit gaming organisation Infinite Frontiers in 2003.*[8] It was also included as part of magnussoft's *C64 Classix* compilation for Windows and Mac CD-ROM.*[9]

A Sinclair Spectrum version was planned but never released.*[10]

4.38.3 Reception

Electronic Games in 1981 warned against beginners to adventure games playing *Mystery Fun House* because of the difficulty, but stated that for others it "should provide a rousing good time".*[11]

4.38.4 References

[1] "Mystery Fun House". *MobyGames*. MobyGames. Retrieved 2008-07-22.

[2] Scoleri III, Joseph. "Adventure 7: Mystery Fun House". *allgame*. All Media Guide. Retrieved 2008-07-22.

[3] Francis, Garry (July–August 1985). "Behind the Scenes". *New Atari User* (16): 20. Retrieved 2008-07-22.

[4] Francis, Garry (July–August 1984). "Scott Adams' Adventureland". *New Atari User* (10): 10. Retrieved 2008-07-22.

[5] Griffin, Brad (March–April 1983). "Scott Adams Adventures 1–12". *A.N.A.L.O.G. Computing* (10).

[6] Matthews, Ken (December 1984). "Scott Adams' Classic Adventures: Mystery Funhouse". *Micro Adventurer* (London: Sunshine Publications) (14): 19.

[7] "Scott Adams Text Adventure Games: Mystery Fun House". *Malinche Entertainment*. Malinche Entertainment Corp. Retrieved 2008-07-22.

[8] Walkland, Nick (2003-06-12). "Classic Scott Adams Adventures for free". *Computer and Video Games* (Future Publishing). Retrieved 2008-07-22.

[9] "C64 Classix". *MobyGames*. MobyGames. Retrieved 2008-07-22.

[10] "Mystery Fun House". *World of Spectrum*. World of Spectrum. Retrieved 2008-12-13.

[11] "Computer Playland". *Electronic Games*. January 1981. p. 61. Retrieved 28 January 2015.

4.38.5 External links

- *Mystery Fun House* at World of Spectrum

4.39 Omega Race

Omega Race is an arcade game programmed by Ron Haliburton*[1] and released in 1981 by Midway. It was the only arcade game with vector graphics Midway created.

4.39.1 Description

In 1982 the editors of *Consumer Guide* magazine published a book entitled *How To Win At Video Games*, which featured detailed strategies for nine of the most popular arcade games of the time. *Omega Race* was chosen as one such game, mostly due to its approachability. The book states that "any unskilled player can pop a quarter into the machine and stay up there for up to 20,000 points." According to the book, more than 35,000 machines were created, with the average machine taking in $181.00 per week at the time of the book's publication.*[2] Frequently, it was one of the top ten money-making arcade machines in any given week in that time period.*[2]

4.39.2 Gameplay

Set in the year 2003, the game involves using a spaceship to destroy enemy droid ships. The player's ship is controlled using a spinner to rotate the ship's direction, a button for thrusting, and a button for firing lasers. The enemies that the player must destroy or avoid are drone ships, commander ships, two types of space mines, and shooting star ships. Extra ships were usually awarded at 40,000 and 100,000 points, but this default setting could be changed by the machine's owner. Its gameplay has been compared to *Asteroids*, in that the game uses black and white vector graphics and the ship is moved using a thrust button. Unlike *Asteroids*, the ship wasn't allowed to warp to the other side of the screen; it would bounce off an invisible barrier on the edges of the screen that would briefly appear when something hit it.

4.39.3   Reception

Compute! called *Omega Race* "a real winner for the VIC" .[3] *BYTE* stated that the VIC-20 version "is fast paced, has colorful graphics, and features good sound effects ... *Omega Race* is a fun game that retains all the best characteristics of the arcade version" .[4] *Ahoy!* called the VIC-20 version "fairly faithful to the arcade game, and very exciting" .[5] The VIC-20 version of *Omega Race* was awarded a Certificate of Merit in the category of "Best Solitaire Computer Game" at the 4th annual Arkie Awards.[6]:33

4.39.4   Legacy

Versions of the game were released for some home video game consoles of the early to mid 80s, including the Atari 2600 and ColecoVision, as well as the VIC-20 and Commodore 64 home computers. The Atari 2600 version came bundled with a special 2-button, 'booster grip,' controller. As of 2007, *Omega Race* remains absent from every collection in the *Midway Arcade Treasures* series.

4.39.5   References

[1] http://www.tomheroes.com/Video%20Games%20FS/Retrotimes/Best%20Of/Many%20Faces/omega_race.htm

[2] Editors of Consumers Guide, The (individuals uncredited). (1982), *How To Win At Video Games*. Publications International, Ltd. ISBN 0-517-38119-2.

[3] Herman, Harvey B. (October 1982). "Four New Cartridges for VIC-20" . *Compute!*. p. 132. Retrieved 30 October 2013.

[4] Wszola, Stan (March 1983). "Omega Race for the VIC-20" . *BYTE*. p. 251. Retrieved 19 October 2013.

[5] Salm, Walter (March 1984). "VIC Game Buyer's Guide" . *Ahoy!*. p. 49. Retrieved 27 June 2014.

[6] Kunkel, Bill; Katz, Arnie (March 1983). "Arcade Alley: The Best Computer Games" . *Video* (Reese Communications) **6** (12): 32–33. ISSN 0147-8907.

4.39.6   External links

- *Omega Race* at the Killer List of Videogames

- *Omega Race* guide at StrategyWiki

- Omega Race played on a Vic-20 (YouTube clip)

4.40   Pac-Man

This article is about the original Pac-man arcade game from 1980. For the video game series, see List of Pac-Man video games. For the video game character, see Pac-Man (character). For other uses, see Pac-Man (disambiguation).

Pac-Man (Japanese: パックマン Hepburn: *Pakkuman*) is an arcade game developed by Namco and first released in Japan on May 22, 1980.[1][2] It was created by Japanese video game designer Toru Iwatani. It was licensed for distribution in the United States by Midway and released in October 1980. Immensely popular from its original release to the present day, *Pac-Man* is considered one of the classics of the medium, virtually synonymous with video games, and an icon of 1980s popular culture.[6][7][8][9] Upon its release, the game—and, subsequently, Pac-Man derivatives—became a social phenomenon[10] that sold a large amount of merchandise and also inspired, among other things, an animated television series and a top-ten hit single.[11]

When *Pac-Man* was released, the most popular arcade video games were space shooters, in particular *Space Invaders* and *Asteroids*. The most visible minority were sports games that were mostly derivatives of *Pong*. *Pac-Man* succeeded by creating a new genre.[12] *Pac-Man* is often credited with being a landmark in video game history, and is among the most famous arcade games of all time.[13] It is also one of the highest-grossing video games of all time,[14] having generated more than $2.5 billion in quarters by the 1990s.[15][16]

The character has appeared in more than 30 officially licensed game spin-offs,[17] as well as in numerous unauthorized clones and bootlegs.[18] According to the Davie-Brown Index, Pac-Man has the highest brand awareness of any video game character among American consumers, recognized by 94 percent of them.[19] Pac-Man is one of the longest running video game franchises from the golden age of video arcade games. It is part of the collection of the Smithsonian Institution in Washington, D.C.[20] and New York's Museum of Modern Art.[21]

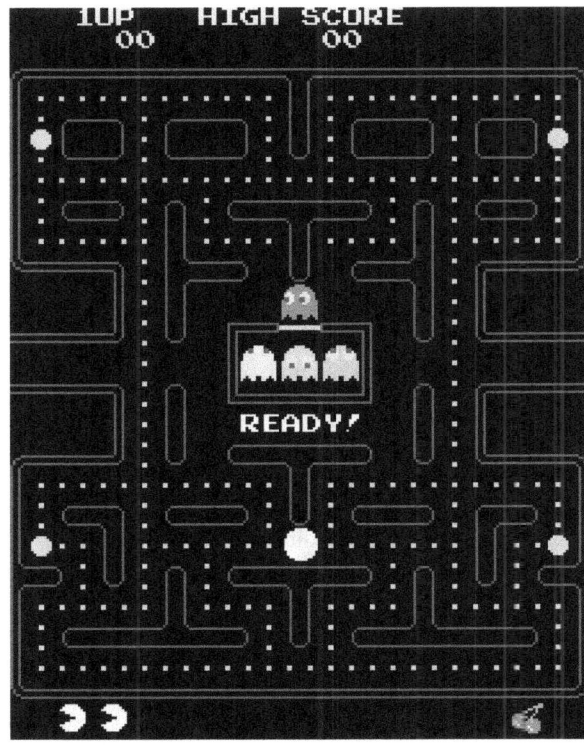

Screenshot of play area

4.40.1 Gameplay

The player controls Pac-Man through a maze, eating pac-dots (also called pellets or just dots). When all pac-dots are eaten, Pac-Man is taken to the next stage. Between some stages one of three intermission animations plays.*[22] Four enemies (Blinky, Pinky, Inky and Clyde) roam the maze, trying to catch Pac-Man. If an enemy touches Pac-Man, a life is lost and Pac-Man himself withers and dies. However, it sometimes happens that Pac-man passes through an enemy unharmed. This is caused by the way the game detects collisions between Pac-Man and an enemy. Whenever Pac-Man occupies the same tile as an enemy, he is considered to have collided with that ghost and a life is lost. This logic proves sufficient for handling collisions more than 99% of the time during gameplay, but does not account for one very special case. Pac-Man's center point moves upwards into the enemy's tile in the same 1/60th of a second that the enemy's center point moves downwards into Pac-Man's tile, resulting in them moving past each other without colliding.*[23] When all lives have been lost, the game ends. Pac-Man is awarded a single bonus life at 10,000 points by default—DIP switches inside the machine can change the required points or disable the bonus life altogether.

Near the corners of the maze are four larger, flashing dots known as power pellets that provide Pac-Man with the temporary ability to eat the enemies. The enemies turn deep blue, reverse direction and usually move more slowly. When an enemy is eaten, its eyes remain and return to the center box where it is regenerated in its normal color. Blue enemies flash white to signal that they are about to become dangerous again and the length of time for which the enemies remain vulnerable varies from one stage to the next, generally becoming shorter as the game progresses. In later stages, the enemies go straight to flashing, bypassing blue, which means that they can only be eaten for a short amount of time, although they still reverse direction when a power pellet is eaten; in even later stages, the ghosts do not become edible (i.e., they do not change color and still make Pac-man lose a life on contact), but they still reverse direction.

Enemies

Main article: Ghosts (Pac-Man)
The enemies in *Pac-Man* are known variously as 'ghosts,"

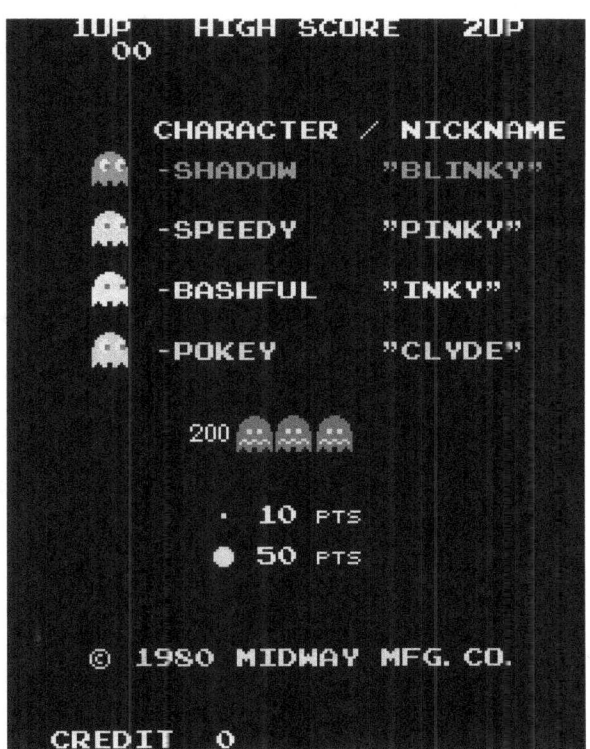

North American Pac-Man *title screen, showing the official enemy names.*

"goblins," "octopi" and "monsters".*[24]*[25]*[26] Despite the seemingly random nature of the enemies, their movements are strictly deterministic, which players have used to their advantage. In an interview, creator Toru Iwatani stated that he had designed each enemy with its own distinct personality in order to keep the game from becoming impossibly difficult or boring to play.*[27] More recently, Iwatani described the enemy behaviors in more detail at the 2011 Game Developers Conference. He stated that the red enemy chases Pac-Man, and the pink and blue enemies try to position themselves in front of Pac-Man's mouth.[28] Although he claimed that the orange enemy's behavior is random, a careful analysis of the game's code reveals that it actually chases Pac-Man most of the time, but also moves toward the lower-left corner of the maze when it gets too close to Pac-Man.

The term "ghosts" originates from the failed Atari 2600 port. Technical limitations caused the villans to flicker, and the game's manual dubbed them "ghosts" so as to cover up the flaw. Although the game was ultimately unsuccessful due to these flaws, the term stuck, and soon spread to all of the bubble gum cards, stickers, and other merchandise released afterwards. In the Japanese cabinet art and flyers, the villains appeared somewhat like sheeted ghosts. These became the basis for most drawings on the various merchandise. Consequently, cabinet artwork for later arcade games depicted the villans as more ghost-like.[29]

Split-screen

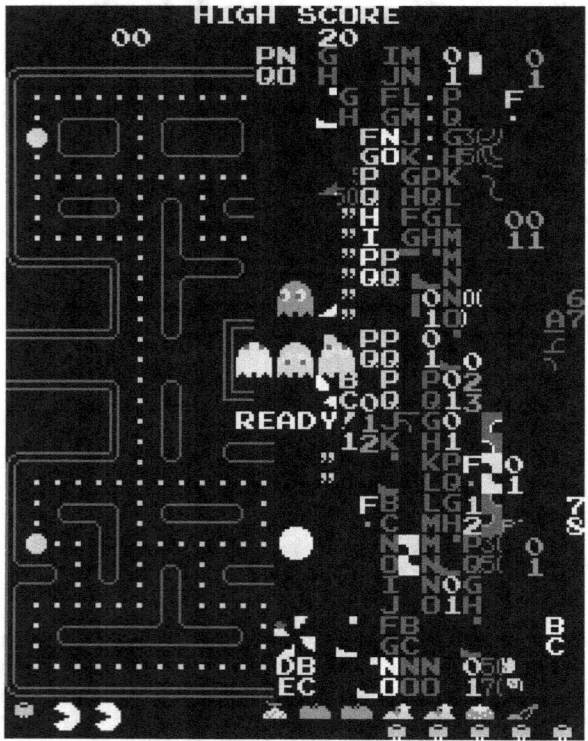

The 256th "Split-Screen" level cannot be completed due to a software bug.

Pac-Man was designed to have no ending – as long as at least one life was left, the game should be able to go on indefi-

nitely. However, a bug keeps this from happening: Normally, no more than seven fruit are displayed on the HUD at the bottom of the screen at any given time. But when the internal level counter, which is stored in a single byte (eight bits), reaches 255, the subroutine that draws the fruit erroneously "rolls over" this number to zero when it is determining the number of fruit to draw, using fruit counter = internal level counter + 1. Normally, when the fruit counter is below eight, the drawing subroutine draws one fruit for each level, decrementing the fruit counter until it reaches 0. When the fruit counter has overflowed to zero, the first decrement sets the fruit counter back to 255, causing the subroutine to draw a total of 256 fruit instead of the maximum of seven.

This corrupts the bottom of the screen and the entire right half of the maze with seemingly random symbols and tiles, overwriting the values of edible dots which makes it impossible to eat enough dots to beat the level. Because this effectively ends the game, this "split-screen" level is often referred to as the "kill screen". There are 114 dots on the left half of the screen, nine dots on the right, and one bonus key, totaling 6,310 points. When all of the dots have been cleared, nothing happens. The game does not consider a level to be completed until 244 dots have been eaten. Each time a life is lost, the nine dots on the right half of the screen get reset and can be eaten again, resulting in an additional 90 points per extra man. In the best-case scenario (five extra men), 6,760 points is the maximum score possible, but only 168 dots can be harvested, and that is not enough to change levels. Four of the nine dots on the right half of the screen are invisible, but can be heard when eaten. Some dots are invisible but the rest can be seen, although some are a different color than normal. One method for safely clearing this round is to trap the ghosts. To trap the three important ghosts, the player must begin by going right until Pac-man reaches a blue letter 'N', then he goes down. He keeps going down until he reaches a blue letter 'F', then he goes right. He keeps going right until he reaches a yellow 'B', then he goes down again. When executed properly, Pac-Man will hit an invisible wall almost immediately after the last turn is made. Eventually, the red ghost will get stuck. The pink ghost follows a few seconds later. The blue ghost will continue to move freely for several moments until the next scatter mode occurs. At that point, it will try to reach some location near the right edge of the screen and get stuck with the pink and red ghost instead. The orange ghost is the only one still on the loose, but he is no real threat since he runs to his corner whenever Pac-Man gets close, so that it is easy to eat all the dots.[31] Emulators and code analysis have revealed what would happen should this 256th level be cleared: The fruit and intermissions would restart at level 1 conditions, but the enemies would retain their higher speed and invulnerability to power pellets from the higher stages.[*][32]

World record and perfect play

As of 2015, the world record according to Twin Galaxies is held by David Race, who in 2013 attained the maximum possible score of 3,333,360 points in 3 hours, 28 minutes and 49 seconds. He has uploaded a recording of his record-breaking game to YouTube.

A perfect *Pac-Man* game occurs when the player achieves the maximum possible score on the first 255 levels (by eating every possible dot, power pellet, fruit, and enemy) without losing a single life, and using all extra lives to score as many points as possible on Level 256.[*][33][*][34] The first person to achieve this score was Billy Mitchell of Hollywood, Florida, who performed the feat in about six hours.[*][34][*][35] Since then, six other players have attained the maximum score in increasingly faster times.

In December 1982, an 8-year-old boy, Jeffrey R. Yee, supposedly received a letter from U.S. President Ronald Reagan congratulating him on a worldwide record of 6,131,940 points, a score only possible if he had passed the unbeatable Split-Screen Level.[*][34] In September 1983, Walter Day, chief scorekeeper at Twin Galaxies, took the US National Video Game Team on a tour of the East Coast to visit video game players who claimed they could get through the Split-Screen Level. No video game player could demonstrate this ability. In 1999, Billy Mitchell offered $100,000 to anyone who could pass through the Split-Screen Level before January 1, 2000. The prize was never claimed.[*][34]

4.40.2 Development

The game was developed primarily by a young Namco employee named Toru Iwatani over the course of a year, beginning in April 1979, employing a nine-man team. It was based on the concept of eating, and the original Japanese title was *Pakkuman* (パックマン), inspired by the Japanese onomatopoeic slang phrase *paku-paku taberu* (パクパク食べる),[*][36][*][37] where *paku-paku* describes (the sound of) the mouth movement when widely opened and then closed in succession.[*][38]

Although Iwatani has repeatedly stated that the character's shape was inspired by a pizza missing a slice,[*][10] he admitted in a 1986 interview that this was a half-truth and the character design also came from simplifying and rounding out the Japanese character for mouth, *kuchi* (口).[*][39] Iwatani attempted to appeal to a wider audience—beyond the typical demographics of young boys and teenagers. His intention was to attract girls to arcades because he found there were very few games that were played by

The North American Pac-Man *cabinet design (left) differs significantly from the Japanese* Puck Man *design (right).*

women at the time.[*][40] This led him to add elements of a maze, as well as cute ghost enemy characters. Eating to gain power, Iwatani has said, was a concept he borrowed from Popeye.[*][41] The result was a game he named *Puck Man*[*][42] as a reference to the main character's hockey puck shape.[*][43]

Later in 1980, the game was picked up for manufacture in the United States by Bally division Midway,[*][39] which changed the game's name from *Puck Man* to *Pac-Man* in an effort to avoid vandalism from people changing the letter 'P' into an 'F' to form the word fuck.[*][42][*][43][*][44] The cabinet artwork was also changed and the pace and level of difficulty increased to appeal to western audiences.[*][42]

4.40.3 Impact and legacy

When first launched in Japan by Namco in 1980, the game received a lukewarm response, as *Space Invaders* and other similar games were more popular at the time.[*][12] However, *Pac-Man*'s success in North America in the same year took competitors and distributors completely by surprise. Marketing executives who saw *Pac-Man* at a trade show prior to release completely overlooked the game (along with the now classic *Defender*), while they looked to a racing car game called *Rally-X* as the game to outdo that year.[*][45] The appeal of *Pac-Man* was such that the game caught on immediately with the public; it quickly became far more popular than anything seen in the game industry up to that point. *Pac-Man* outstripped *Asteroids* as the best-selling arcade game in North America,[*][46][*][47] grossing over $1 billion in quarters within a decade,[*][15][*][48] by the end of the 1980s,[*][49] surpassing the revenues grossed by the then highest-grossing film *Star Wars*.[*][50]

It sold more than 350,000 arcade cabinets[*][51][*][52] (retailing at around $2400 each)[*][53] for $1 billion within 18 months[*][51] (inflation adjusted: $2.4 billion in 2011).[*][54] By 1982, the game had sold 400,000 arcade machines worldwide and an estimated 7 billion coins had been inserted into *Pac-Man* machines.[*][55] In addition, United States revenues from *Pac-Man* licensed products (games, T-shirts, pop songs, wastepaper baskets, etc.) exceeded $1 billion[*][55] (inflation adjusted: $2.33 billion in 2011).[*][54] The game was also estimated to have had 30 million active players across the United States in 1982.[*][56]

Towards the end of the 20th century, the game's total gross in quarters had been estimated by Twin Galaxies at more than 10 billion quarters ($2.5 billion),[*][15][*][16] making it the highest-grossing video game of all time.[*][14] In January 1982, the game won the overall Best Commercial Arcade Game award at the 1981 Arcade Awards.[*][57] In 2001, it was voted the greatest video game of all time by a Dixons poll in the UK.[*][58]

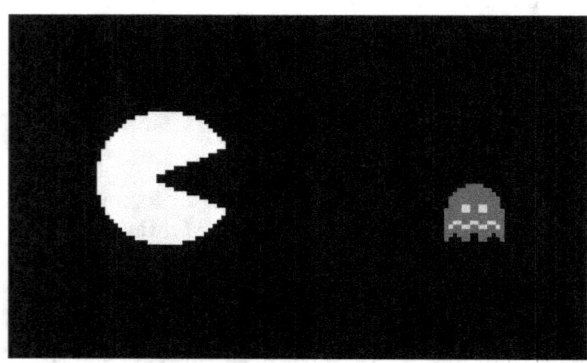

Pac-Man intermission cutscene. It exaggerated the effect of the power pellet power-up showing a comically large Pac-Man.[*][59]

The game is regarded as one of the most influential video games of all time,[*][60][*][61][*][62] for a number of reasons: its titular character was the first original gaming mascot, the game established the maze chase game genre, it demonstrated the potential of characters in video games, it opened gaming to female audiences, and it was gaming's first licensing success.[*][60] In addition, it was the first video game to feature power-ups,[*][63] and it is frequently credited as the first game to feature cut scenes, in the form of brief comical interludes about Pac-Man and Blinky chasing each other around during those interludes,[*][64] though *Space Invaders Part II* employed a similar technique that same year.[*][65] *Pac-Man* is also credited for laying the foundations for the stealth game genre, as it emphasized avoiding enemies rather than fighting them,[*][66] and had an influence on the early stealth game *Metal Gear*, where guards

chase Solid Snake in a similar manner to *Pac-Man* when he is spotted.[*][67]

Pac-Man has also influenced many other games, ranging from the sandbox game *Grand Theft Auto* (where the player runs over pedestrians and gets chased by police in a similar manner)[*][68] to early first-person shooters such as *MIDI Maze* (which had similar maze-based gameplay and character designs).[*][69][*][70] Game designer John Romero credited *Pac-Man* as the game that had the biggest influence on his career;[*][71] *Wolfenstein 3D* was similar in level design[*][72] and featured a *Pac-Man* level from a first-person perspective,[*][73][*][74] while *Doom* had a similar emphasis on mazes, power-ups, killing monsters, and reaching the next level.[*][75] *Pac-Man* also influenced the use of power-ups in later games such as *Arkanoid*.[*][76]

Remakes and sequels

See also: List of Pac-Man video games and Pac-Man clones

Pac-Man is one of the few games to have been consistently published for over three decades, having been remade on numerous platforms and spawned many sequels. Re-releases include ported and updated versions of the original arcade game. Numerous unauthorized Pac-Man clones appeared soon after its release. The combined sales of counterfeit arcade machines sold nearly as many units as the original *Pac-Man*, which had sold more than 300,000 machines.[*][77]

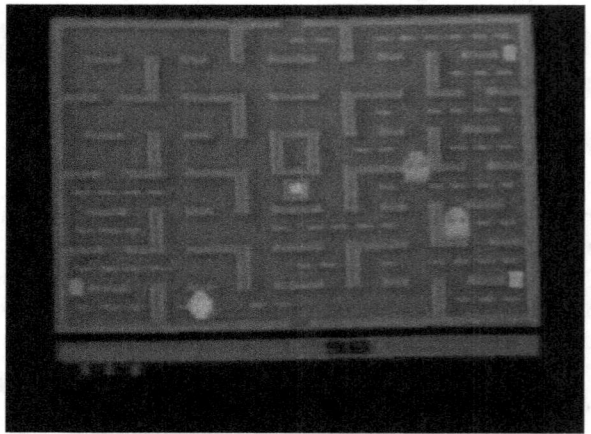

The Atari 2600 version of Pac-Man *only somewhat resembled the original and had the ghosts take turns appearing on the screen, creating a flicker effect.*

One of the first ports to be released was the much-maligned port for the Atari 2600, which only somewhat resembled the original and was widely criticized for its flickering ghosts, due to the 2600's limited memory and hardware compared to the arcade machine.*[78]*[79]*[80] Despite the criticism, this version of *Pac-Man* sold seven million units*[81] at \$37.95 per copy,*[11]*[82] and became the best-selling game of all time on the Atari 2600 console. While enjoying initial sales success, Atari had overestimated demand by producing 12 million cartridges, of which 5 million went unsold.*[81]*[83]*[84] The port's poor quality damaged the company's reputation among consumers and retailers, which would eventually become one of the contributing factors to Atari's decline and the North American video game crash of 1983, alongside Atari's *E.T. the Extra-Terrestrial.**[81]

Meanwhile, Coleco's tabletop Mini-Arcade versions of the game sold 1.5 million units in 1982.*[85]*[86]

II Computing listed it tenth on the magazine's list of top Apple II series games as of late 1985, based on sales and market-share data,*[87] and in December 1987 alone Mindscape's IBM PC version of *Pac-Man* sold over 100,000 copies.*[88] The game was also released for Atari's 5200 and 8-bit computers, Intellivision, the Commodore 64 and VIC-20, and the Nintendo Entertainment System. For handheld game consoles, it was released on the Game Boy, Sega Game Gear, Game Boy Color, and the Neo Geo Pocket Color

The game has also been featured in Namco's long-running *Namco Museum* video game compilations. Downloads of the game have been made available on game services such as Xbox Live Arcade, GameTap and Virtual Console. Namco has also released mobile versions for BREW, Java, and iOS, as well as Palm PDAs and Windows Mobile-based devices. A port of *Pac-Man* for Android*[89] can be controlled not only through an Android phone's trackball but through touch gestures or its on-board accelerometer. As of 2010, Namco had sold over 30 million paid downloads of *Pac-Man* on BREW in the United States alone.*[90]

Microsoft released *Microsoft Return of Arcade* in 1996 and *Microsoft Return of Arcade: Anniversary Edition* in 2000 and includes Pac-Man as one of its bundled arcade games.*[91]

In addition, Namco has repeatedly re-released the game to arcades. In 2001, Namco released a *Ms. Pac-Man/Galaga* "Class of 1981 Reunion Edition" cabinet with *Pac-Man* available for play as a hidden game. To commemorate *Pac-Man*'s 25th anniversary in 2005, Namco released a revision that officially featured all three games.

Namco Networks ported Pac-Man to the PC (bought online) in 2009 which also includes an "Enhanced" mode which replaces all of the original sprites with the sprites from *Pac-Man Championship Edition* but it's still the original Pac-Man otherwise, Namco Networks also made a bundle (also bought online) which includes their PC version of

Pac-Man as well as their port of *Dig Dug* called *Namco All-Stars: Pac-Man and Dig Dug.*

In 2010 Namco Bandai announced the release of the game on Windows Phone 7 as an Xbox Live game.*[92]

Pac-Man's spawned sequels and spin-offs includes only one which was designed by Tōru Iwatani . Some of the follow-ups were not developed by Namco either – including the most significant, *Ms. Pac-Man*, released in the United States in 1981. Originally called *Crazy Otto*, this unauthorized hack of *Pac-Man* was created by General Computer Corporation and sold to Midway without Namco's permission. The game features several changes from the original *Pac-Man*, including faster gameplay, more mazes, new intermissions, and moving bonus items. Some consider *Ms. Pac-Man* to be superior to the original or even the best in the entire series.*[13] Stan Jarocki of Midway stated that *Ms. Pac-Man* was conceived in response to the original *Pac-Man* being "the first commercial videogame to involve large numbers of women as players" and that it is "our way of thanking all those lady arcaders who have played and enjoyed *Pac-Man.*" *[93] Namco sued Midway for exceeding their license. Eventually, Bally Midway struck a deal with Namco to officially license *Ms. Pac-Man* as a sequel. Namco today officially owns Ms. Pac-Man in its other releases.

Following *Ms. Pac-Man*, Bally Midway released several other unauthorized spin-offs, such as *Pac-Man Plus*, *Jr. Pac-Man*, *Baby Pac-Man* and *Professor Pac-Man*, resulting in Namco severing business relations with Midway.*[10]*[94]

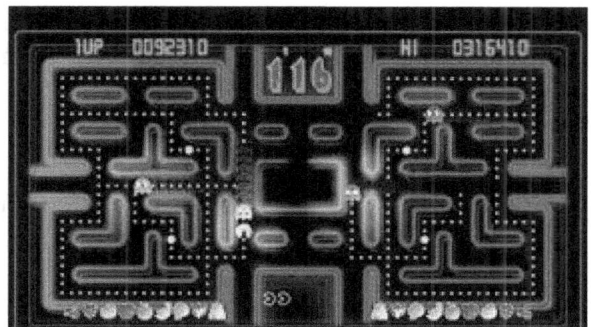

Pac-Man Championship Edition *(2007), commemorating the first World Championship*

Various platform games based on the series have also been released by Namco, such as 1984's *Pac-Land* and the *Pac-Man World* series, which features Pac-Man in a 3-D world. More modern versions of the original game have also been developed, such as the multiplayer *Pac-Man Vs.* for the Nintendo GameCube.

On June 5, 2007, the first Pac-Man World Championship

was held in New York City, which brought together ten competitors from eight countries to play the new *Pac-Man Championship Edition* developed by Tōru Iwatani.[*][95] Its sequel was released on November 2010.

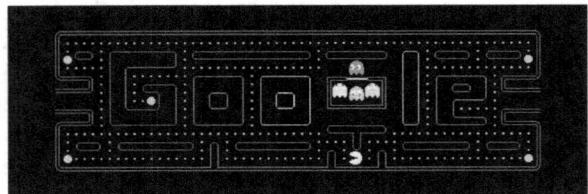

The initial configuration of the Google Pac-Man banner

For the weekend of May 21–23, 2010, Google changed the Google logo on its homepage to a Google Doodle of a fully playable version of the game[*][96] in recognition of the 30th anniversary of the game's release. The game featured the ability to play both Pac-Man and Ms. Pac-Man simultaneously.[*][97] After finishing the game, the website automatically redirected the user to a search of *Pac-Man 30th Anniversary*.[*][98] Companies across the world experienced slight drops in productivity due to the game, estimated to be valued at the time as $120,000,000 (approximately €95,400,000; £83,000,000). However, The Official ASTD Blog noted that the total loss, "spread out across the entire world isn't a huge loss, comparatively speaking" .[*][99] In total, the game devoured around 4.8 million hours of work productivity that day.[*][61] Some organizations even temporarily blocked Google's website from workplace computers on the Friday it was uploaded, particularly where it violated regulations against recreational games.[*][100][*][101][*][102] Because of the popularity of the Pac-Man doodle, Google decided to allow access to the game through a separate web page; doing a Google search for the phrase "pac man" yields this doodle as the first result. On March 31, 2015, Google Maps added an option allowing a Pac-Man style game to be played using streets on the map as the maze.[*][103]

In 2011, Namco sent a DMCA notice to the team that made the programming language Scratch saying that a programmer had infringed copyright by making a Pac-Man game using the language and uploading it to Scratch's official website.[*][104]

In April 2011, Soap Creative published World's Biggest Pac-Man working together with Microsoft and Namco-Bandai to celebrate Pac-Man's 30th anniversary. It is a multiplayer browser-based game with user-created, interlocking mazes.[*][105]

Non video game versions

In 1982, Milton Bradley released a board game based on Pac-Man.[*][106] In this game, players move up to four Pac-Mans (traditional yellow plus red, green and blue) plus two ghosts as per the throws of a pair of dice. Each Pac-Man is assigned to a player while the ghosts are neutral and controlled by all players. Each player moves their Pac-Man the number of spaces on either die and a ghost the number of spaces on the other die, the Pac-Man consuming any white marbles (equivalent of pac-dots) and yellow marbles (equivalent of power pellets) in its path. Players can move a ghost onto a Pac-Man and claim two white marbles from its player. They can also move a Pac-Man with a yellow marble inside it onto a ghost and claim two white marbles from any other player (following which the yellow marble is placed back on the maze. The game ends when all white marbles have been cleared from the board and the player with the largest number of white marbles is then declared the winner.[*][107]

Sticker manufacturerer Fleer included Pac-Man Rub Off Game cards with their Pac-Man stickers. The cards contained a Pac-Man style maze with all points along the path covered with opaque coverings. Starting from the lower board Pac-Man starting position, the player moved around the maze scratching off the coverings to score points. A white dot scored one point, a Blue Monster scored ten points and a cherry scored 50 points. Uncovering a red, orange or pink monster scored no points but the game ended when a third such monster had been uncovered. An Ms Pac-Man version of the game also included pretzels (100 points) and bananas (200 points) [*][108]

Nelsonic Industries produced a Pac-Man game LCD wristwatch. This followed essentially the same rules as the video version albeit with a simplified maze.[*][109]

A pinball version, Mr. & Mrs. Pac-Man was designed by George Christian and released by Bally/Midway in 1982.[*][110] The spin-off arcade game Baby Pac Man also contains a non-video pinball element.[*][111]

In popular culture

Pac-Man went on to become an icon of video game culture during the 1980s, and a wide variety of *Pac-Man* merchandise was marketed with the character's image, from t-shirts and toys to hand-held video game imitations and even specially shaped pasta. An animated TV series produced by Hanna–Barbera aired on ABC from 1982 to 1983.[*][112] The Killer List of Videogames lists *Pac-Man* as the No. 1 video game on its "Top 10 Most Popular Video games" list.[*][113] At one time, a feature film based on the game was also in development.[*][114][*][115] In 2010, a computer-generated animated series titled *Pac-Man and the Ghostly Adventures*, was reported to be in the works. The show was released on Disney XD in June, 2013.[116][117] *Pac-Man* has also been referenced in the 2010 film *Scott Pilgrim vs. the World*, where the game's origins as *Puck-Man* is mentioned several times.[118] Clyde appears in the Disney animated film *Wreck-It Ralph* as one of several villains participating in a group therapy session, voiced by Kevin Deters. His cohorts, Inky, Blinky, and Pinky, appear together in Game Central Station for a few scenes, and Pac-Man makes a cameo appearance during the Fix-It Felix Jr. 30th Anniversary party. General Mills manufactured a cereal by the Pac-Man name in 1983. Over the cereal's lifespan, characters from sequels *Super Pac-Man* and *Ms. Pac-Man* were also added.

Guinness World Records has awarded the *Pac-Man* series eight records in *Guinness World Records: Gamer's Edition 2008*, including First Perfect Pac-Man Game for Billy Mitchell's July 3, 1999 score and "Most Successful Coin-Operated Game". On June 3, 2010, at the NLGD Festival of Games, the game's creator Toru Iwatani officially received the certificate from Guinness World Records for Pac-Man having had the most "coin-operated arcade machines" installed worldwide: 293,822. The record was set and recognized in 2005 and mentioned in the *Guinness World Records: Gamer's Edition 2008*, but finally actually awarded in 2010.[26]

Pac-Man has been referenced in numerous other media. In music, the Buckner & Garcia song "Pac-Man Fever" (1981) went to No. 9 on the Billboard Hot 100 charts,[11] and received a Gold certification with over a million records sold by 1982,[119] and a total of 2.5 million copies sold as of 2008.[120] Their *Pac-Man Fever* album (1982) also received a Gold certification for selling over a million records.[121] "Weird Al" Yankovic recorded a song titled "Pac-Man" that was a parody of The Beatles' "Taxman", in 1981.[122] Jonzun Crew's "Pack Jam" (1983) was inspired by Michael Jonzun's distaste towards the popular *Pac-Man* game.[123] Hip hop emcee Lil' Flip sampled sounds from the game *Pac-Man* and *Ms. Pac-Man* to make his top-20 single "Game Over" (2004). Namco America filed a lawsuit against Sony Music Entertainment for unauthorized use of these samples. The suit was eventually settled out of court.[124][125] Aphex Twin released an EP dedicated to the game, *Pac-Man EP*, in 1992.

Ken Uston's strategy guide *Mastering Pac-Man* sold 750,000 copies, reaching No. 5 on B. Dalton's mass-market bestseller list.[126] By 1983, 1.7 million copies of *Mastering Pac-Man* had been printed.[127] In comedy, there is a popular Pac-Man joke on the controversy regarding the influence of video games on children.[128]

The game has also inspired various real-life recreations, in-

Students' Spring Days in Tartu: runners in Pac-Man costume.

volving either real people or robots. One event called Pac-Manhattan set a Guinness World Record for "Largest Pac-Man Game" in 2004.[129][130][131] The term Pac-Man defense in mergers and acquisitions refers to a hostile takeover target that attempts to reverse the situation and take over its would-be acquirer instead, a reference to Pac-Man's power pellets.[132] The game's popularity has led to "Pac-Man" being adopted as a nickname, most notably by boxer Manny Pacquiao,[133] as well as the American football player Adam Jones.

Pac-Man has also found its position beyond the game world. Under a National Science Foundation funded project, the computer science department at UC Berkeley has developed a custom version of the Pac-Man in Python to teach students basic Artificial Intelligence concepts, such as informed state-space search, probabilistic inference, and reinforcement learning.[134] Students are asked to complete a series of problems from simple to difficult, to eventually design a Pac-Man agent that automatically eats all the dots on the map. The concepts learned during these problems underly many real-world AI application areas, such as natural language processing, computer vision, and robotics.

In January 2013, Pac-Man and Blinky appeared on the top Massachusetts Institute of Technology's Great Dome as part of a traditional hack or prank used to demonstrate the technical aptitude and cleverness of the students. According to the MIT alumni blog, Slice of MIT, the Pac-Man, Blinky battle was intended to serve as a metaphor for the semester. "Pac-Man represents the unquenchable search for knowledge, while Blinky represents the unforeseen distractions that may occur." [135][136]

On June 10, 2014, Pac-Man was confirmed to appear as a playable character in the game *Super Smash Bros. for Nintendo 3DS and Wii U*. The 3DS version also has a stage based on the original arcade game, called *Pac-Maze*.[137]

A Pac-Man Amiibo figurine was also released by Nintendo on May 29, 2015.

On February 1, 2015, as part of the Super Bowl, a Bud Light commercial featured a real-life *Pac-Man* board with one of the players as Pac-Man, and using computer graphic ghosts.

Pac-Man appears in the film *Pixels* (2015), with Denis Akiyama playing series creator Toru Iwatani.[*][138][*][139] Iwatani himself makes a cameo at the beginning of the film as an arcade technician.

4.40.4 References

[1] Namco Bandai Games Inc. (June 2, 2005). "Bandai Namco press release for 25th Anniversary Edition" (in Japanese). bandainamcogames.co.jp/. Archived from the original on December 30, 2007. Retrieved October 10, 2007. 2005 年 5 月 22 日で生誕 25 周年を迎えた『パックマン』。 ("Pac-Man celebrates his 25th anniversary on May 22, 2005", seen in image caption)

[2] Long, Tony (October 10, 2007). "Oct. 10, 1979: Pac-Man Brings Gaming Into Pleistocene Era". *Wired*. Archived from the original on September 11, 2014. Retrieved October 10, 2007. [Bandai Namco] puts the date at May 22, 1980 and is planning an official 25th anniversary celebration next year.

[3] Date shown in January 1982 article "Midway celebrates Pac-Man".

[4] "Game Board Schematic". *Midway Pac-Man Parts and Operating Manual* (PDF). Chicago, Illinois: Midway Games. December 1980. Retrieved July 20, 2009.

[5] Nitsche, Michael (March 31, 2009). "Games and Rules". *Video Game Spaces: Image, Play, and Structure in 3D Worlds*. Cambridge, Massachusetts: MIT Press. p. 26. ISBN 0-262-14101-9. [...] they would not realize the fundamental logical difference between a version of Pac-Man (Iwatani 1980) running on the original Z80 [...]

[6] "Pac-Man still going strong at 30". UPI.com. May 22, 2010. Retrieved 2012-05-22.

[7] "Oct. 10, 1979: Pac-Man Brings Gaming Into Pleistocene Era". *Wired*. October 10, 2007.

[8] "Pac 'n Roll Review". GameSpot.com. August 23, 2005. Retrieved 2012-05-22.

[9] Wolf, Mark J. P. (2008). "The video game explosion: A history from PONG to PlayStation and beyond". ISBN 978-0-313-33868-7.

[10] Green, Chris (June 17, 2002). "Pac-Man". Salon.com. Retrieved February 12, 2006.

[11] "Pac-Man Fever". *Time Magazine*. April 5, 1982. Archived from the original on January 22, 2011. Retrieved October 15, 2009. Columbia Records' Pac-Man Fever ... was No. 9 on the Billboard Hot 100 last week.

[12] Goldberg, Marty (January 31, 2002). "Pac-Man: The Phenomenon: Part 1". Arcadegaming.us. Retrieved July 31, 2006.

[13] Parish, Jeremy (2004). "The Essential 50: Part 10 – Pac Man". 1UP.com. Retrieved July 31, 2006.

[14] Steve L. Kent (2001). *The ultimate history of video games: from Pong to Pokémon and beyond : the story behind the craze that touched our lives and changed the world*. Prima. p. 143. ISBN 0-7615-3643-4. Retrieved May 1, 2011. Despite the success of his game, Iwatani never received much attention. Rumors emerged that the unknown creator of Pac-Man had left the industry when he received only a $3500 bonus for creating the highest-grossing video game of all time.

[15] Mark J. P. Wolf (2008). *The video game explosion: a history from PONG to PlayStation and beyond*. ABC-CLIO. p. 73. ISBN 0-313-33868-X. Retrieved April 10, 2011. It would go on to become arguably the most famous video game of all time, with the arcade game alone taking in more than a billion dollars, and one study estimated that it had been played more than 10 billion times during the twentieth century.

[16] Chris Morris (May 10, 2005). "Pac Man turns 25: A pizza dinner yields a cultural phenomenon – and millions of dollars in quarters". CNN. Retrieved April 23, 2011. In the late 1990s, Twin Galaxies, which tracks video game world record scores, visited used game auctions and counted how many times the average Pac Man machine had been played. Based on those findings and the total number of machines that were manufactured, the organization said it believed the game had been played more than 10 billion times in the 20th century.

[17] "The Legacy of Pac-Man". Archived from the original on January 21, 1998.

[18] "Pac Man Bootleg Board Information". Archived from the original on July 2, 2007.

[19] "Davie Brown Celebrity Index: Mario, Pac-Man Most Appealing Video Game Characters Among Consumers". PR Newswire. Archived from the original on June 27, 2009. Retrieved May 6, 2011.

[20] "History of Computing: Video games – Golden Age". Thocp.net. Retrieved 2012-05-22.

[21] Antonelli, Paola (29 November 2012). "Video Games: 14 in the Collection, for Starters". MoMA. Retrieved 30 November 2012.

[22] "Pacman Game". Retrieved 13 November 2012.

[23] http://home.comcast.net/~{}jpittman2/pacman/ pacmandossier.html#CH3_Just_Passing_Through

[24] *Pac-Man*, The Arcade Flyer Archive, 1980, archived from the original on November 30, 2013, retrieved May 23, 2012

[25] "What is Pacman?". *Pacman.com*. Namco. Archived from the original on 2010-11-28. Retrieved July 14, 2010.

[26] Martijn Müller (June 3, 2010). "Pac-Man wereldrecord beklonken en het hele verhaal" (in Dutch). NG-Gamer.

[27] Mateas, Michael (2003). "Expressive AI: Games and Artificial Intelligence" (PDF). *Proceedings of Level Up: Digital Games Research Conference, Utrecht, Netherlands.*

[28] "News Headlines". Cnbc.com. March 3, 2011. Retrieved 2012-05-22.

[29] http://pacmanmuseum.com/history/nomenclatureconflicts.php

[30] DeMaria, Rusel; Wilson, Johnny L. (December 18, 2003). *High Score!: The Illustrated History of Electronic Games* (2nd ed.). McGraw-Hill Osborne Media. ISBN 9780072224283.

[31] http://home.comcast.net/~{}jpittman2/pacman/pacmandossier.html#CH5_Playing_The_Level

[32] Don Hodges. "Pac-Man's Split-screen level analyzed and fixed". Retrieved April 29, 2008.

[33] "Pac-Man review at OAFE". Oafe.net. Retrieved 2012-09-15.

[34] Ramsey, David. "The Perfect Man". Oxford American. Archived from the original on 2008-02-29. Retrieved 13 November 2012.

[35] "Pac-Man at the Twin Galaxies Official Scoreboard". Twin Galaxies. Archived from the original on May 23, 2006. Retrieved July 22, 2006.

[36] "Top 25 Smartest Moves in Gaming". Gamespy.com. Archived from the original on 2009-02-18. Retrieved July 26, 2010.

[37] Kohler, Chris (2005). *Power-Up: How Japanese Video Games Gave the World an Extra Life*. Brady Games. ISBN 0-7440-0424-1. External link in |title= (help)

[38] "Daijisen Dictionary entry for ぱくぱく (*paku-paku*), in Japanese". Retrieved January 27, 2007.

[39] Lammers, Susan M. (1986). *Programmers at Work: Interviews*. New York: Microsoft Press. p. 266. ISBN 0-914845-71-3.

[40] http://www.cnbc.com/id/41888021

[41] "The Collection: Selected Works from Applied Design; Pac-Man". MoMA. Retrieved March 4, 2013.

[42] Kohler, Chris (May 21, 2010). "Q&A: Pac-Man Creator Reflects on 30 Years of Dot-Eating". *Wired*. Retrieved July 15, 2010.

[43] Kent, Steve. *Ultimate History of Video Games*, p.142. "Before Namco showed Pac-Man to Midway, one change was made to the game. Pac-Man was originally named Puck-Man, a reference to the puck-like shape of the main character. Nakamura worried about American vandals changing the "P" to an "F." To prevent any such occurrence, he changed the name of the game."

[44] Brian Ashcraft. "This Guy Has a Rare Arcade Cabinet. Is It Real?". Kotaku.

[45] Bowen, Kevin (2001). "Game of the Week: *Defender*". ClassicGaming.com. Retrieved August 17, 2006.

[46] "Pac-Man – The Dot Eaters". The Dot Eaters. Retrieved August 17, 2006.

[47] Mark J. P. Wolf (2001). *The medium of the video game.* University of Texas Press. p. 44. ISBN 0-292-79150-X. Retrieved April 9, 2011

[48] Bill Loguidice & Matt Barton (2009). *Vintage games: an insider look at the history of Grand Theft Auto, Super Mario, and the most influential games of all time.* Focal Press. p. 181. ISBN 0-240-81146-1. Retrieved April 23, 2011. The machines were well worth the investment; in total they raked in over a billion dollars worth of quarters in the first year alone.

[49] Kline, Stephen; Nick Dyer-Witheford; Greig de Peuter (2003). *Digital play: the interaction of technology, culture, and marketing* (Reprint ed.). Montréal, Quebec: McGill-Queen's University Press. p. 96. ISBN 0-7735-2591-2. Retrieved February 25, 2012. The game produced one billion dollars in 1980 alone

[50] "Electronic and Computer Games: The History of an Interactive Medium". *Screen* **29** (2): 52–73 [53]. 1988. doi:10.1093/screen/29.2.52. Retrieved January 25 2012. Revenue from the game Pac-Man alone was estimated to exceed that from the cinema box-office success Star Wars.

[51] Marlene Targ Brill (2009). *America in the 1980s.* Twenty-First Century Books. p. 120. ISBN 0-8225-7602-3. Retrieved May 1, 2011

[52] Kevin "Fragmaster" Bowen (2001). "Game of the Week: Pac-Man". GameSpy. Retrieved April 9, 2011.

[53] Infoworld Media Group, Inc (April 12, 1982). "Video arcades rival Broadway theatre and girlie shows in NY". *InfoWorld* **4** (14). p. 15. ISSN 0199-6649. Retrieved May 1, 2011

[54] "CPI Inflation Calculator". Bureau of Labor Statistics. Retrieved March 22, 2011.

[55] Kao, John J. (1989). *Entrepreneurship, creativity & organization: text cases & readings.* Englewood Cliffs, NJ: Prentice Hall p. 45. ISBN 0-13-283011-6. Retrieved February 12, 2012. Estimates counted 7 billion coins that by 1982 had been inserted into some 400,000 Pac Man machines worldwide, equal to one game of Pac Man for every

person on earth. US domestic revenues from games and licensing of the Pac Man image for T-shirts, pop songs, to wastepaper baskets, etc. exceeded \$1 billion.

[56] "Men's wear, Volume 185". *Men's wear* (Fairchild Publications) **185**. 1982. Retrieved February 28, 2012.

[57] "Electronic Games Magazine". Internet Archive. Retrieved February 1, 2012.

[58] "Pac Man 'greatest video game'". BBC News. November 13, 2001. Retrieved March 13, 2012.

[59] Aaron Matteson. "Five Things We Learned From Pac-Man". http://joystickdivision.com. "This cutscene furthers the plot by depicting a comically large Pac-Man".

[60] The Essential 50 – Pac-Man, 1UP

[61] Wilson, Jeffrey L. (June 11, 2010). "The 10 Most Influential Video Games of All Time". *PC Magazine*. 1. Pac-Man (1980). Retrieved April 19, 2012.

[62] The ten most influential video games ever, *The Times*, September 20, 2007

[63] Playing With Power: Great Ideas That Have Changed Gaming Forever, 1UP

[64] Gaming's Most Important Evolutions, GamesRadar

[65] "Space Invaders Deluxe". *klov.com*. Retrieved March 28, 2011.

[66] Al-Kaisy, Muhammad (June 10, 2011). "The history and meaning behind the 'Stealth genre'". Gamasutra. Retrieved September 15, 2011.

[67] David Low (April 2, 2007). "GO3: Kojima Talks Metal Gear History, Future". Gamasutra. Retrieved August 3, 2011.

[68] Brian Ashcraft (July 16, 2009). "Grand Theft Auto And Pac-Man? "The Same"". Retrieved March 8, 2011.

[69] "25 years of Pac-Man". MeriStation. July 4, 2005. Retrieved May 6, 2011. (Translation)

[70] "Gaming's Most Important Evolutions". GamesRadar. October 8, 2010. p. 5. Retrieved April 27, 2011.

[71] Bailey, Kat (March 9, 2012). "These games inspired Cliff Bleszinski, John Romero, Will Wright, and Sid Meier". Joystiq. Retrieved April 2, 2012.

[72] Stephan Günzel, Michael Liebe, Dieter Mersch (2008). Sebastian Möring, ed. *Conference Proceedings of The Philosophy of Computer Games 2008*. Potsdam University Press. pp. 191–2. ISBN 3-940793-49-3. Retrieved May 6, 2011

[73] *Book of Games: The Ultimate Reference on PC & Video Games*. Book of Games. 2006. p. 24. ISBN 82-997378-0-X. Retrieved May 6, 2011

[74] "Game developer". 2 & 5. Miller Freeman. 1995. p. 62. Retrieved June 6, 2011. If you made it to the secret Pac-Man level in Castle Wolfenstein, you know what I mean (Pac-Man never would have made it as a three-dimensional game). Though it may be less of a visual feast, two dimensions have a well-established place as an electronic gaming format.

[75] Media, Spin L.L.C. (September 1995). "Children of Doom". *Spin* **11** (6). p. 118. ISSN 0886-3032. Retrieved May 6, 2011

[76] Gutman, Dan (July 1989). "Nine for '89". *Compute!*. p. 19. Retrieved 11 November 2013.

[77] Leonard Herman, Jer Horwitz, Steve Kent, Skyler Miller (2002). "The History of Video Games" (PDF). GameSpot. p. 7. Retrieved March 14, 2012.

[78] "Creating a World of Clones". *Philadelphia Inquirer*. October 9, 1983. p. 16.

[79] Thompson, Adam (Fall 1983). "The King of Video Games is a Woman". *Creative Computing Video and Arcade Games* **1** (2): 65.

[80] Ratcliff, Matthew (August 1988). "Classic Cartridges II". *Antic* **7** (4): 24.

[81] Buchanan, Levi (August 26, 2008). "Top 10 Best-Selling Atari 2600 Games". IGN. Retrieved July 15, 2009.

[82] "The A-Maze-ing World of Gobble Games". *Electronic Games* **1** (3): 62–63 [63]. May 1982. Retrieved February 3, 2012.

[83] Ellis, David (2004). "The Atari VCS (2000)". *Official Price Guide to Classic Video Games*. Random House. pp. 98–99. ISBN 0-375-72038-3.

[84] Buchanan, Levi (2008-11-26). "Top 10 Videogame Turkeys". IGN. Retrieved 2009-07-15.

[85] "Mini-Arcades 'Go Gold'". *Electronic Games* **1** (9): 13. November 1982. Retrieved February 5, 2012.

[86] "Coleco Mini-Arcades Go Gold" (PDF). *Arcade Express* **1** (1): 4. August 15, 1982. Retrieved February 3, 2012.

[87] Ciraolo, Michael (Oct–Nov 1985). "Top Software / A List of Favorites". *II Computing*. p. 51. Retrieved 28 January 2015.

[88] J.F. Archibald, J. Haynes, ed. (1988). "Video Games Are Back". *The Bulletin* (5609–5616): 134. Retrieved January 29, 2012. Mindscape, a software company based in Northbrook, sold more than 100,000 copies of Pac Man for the PC last December alone.

[89] Nguyen, Vincent (May 28, 2008). "First LIVE images and videos of fullscreen Android demos!". Retrieved July 5, 2008.

[90] "Namco Networks' Pac-Man Franchise Surpasses 30 Million Paid Transactions in the United States on Brew". AllBusiness.com. 2010. Retrieved February 22, 2012.

[91] Microsoft Arcade.

[92] "A quick look at some of the new WP7 games from Namco". BestWP7Games. November 9, 2010.

[93] Worley, Joyce (May 1982). "Women Join the Arcade Revolution". Electronic Games 1 (3): 30–33 [33]. Retrieved February 3, 2012.

[94] "Ms. Pac-Man". Killer List of Videogames. Retrieved July 31, 2006.

[95] Schiesel, Seth (2007-06-06). "Run, Gobble, Gobble, Run: Vying for Pac-Man Acclaim". The New York Times. Retrieved 2010-05-20.

[96] "Google gets Pac-Man fever". cnet. May 21, 2010.

[97] Terdiman, Daniel (May 21, 2010). "Google gets Pac-Man fever". News.cnet.com. Retrieved 2012-05-22.

[98] "'Insert Coin': Google Doodle Celebrates Pac-Man's 30th Anniversary". ABC. ABC. May 21, 2010. Retrieved May 21, 2010.

[99] "Pac-Man gobbles up $120M in workplace productivity". .astd.org. May 26, 2010. Archived from the original on January 10, 2015. Retrieved 2015-03-09.

[100] "CANOE – Technology: Pac-Man gobbles up $120M in workplace productivity". Technology canoe.ca. Retrieved 2012-05-22.

[101] "Quit playing Google Pac Man and get back to work, everyone!". Inquisitr.com. May 21, 2010. Retrieved 2012-05-22.

[102] Terdiman, Daniel (May 21, 2010). "Is playable Pac-Man getting Google's home page banned?". News.cnet.com. Retrieved 2012-05-22.

[103] Michael Calia. "You Can Play Pac-Man on Your City's Streets". Wall Street Journal (blogs). Retrieved March 31, 2015.

[104] Tom Goldman. "The Escapist : News : Namco Shuts Down Student's Pac-Man Project". Escapistmagazine com. Retrieved 2012-05-22.

[105] Ki Mae Huessner. "World's Biggest Pac-Man Is Web Sensation". ABC News Internet Ventures Retrieved April 13, 2013.

[106] Coopee, Todd. "Pac-Man Turns 35!". ToyTales.ca.

[107] http://www.noisetosignal.org/2009/06/the-mb-official-pac-man-board-game.html

[108] http://pacstar.mycoldwater.com/zindex.htm

[109] http://www.handheldmuseum.com/Manuals/Nelsonic-PacManWatch.pdf

[110] http://www.ipdb.org/machine.cgi?id=1639

[111] http://www.ipdb.org/machine.cgi?id=125

[112] "The Pac-Page (including database of Pac-Man merchandise and TV show reference)". GameSpy. Archived from the original on February 18, 2009. Retrieved May 7, 2011.

[113] McLemore, Greg. "The Top Coin-Operated Videogames of All Times". Killer List of Videogames. Archived from the original on July 17, 2006. Retrieved July 22, 2006.

[114] "Crystal Sky, Namco & Gaga are game again". Crystalsky.com. Retrieved August 11, 2008.

[115] Jaafar, Ali (May 19, 2008) "Crystal Sky signs $200 million deal". Variety.com. Retrieved September 4, 2008.

[116] White, Cindy. (June 17, 2010) "E3 2010: Pac-Man Back on TV?". IGN.com. Retrieved July 7, 2010.

[117] Morris, Chris. (June 17, 2010) "Pac-Man chomps at 3D TV. Variety.com. Retrieved July 7, 2010.

[118] Ivan-Zadeh, Larushka (August 26, 2010). "Scott Pilgrim Vs The World is almost Spaced in Toronto". Metro. Retrieved March 15, 2012.

[119] "Popular Computing". McGraw-Hill. 1982. Retrieved August 14, 2010. Pac-Man Fever went gold almost instantly with 1 million records sold.

[120] Turow, Joseph (2008). Media Today: An Introduction to Mass Communication (3rd ed.). Taylor & Francis. p. 554. ISBN 0-415-96058-4. Retrieved January 29, 2012.

[121] RIAA Gold & Platinum Searchable Database – Pac-Man Fever. RIAA.com. Retrieved November 1, 2009.

[122] Dr. Demento's Basement Tapes #4, a Demento Society members-only compilation from 1994, contains the demo. It was never commercially recorded or released.

[123] "The Vocoder: From Speech-Scrambling To Robot Rock". NPR. May 13, 2010. Retrieved February 4, 2012

[124] Carless, Simon (August 29, 2005). "Namco, Sony Music Settle Over Pac-Man Samples". Gamasutra.com. Retrieved 2012-09-15.

[125] Lai, Marcus (August 29, 2005). "Namco and Sony settle Pac-Man lawsuit". News.punchjump.com. Retrieved 2012-09-15.

[126] "Learn The Code Book And Beat Video Games". Ludington Daily News. March 1, 1982. p. 25. Retrieved April 30, 2011.

[127] Uston, Ken (Fall 1983). "Mastering Pac-Man Plus and Super Pac-Man". Creative Computing Video & Arcade Games 1 (2): 32. Retrieved February 22, 2012.

[128] "Official site for the stand-up comic, writer, presenter & actor" . Marcus Brigstocke. Archived from the original on 2008-11-20. Retrieved March 13, 2009. "If Pacman had affected us as kids we'd be running around in dark rooms, munching pills and listening to repetitive music."

[129] "About Pac-Manhattan". Pac-Manhattan. 2004. Retrieved July 3, 2009.

[130] "Roomba Pac-Man Web Site". Retrieved October 10, 2009.

[131] Lau, Dominic. "Pacman in Vancouver" . SFU Computing Science. Archived from the original on 2009-05-30. Retrieved July 3, 2009.

[132] "Origins of the 'Pac-Man' Defense" . The New York Times. January 23, 1988. Retrieved November 20, 2010.

[133] Brunell, Evan (May 22, 2010). "Popular Video Game Pac-Man Celebrates 30th Anniversary" . New England Sports Network. Retrieved April 11, 2012.

[134] "The Pac-Man Project" . UC Berkeley. Retrieved July 19, 2012.

[135] Landry, Lauren (January 11, 2013). "New Year, New Hack: MIT Students Place Pac-Man On Top of the Great Dome" . BostInno (Streetwise Media). Retrieved 2013-01-11.

[136] Dezenski, Lauren (January 11, 2013). "In whimsical retro tribute, Pac-Man appears on MIT's Great Dome" . Boston.com (NY Times Co). Retrieved 2013-01-11.

[137] "Pac-Man" .

[138] "Classic video game characters unite via film 'Pixels'". Philstar. July 23, 2014. Retrieved July 23, 2014.

[139] Tarek Bazley: Pac-man at 35: the video game that changed the world. Al Jazeera English, 2015-05-25

4.40.5 Further reading

- Trueman, Doug (November 10, 1999). "The History of Pac-Man". *GameSpot*. Comprehensive coverage on the history of the entire series up through 1999.

- Müller, Martijn (June 3, 2010). "Tōru Iwatani on how Pac-Man came to be". *NG-Gamer*.

- Morris, Chris (May 10, 2005). "Pac Man Turns 25". *CNN Money*.

- Vargas, Jose Antonio (June 22, 2005). "Still Love at First Bite: At 25, Pac-Man Remains a Hot Pursuit". *The Washington Post*.

- Hirschfeld, Tom. *How to Master the Video Games*, Bantam Books, 1981. ISBN 0-553-20164-6 Arcade strategy guide to several games including incarnations of Pac-Man. Includes hand drawings of some of the common patterns for use in the arcade Pac-Man. 1982 edition ISBN 0-553-20195-6 covers home versions.

4.40.6 External links

- *Pac-Man* at the Killer List of Videogames

- *Pac-Man* at the Arcade History database

- Pac-man 30th Anniversary website

- *Pac-Man* guide at StrategyWiki

- Twin Galaxies' High-Score Rankings for Pac-Man

4.41 Pirate Adventure

Pirate Adventure or *Pirate Cove* was a text-based adventure program written by Scott Adams.

4.41.1 Description

Published by Adventure International and the second game of the series, after *Adventureland*, this text-based adventure game was one of many adventure games created by Scott Adams,[1] in this case based on his wife Alexis' ideas.[2] The setting was inspired by the novel Treasure Island and involved a quest to retrieve Long John Silver's lost treasures.[3]

Gameplay involved moving from location to location, picking up any objects found there, and using them somewhere else to unlock puzzles. Commands took the form of verb and noun, e.g. "Climb Tree" .

The player started the game in a flat, and progressed via a bit of magic to Pirates Island. Here, the player had to build a ship to reach Treasure Island and there find two pieces of treasure. The player also had to contend with an unpredictable pirate ally; it was the first text adventure game of the Adams series in which the player shared the adventure with a second character.

The magic phrase to reach the island in this game, 'Say Yoho', was the name of a long-running column in SoftSide magazine by Scott Adams.

4.41.2 Source code

The source code for *Pirate Adventure* was printed in the December 1980 issue of *BYTE*, with an addendum in April 1981.[2][4] This enabled others to discover how the engine worked and to create their own adventures using this or a similar design.

4.41.3 Reception

Pirate Adventure was reviewed in issue #42 of *The Dragon* magazine. The reviewer, Mark Herro, commented on the difficulty of the game: "Supposedly one of the "easier" programs of the series, I'm embarrassed to say that I have yet to find a treasure in *Pirate Adventure...* This Program has been sending me in circles for weeks." *[1]

4.41.4 References

[1] Herro, Mark (October 1980). "The Electric Eye". *The Dragon* (42): 42–43.

[2] Adams, Scott (December 1980). "Pirate's Adventure". *BYTE*. p. 192. Retrieved 18 October 2013.

[3] "Scott Adams Classic Adventures".

[4] "Adventurous Bugs". *BYTE*. April 1981. p. 302. Retrieved 18 October 2013.

4.41.5 External links

- *Pirate Adventure* at World of Spectrum

- Pirate Adventure at MobyGames

4.42 Pyramid of Doom

Not to be confused with pyramid of doom (programming).

Pyramid of Doom is a text-based adventure program written by Scott Adams.

4.42.1 Description

Published by Adventure International, this text-based adventure game was one of many from Scott Adams, co-written by Alvin Files.

Gameplay involved moving from location to location, picking up any objects found there, and using them somewhere else to unlock puzzles. Commands took the form of verb and noun, e.g. "Take Shovel". Movement from location to location was limited to North, South, East, West, Up and Down.

The object of the game was to enter an Egyptian pyramid and plunder its treasures. The player must face a variety of challenges, such as an angry mummy, a purple worm, and an irate desert nomad.

4.43 Q*bert

"Q-bert" and "Q*bert's Qubes" redirect here. For the disc jockey, see DJ Qbert.

Q*bert /ˈkjuːbərt/ is an arcade video game developed and published by Gottlieb in 1982. It is a 2D action game with puzzle elements that uses "isometric" graphics to create a pseudo-3D effect, and serves as a precursor to the isometric platformer genre. The objective is to change the color of every cube in a pyramid by making the on-screen character hop on top of the cube while avoiding obstacles and enemies. Players use a joystick to control the character.

The game was conceived by Warren Davis and Jeff Lee. Lee designed the title character and original concept, which was then further developed and implemented by Davis. *Q*bert* was developed under the project name *Cubes*, but was briefly named *Snots And Boogers* and *@!#?@!* during development.

Q*bert was well received in arcades and among critics. The game was Gottlieb's most successful video game and among the most recognized brands from the golden age of arcade video games. It has been ported to numerous platforms. The success resulted in sequels and the use of the character's likeness in merchandising, such as appearances on lunch boxes, toys, and an animated television show. The character Q*bert became known for his "swearing", an incoherent phrase of synthesized speech generated by the sound chip and a speech balloon of nonsensical characters that appear when he collides with an enemy.

Because the game was developed during the period when Columbia Pictures owned Gottlieb, the intellectual rights to *Q*bert* stayed with Columbia even after they divested themselves of Gottlieb's assets in 1984. Therefore, it is currently a property of Sony Pictures Entertainment when its parent Sony acquired Columbia in 1989. Q*bert appeared in Disney's computer-animated film *Wreck-It Ralph*, under a license from Sony, and later appeared in the live-action comedy *Pixels* in 2015.

4.43.1 Gameplay

Q*bert is an action game with puzzle elements played from an axonometric third-person perspective to convey a three-dimensional look.*[1] The game is played using a single, diagonally mounted four-way joystick.*[2] The player controls Q*bert, who starts each game at the top of a pyramid made of 28 cubes, and moves by hopping diagonally from cube to cube. Landing on a cube causes it to change color, and changing every cube to the target color allows the player to progress to the next stage.*[3]

At the beginning, jumping on every cube once is enough to advance. In later stages, each cube must be hit twice to reach the target color. Other times, cubes change color every time Q*bert lands on them, instead of remaining on the target color once they reach it. Both elements are then combined in subsequent stages. Jumping off the pyramid results in the character's death.[4]

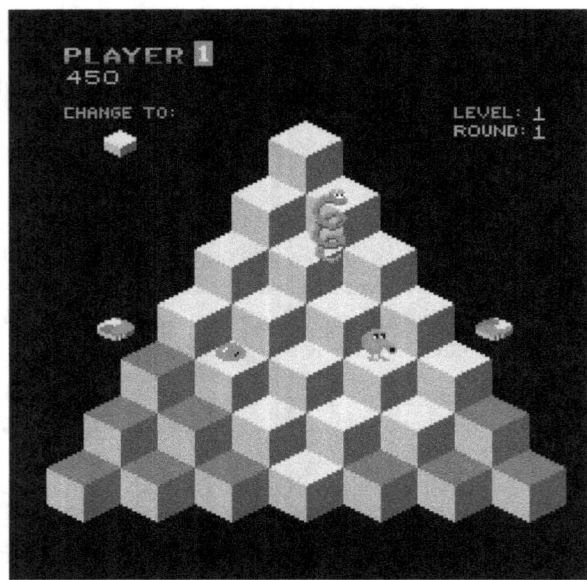

*The eponymous Q*bert hops diagonally down the pyramid to avoid Coily, who is pursuing him. The game tracks the player's progress above the pyramid.*

The player is impeded by several enemies, introduced gradually to the game:

- **Coily** - Coily first appears as a purple egg that bounces to the bottom of the pyramid and then transforms into a snake that chases after Q*bert.[2]

- **Ugg and Wrongway** - Two purple creatures that hop along the sides of the cubes in an Escheresque manner. Starting at either the bottom left or bottom right corner, they keep moving toward the top right or top left side of the pyramid respectively, and fall off the pyramid when they reach the end.[2]

- **Slick and Sam** - Two green creatures that descend down the pyramid and revert cubes whose color has already been changed.[4]

A collision with purple enemies is fatal to the character, whereas the green enemies are removed from the board upon contact.[2] Colored balls occasionally appear at the second row of cubes and bounce downward; contact with a red ball is lethal to Q*bert, while contact with a green one immobilizes the on-screen enemies for a limited time.[4]

Multicolored floating discs on either side of the pyramid serve as an escape from danger, particularly Coily. When Q*bert jumps on a disc, it transports him to the top of the pyramid. If Coily is in close pursuit of the character, he will jump after Q*bert and fall to his death, awarding bonus points.[2] This causes all enemies and balls on the screen to disappear, though they start to return after a few seconds.

Points are awarded for each color change (25), defeating Coily with a flying disc (500), remaining discs at the end of a stage (at higher stages, 50 or 100) and catching green balls (100) or Slick and Sam (300 each).[4] Extra lives are granted for reaching certain scores, which are set by the machine operator.[5]

4.43.2 Development

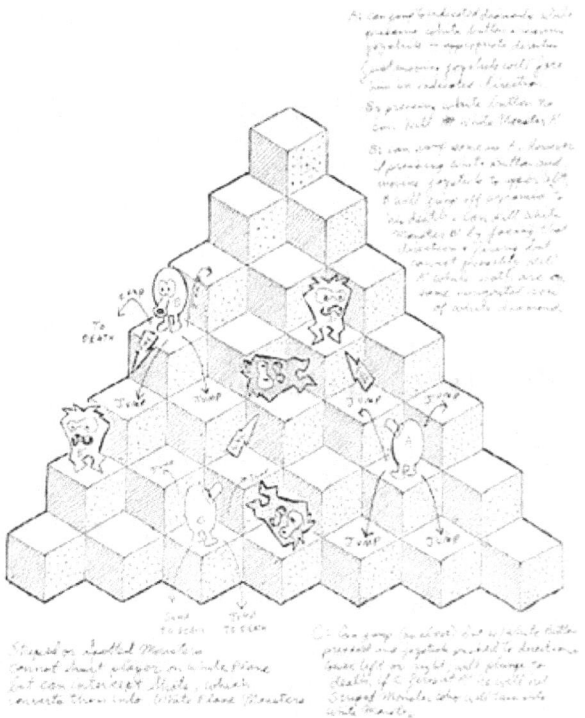

*In this concept sketch, Q*bert is still depicted shooting his foes. The sole enemy type is also not included in the final game.*

Concept

The basic ideas for the game were thought up by Warren Davis and Jeff Lee. The initial concept began when artist Jeff Lee drew a pyramid of cubes inspired by M. C. Escher.[6] Lee felt a game could be derived from the artwork, and created an orange, armless main character. The character jumped along the cubes and shot projectiles, called "mucus bombs", from a tubular nose at enemies.[*][7] Enemies included a blue creature, later changed purple and named Wrong Way, and an orange creature, later changed green and named Sam.[*][8] Lee had drawn similar characters since childhood, inspired by characters from comics, cartoons, *Mad* magazine and by artist Ed "Big Daddy" Roth.[*][9] Q*bert's design later included a speech balloon with a string of nonsensical characters, "@!#?@!",[*][Note 1] which Lee originally presented as joke.[*][8]

Implementation

Warren Davis, a programmer hired to work on the action game *Protector*,[*][6] noticed Lee's ideas, and asked if he could use them to practice programming randomness and gravity as game mechanic. Thus, he added balls that bounced from the pyramid's top to bottom.[*][8] Because Davis was still learning how to program game mechanics, he wanted to keep the design simple. He also felt games with complex control schemes were frustrating and wanted something that could be played with one hand. To accomplish this, Davis removed the shooting and changed the objective to saving the protagonist from danger.[*][9] As Davis worked on the game one night, Gottlieb's vice president of engineering, Ron Waxman, noticed him and suggested to change the color of the cubes after the game's character has landed on them.[*][6][*][8][*][9] Davis implemented a unique control scheme; a four-way joystick was rotated 45° to match the directions of Q*bert's jumping. Staff members at Gottlieb urged for a more conventional orientation, but Davis stuck to his decision.[*][6][*][8] Davis remembered to have started programming in April 1982,[*][10] but the project was only put on schedule as an actual product several months later.[*][Note 2]

Audio

We wanted the game to say, 'You have gotten 10,000 bonus points', and the closest I came to it after an entire day would be "bogus points". Being very frustrated with this, I said, "Well, screw it. What if I just stick random numbers in the chip instead of all this highly authored stuff, what happens?"

David Thiel on the creation of Q*bert's incoherent swearing.[*][6]

A MOS Technology 6502 chip that operates at 894 kHz generates the sound effects, and a speech synthesizer by Votrax generates Q*bert's incoherent expressions.[*][11] The audio system uses 128B of random-access memory and 4KB of erasable programmable read only memory to store the sound data and code to implement it. Like other Gottlieb games, the sound system was thoroughly tested to ensure it would handle daily usage. In retrospect, audio engineer David Thiel commented that such testing minimized time available for creative designing.[*][12]

Thiel was tasked with using the synthesizer to produce English phrases for the game. However, he was unable to create coherent phrases and eventually chose to string together random phonemes instead. Thiel also felt the incoherent speech was a good fit for the "@!#?@!" in Q*bert's speech balloon. Following a suggestion from technician Rick Tighe, a pinball machine component was included to make a loud sound when a character falls off the pyramid.[*][6][*][8] The sound is generated by an internal coil that hits the interior of a cabinet wall. Foam padding was added to the area of contact on the cabinet; the developers felt the softer sound better matched a fall rather than a loud knocking sound. The cost of installing foam, however, was too expensive and the padding was omitted.[*][9]

Title

The Gottlieb staff had difficulty naming the game. Aside from the project name "*Cubes*", it was untitled for most of the development process. The staff agreed the game should be named after the main character, but disagreed on the name.[*][8] Lee's title for the initial concept—*Snots And Boogers*—was rejected, as was a list of suggestions compiled from company employees.[*][8][*][13] According to Davis, vice president of marketing Howie Rubin championed @!#?@! as the title. Although staff members argued it was silly and would be impossible to pronounce a few early test models were produced with @!#?@! as the title on the units' artwork.[*][8][*][13] During a meeting, "Hubert" was suggested, and a staff member thought of combining "Cubes" and "Hubert" into "Cubert".[*][8][*][13] Art director Richard Tracy changed the name to "Q-bert", and the dash was later changed to an asterisk. In retrospect, Davis expressed regret for the asterisk, because he felt it prevented the name from becoming a common crossword term and it is a wildcard character for search engines.[*][8]

Testing

As development neared the production stage, *Q*bert* underwent location tests in local arcades under its preliminary title @!#?@!, before being widely distributed. According to Jeff Lee, his oldest written record attesting to the game being playable as @!#?@! in a public location, a Brunswick bowling alley, dates back to September 11, 1982.[*][8] Gottlieb also conducted focus groups, in which the designers observed players through a two-way mirror.[*][8] The control scheme received a mixed reaction during play testing; some players adapted quickly while others found it frustrating.[8][9] Initially, Davis was worried players would not adjust to the different controls; some players would unintentionally jump off the pyramid several times, reaching a game over in about ten seconds. Players, however, became accustomed to the controls after playing several rounds of the game.[8] The different responses to the controls prompted Davis to reduce the game's level of difficulty —a decision that he would later regret.[9]

Release

A copyright claim registered with the United States Copyright Office by Gottlieb on February 10, 1983 cites the date of publication of Q*bert as October 18, 1982.[14] Video Games reported that the game was sold directly to arcade operators at its public showing at the AMOA show held November 18–20, 1982.[15] Gottlieb offered the machines for $2600 per unit.[16] Q*bert is Gottlieb's fourth video game.[17]

4.43.3 Reception

Q*bert was Gottlieb's only video game that gathered huge critical and commercial success, selling around 25,000 arcade cabinets.[6] Cabaret and cocktail versions of the game were later produced. The machines have since become collector's items; the rarest of them are the cocktail versions.[18]

When the game was first introduced to a wider industry audience at the November 1982 AMOA show, it was immediately received favorably by the press. Video Games placed Q*bert first in its list of Top Ten Hits, describing it as "the most unusual and exciting game of the show" and stating that "no operator dared to walk away without buying at least one".[15] The Coin Slot reported "Gottlieb's game, Q*BERT, was one of the stars of the show", and predicted that "The game should do very well."[19]

Contemporaneous reviews were equally enthusiastic, and focused on the uniqueness of the gameplay and audiovisual presentation. Roger C. Sharpe of Electronic Games considered it "a potential Arcade Award winner for coin-op game of the year", praising innovative gameplay and outstanding graphics.[3] William Brohaugh of Creative Computing Video & Arcade Games described the game as an "all-round winner" that had many strong points. He praised the variety of sound effects and the graphics, calling the colors vibrant. Brohaugh lauded Q*bert's inventiveness and appeal, stating that the objective was interesting and unique.[17] Michael Blanchet of Electronic Fun suggested the game might push Pac-Man out of the spotlight in 1983.[2] Neil Tesser of Video Games also likened Q*bert to Japanese games like Pac-Man and Donkey Kong, due to the focus on characters,

animation and story lines, as well as the "absence of violence".[7] Computer and Video Games magazine praised the game's graphics and colors.[5]

Electronic Games awarded Q*bert "Most Innovative Coin-op Game" of the year.[20] Video Game Player called it the "Funniest Game of the Year" among arcade games in 1983.[21]

Q*bert continues to be widely recognized as a significant part of video game history. Author Steven Kent and GameSpy's William Cassidy considered Q*bert one of the more memorable games of its time.[22][23] Author David Ellis echoed similar statements, calling it a "classic favorite".[24] 1UP.com's Jeremy Parish and Kim Wild of Retro Gamer magazine described the game as difficult yet addictive.[8][25] Author John Sellers also called Q*bert addictive, and praised the sound effects and three-dimensional appearance of the graphics.[13] Cassidy called the game unique and challenging; he attributed the challenge in part to the control scheme.[23] IGN's Jeremy Dunham felt the controls were poorly designed, describing them as "unresponsive" and "a struggle". He nonetheless commented that the game was addictive despite the controls.[26]

The main character also received positive press coverage. Edge magazine attributed the success of the game to the title character. They stated that players could easily relate to Q*bert, particularly because he swore.[9] Computer and Video Games, however, considered the swearing a negative, but still felt the character was appealing.[5] Cassidy believed the game's appeal lay in the main character. He described Q*bert as cute and having a personality that made him stand out in comparison to other popular video game characters.[23] The authors of High Score! referred to Q*bert as "ultra-endearing alien hopmeister", and the cutest game character of 1982.[27]

4.43.4 Ports

At the 1982 AMOA Show, Parker Brothers secured the license to publish home conversions of the Q*bert arcade game.[28] Parker first published a port to the Atari 2600,[29] and by the end of 1983, the company also advertised versions for Atari 5200, Intellivision, ColecoVision, the Atari 8-bit computer family, Commodore VIC-20, Texas Instruments TI-99/4A and Commodore 64.[30] The release of the Commodore 64 version was noted to lack behind the others[29] but appeared in 1984.[31] Parker Brothers also translated the game into a stand-alone tabletop electronic game.[32] It uses a VFD screen, and has since become a rare collector's item.[33] Q*bert was also published by Parker Brothers for the Philips Videopac in Europe,[34] by Tsukuda Original for the Othello Multi-

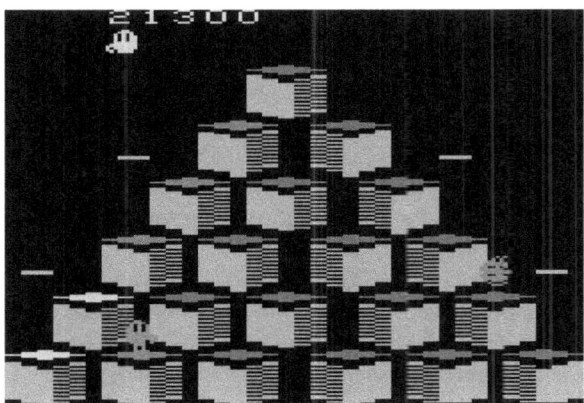

*A screenshot of the Atari 2600 version by Parker Brothers. The Escher-inspired visual style of the pyramid could not be preserved and the pyramid was shortened by one row. The discs that transport Q*bert to the top of the screen are represented as simple horizontal lines.*

vision in Japan,[35] and by Ultra Games for the NES in North America.[36]

The initial home port for the Atari 2600, the most widespread system at the time, was met with mixed reactions. *Video Games* warned that buyers of the Atari 2600 version "may find themselves just a little disappointed." They criticized the lack of music, the removing of the characters Ugg and Wrongway, and the system's troubles to handle the character sprites on screen at a steady performance.[37] Later Mark Brownstein of the same magazine was more in favor of the game, but still cited the presence of fewer cubes in the game's pyramidal layout and "pretty poor control" as negatives.[29] Will Richardson of *Electronic Games* noted a lack in audiovisual qualities and counter-intuitive controls, but commended the gameplay, stating that the game "comes much closer to its source of inspiration than a surface evaluation indicates".[38] Randi Hacker of *Electronic Fun with Computers & Games* called it a "sterling adaption [*sic*]"[39] In 2008, however, IGN's Levi Buchanan rated it the fourth worst arcade port for the Atari 2600, mostly due to a lack of jumping animations for enemies, which instead pop up instantly on the adjacent cube, making it impossible to know which direction they travel before they land.[40]

Other home versions were well-received for the most part, with some exceptions. Of the ColecoVision version, Electronic Fun with Computers & Games noted that "Q*bert aficionados will not be disappointed".[41] Marc Brownstein of Video Games called it one of the best of the authorized versions.[29] Warren Davis also considered the ColecoVision version the most accurate port of the arcade.[8] Mark Brownstein judged the Atari 5200 version inferior to the ColecoVision, due to the imprecision of the

Atari 5200 controller, but noted that "it does tend to grow on you."[29] Video Games determined the Intellivision version as the worst of the available ports, criticizing the system's controller for being inadequate for the game.[42] *Antic* magazine's David Duberman called the Atari 8-bit version "one of the finest translations of an arcade game for the home computer format",[43] and Arthur Leyenberger of Creative Computing listed it as a runner-up for Best Arcade Adaptation to the system, praising its faithful graphics, sound, movement and playability.[44] *Softline* was more critical, criticizing the Atari version's controls and lack of swearing. The magazine concluded that "the home computer game doesn't have the sense of style of the one in the arcades ... the execution just isn't there".[45] In 1984 the magazine's readers named the game the fifth-worst Atari program of 1983.[46] *Computer Games* called the C64 version an "absolutely terrific translation" that "almost totally duplicates the arcade game," aside from its lack of synthesized speech.[31] The stand-alone tabletop was awarded Stand-Alone Game of the Year in *Electronic Games.*[20]

In 2003, a version for Java-based mobile phones was announced by Sony Pictures Mobile.[47] Reviewers generally acknowledged it as a faithful port of the arcade original, but criticized the controls. Modojo's Robert Falcon stated that the diagonal controls take time to adapt to on a cell phone with traditional directions.[48] Michael French of Pocket Gamer concluded: "You can't escape the fact it doesn't exactly fit on mobile. The graphics certainly do, and the spruced-up sound effects are timeless···but really, it's a little too perfect a conversion."[49] Airgamer criticized the gameplay as monotonous and the difficulty as frustrating.[50] By contrast, Wireless Gaming Review called it "one of the best of mobile's retro roundup".[51]

On February 22, 2007, *Q*bert* was released on the PlayStation 3's PlayStation Network.[52] It features upscaled and filtered graphics,[25] an online leaderboard for players to post high-scores, and Sixaxis motion controls.[26] The game received a mixed reception. Dunham and Gerstmann did not enjoy the motion controls and felt it was a title only for nostalgic players.[26][53] Eurogamer.net's Richard Leadbetter judged the game's elements "too simplistic and repetitive to make them worthwhile in 2007".[54] In contrast, Parish considered the title worth purchasing, citing its addictive gameplay.[25]

4.43.5 Legacy

According to Jeremy Parish, *Q*bert* was "one of the higher-profile titles of the classic era".[25] In describing *Q*bert*'s legacy, Jeff Gerstmann of GameSpot referred to the game as a "rare arcade success".[53] In 2008, Guinness World

Records ranked it behind 16 other arcade games in terms of their technical, creative and cultural impact.[*][55] Despite its success, the creators of the game did not receive royalties, as Gottlieb had no such program in place at the time.[*][8] Davis and Lee nonetheless expressed pride about the game continuing to be remembered fondly.[*][8]

Market impact

An advertisement flyer by Gottlieb showcasing several of the licensed tie-in products by Parker Brothers, Kenner and others. The character's likeness was often slightly adjusted to serve the specific application.

*Q*bert* became one of the most merchandised arcade games behind *Pac-Man*,[*][8] although according to John Sellers it was not nearly as successful as that franchise or *Donkey Kong*.[*][13] The character's likeness appears on various items including coloring books, sleeping bags, frisbees, board games, wind-up toys, and stuffed animals.[*][8][*][13][*][23] In a flyer distributed in 1983, Gottlieb claimed over 125 licensed products.[*][16] However, the North American video game crash of 1983 depressed the market, and the game's popularity began to decline by 1984.[*][8][*][23]

In the years following its release, *Q*bert* inspired many other games with similar concepts. The magazines *Video Games* and *Computer Games* both commented on the trend with features about *Q*bert*-like games in 1984. They listed *Mr. Cool* by Sierra On-Line, *Frostbite* by Activision, *Q-Bopper* by Accelerated Software, *Juice* by Tronix, *Quick Step* by Imagic, *Flip & Flop* and *Boing* by First Star Software, *Pharaoh's Pyramid* by Master Control Software, *Pogo Joe* by Screenplay, *Rabbit Transit* by Starpath, as games which had been inspired by *Q*bert*.[*][29][*][56] Further titles that have been identified as *Q*bert*-like games include *J-bird* by Orion Software,[*][57] *Cubit* by Micromax,[*][58] and in the UK *Pogo* by Ocean,[*][59] *Spellbound* by Beyond[*][60] and *Hubert* by Blaby Computer Games.[*][61]

Other media appearances

In 1983, *Q*bert* was adapted into an animated cartoon as part of CBS's *Saturday Supercade*, which features segments based on video game characters from the golden age of video arcade games. *Saturday Supercade* was produced by Ruby-Spears Productions, the *Q*bert* segments between 1983 and 1984.[*][62] The show is set in a United States, 1950s era town called "Q-Burg",[*][63] and stars Q*bert as a high school student, altered to include arms and hands.[*][23] He also has the ability to shoot black projectiles from his nose. Characters frequently say puns that add the letter "Q" to words.[*][63] Aside from Q*bert and the known game villains, the cartoon also includes new characters similar to Q*bert in appearance and naming.[*][64]

Q*bert, Coily, Ugg, Slick, and Sam appear in the 2012 Disney animated film *Wreck-It Ralph*.[*][65] They start out as "homeless" video game characters living in Game Central Station after their game was unplugged and taken out of Litwak's Arcade. Ralph gives them a cherry from *Pac-Man* as a gesture of kindness. After Ralph takes Markowski's uniform in *Tapper's*, he accidentally trips over Q*bert on his way to *Hero's Duty*. This leads Q*bert to go to *Fix-It Felix Jr.* to warn Felix that Ralph has "gone Turbo." In that scene, Felix apparently speaks "Q*bert-ese." At the end of the film, Ralph and Felix decide to let Q*bert, Coily, Ugg, Slick, Sam, and the generic homeless video game characters into *Fix-It Felix Jr.*, suggesting that they help out in the bonus levels where Coily, Ugg, Slick, Sam, and the generic video game characters assist Ralph in wrecking the building while Q*bert assists Felix in fixing it.[*][66]

Q*bert makes another appearance in Sony film *Pixels*, which was released on July 24, 2015.[*][67] In the movie, Q*bert is given to the main characters as a "trophy" by the aliens for defeating *Pac-Man*. He then accompanies the team on its last mission. In the end, he randomly transforms into the fictional female character *Lady Lisa* (of the video game *Dojo Quest*), after victory against the aliens.[*][68]

In popular culture

*Q*bert* is seen being played in the 1984 film *Moscow on the Hudson* starring Robin Williams.[*][10] The 1993 IBM PC role-playing game *Ultima Underworld II: Labyrinth of Worlds* features a segment where the player has to solve a pyramid puzzle as an homage to *Q*bert*.[*][69] In the 2009 action-adventure game *Ghostbusters: The Video Game*, a *Q*bert* arcade cabinet can be seen in the Ghostbusters HQ. However, the game is merely decoration and not playable.[*][70]

More recently, the game or its characters have been referenced in several animated television series. In the *Family*

Guy episode "Chick Cancer", Stewie reflects on how it was easier being Q*bert's room mate and an animation of him on the game board is shown.[*][71] In "Anthology of Interest II" of *Futurama*, he is one of the aliens that attack to invade earth in a segment of video game parodies.[*][72] In *The Simpsons* episode "In the Name of the Grandfather" Marge, Bart and Lisa hop around the stones of the Giants Causeway in a game of *Q*bert*.[*][73][*][74] The *Robot Chicken* episode "Sushi Rolls" is in general a *Street Fighter* parody, but in the end M. Bison is shown inside the game *Q*bert*.[*][75] In *Mad*: "James Bond: Reply All", *Q*bert* is seen at the MI6 lab.[*][76] Q*bert also appeared on the battlefield in *South Park*: "Imaginationland: Episode III".[*][77]

High score records

On November 28, 1983, Rob Gerhardt reached a record score of 33,273,520 points in a *Q*bert* marathon.[*][78] He held it for almost 30 years, until George Leutz from Brooklyn, NY played one game of *Q*bert* for eighty-four hours and forty-eight minutes on February 14–18, 2013 at Richie Knucklez' Arcade in Flemington, NJ.[*][79] He scored 37,163,080 points.[*][80]

Doris Self, credited by Guinness World Records as the "oldest competitive female gamer",[*][81] set the tournament record score of 1,112,300 for *Q*bert* in 1984 at the age of 58. Her record was surpassed by Drew Goins on June 27, 1987 with a score of 2,222,220.[*][82] Self continuously attempted to regain the record until her death in 2006.[*][8]

On November 18, 2012, George Leutz broke the *Q*Bert* tournament world record live at the Kong Off 2 event at the 1up Barcade in Denver, Co. George scored 3,930,990 points in just under 8 hours, earning 1.5 million points on his first life, beating Self's score using a single life. Leutz's score was verified by Twin Galaxies.[*][83] The video ends at a score of 3.7 Million points, 1,500,000 points over the previous record.[*][84]

Updates, remakes, and sequels

Faster Harder More Challenging Q*bert   Believing that the original game was too easy, Davis initiated development of *Faster Harder More Challenging Q*bert* (also known as *FHMC Q*bert*) in 1983,[*][9] which increases the difficulty, introduces Q*bertha and adds a bonus round.[*][85] Finally, the project was canceled and the game never entered production.[*][8] Davis later released *FHMC Q*bert*'s ROM image onto the web.[*][8]

Q*bert's Quest   Gottlieb also released a pinball game, *Q*bert's Quest*, based on the arcade version. It features two pairs of flippers in an "X" formation and audio from the arcade.[*][8][*][86] Gottlieb produced less than 900 units.[*][86]

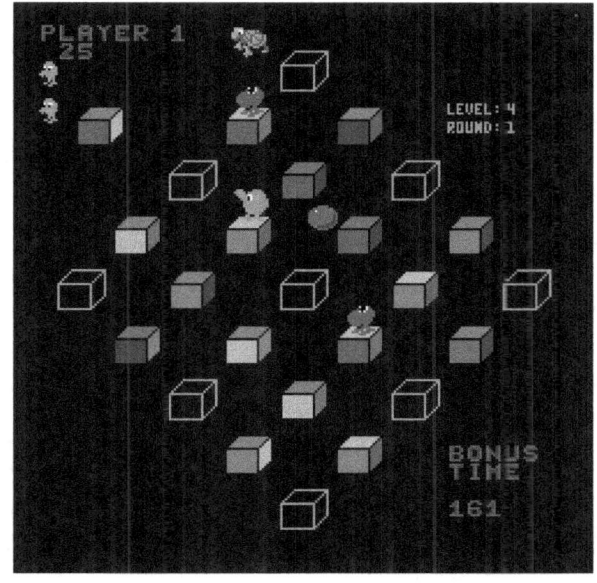

The sequel Q*bert's Qubes *involves matching the color of cubes to a target sample. The player must match the cubes on the field to the cube in the top left corner.*

Q*bert's Qubes   Several video game sequels were released over the years, but did not reach the same level of success as the original.[*][8][*][23] The first, titled *Q*bert's Qubes*, shows a copyright for 1983 on its title screen,[*][13] whereas the instruction manual cites a 1984 copyright.[*][87] It was manufactured by Mylstar Electronics,[*][Note 3] and uses the same hardware as the original.[*][88] The game features Q*bert, but introduces new enemies: Meltniks, Soobops, and Rat-A-Tat-Tat.[*][89] The player navigates the protagonist around a plane of cubes while avoiding enemies. Jumping on a cube causes it to rotate, changing the color of the visible sides of the cube.[*][13] The goal is to match a line of cubes to a target sample; later levels require multiple rows to match.[*][90] Despite the popularity of the franchise, the game's release was hardly noticed.[*][13] Parker Brothers showcased home versions of *Q*bert's Qubes* at the Winter Consumer Electronics Show in January 1985.[*][89] *Q*bert's Qubes* was ported to the Colecovision and Atari 2600.[*][91][*][92]

Q*bert (1986)   Konami, who had distributed the original *Q*bert* to Japanese arcades in 1983,[*][93] produced a game with the title *Q*bert* for MSX computers in 1986, released in Japan and Europe. However, the main character is a little dragon, and the mechanics are based on *Q*bert's Qubes*.

The player once again turns around colored cubes by jumping from cube to cube, trying to reach the displayed target pattern. Contrary to Mylstar's arcade game, each of the 50 stages has a different pattern of cubes, in addition to the known rule extensions in later stages. The game also features a competitive 2-player mode, where each side is assigned a different pattern, and the players can score points either by completing their pattern first, or by pushing the other off the board.[*][94]

Q*bert for Game Boy   In 1992, this handheld game was developed by Realtime Associates and published by Jaleco in 1992. It features 64 boards in different shapes.[*][95]

Q*bert 3   *Q*bert 3* for the SNES was also developed by Realtime Associates and released in 1992 [*][96] Jeff Lee, creator of the Q*bert character, also worked on the graphics for this game.[*][10] *Q*bert 3* features gameplay similar to the original, but like the Game Boy game, it has larger levels of varying shapes. In addition to enemies from the first game, it introduces several new enemies (Frogg, Top Hat, and Derby).[*][97][*][98]

Q*bert (1999)   A remake with three-dimensional (3D) graphics was developed by Artech Studios and released by Hasbro Interactive on the PlayStation in 1999 and on the Dreamcast the following year. It features three modes of play: classic, adventure, and competitive multiplayer.[*][99][*][100] Allgame's Brett Weiss praised all aspects of the game,[*][99] while Parish called it a poor adaptation.[*][25] Kevin Rice of *Next Generation Magazine* praised the game's graphics, but criticized the new level designs. He further commented that adventure mode was not enjoyable.[*][100] The game was the winner of Electronic Gaming Monthly's "Puzzle Game of the Year" award.[*][101]

Q*bert 2004   In 2004, Sony Pictures released a remake/sequel for Adobe Flash titled *Q*bert 2004*, containing a faithful rendition of the original arcade game, along with 50 levels that use new board layouts and six new visual themes.[*][102] *Q*Bert Deluxe* for iOS devices was initially released as a rendition of the arcade game, but later received updates with the themes and stages from *Q*Bert 2004*.[*][103]

Q*bert 2005   In 2005, Sony Pictures released *Q*bert 2005* as a download for Windows[*][104] and as a Flash browser applet, featuring 50 different levels.[*][104]

Q*bert Rebooted   On July 2, 2014 Gonzo Games and Sideline amusement announced *Q*bert Rebooted* to be released on Steam, iOS and Android.[*][105] Versions for PlayStation 3, PlayStation 4 and PlayStation Vita were released 2015 on February 17 in North America and February 18 in Europe.[*][106] According to Mark Caplan, Vice President, Consumer Products, Worldwide Marketing & Distribution at Sony Pictures Entertainment, the release was motivated by "renewed interest in *Q*bert*, in part due to the cameo in the recent *Wreck It Ralph* animated feature film".[*][107]

Q*bert Rebooted contains a port of the classic arcade game alongside a new playing mode that uses hexagonal shapes, increasing the number of possible movement directions to six.[*][108] Additionally, the 'Rebooted' mode features new enemy types, including a boxing glove that punches Q*bert off the levels[*][108] and a treasure chest that tries to avoid him.[*][109] The game has 5 different stage designs spread across 40 levels,[*][108] which contain three rounds and a bonus round and have to be completed with 5 lives.[*][109] Gems are collected to unlock different skins for the Q*bert character, and completing levels multiple times while reaching specific time and score goals is awarded with stars that enable access to more levels.[*][108]

4.43.6   Notes

[1] The original artwork displays the first and fifth character as spirals. The at sign ("@") is used in its place in the text of the references.

[2] Davis stated that this happened "by June or July", whereas Howie Rubin, vice president of Gottlieb, claimed in an early 1983 interview with Video Games that the game was not yet on a list of games voted for in a brainstorming session in August. Tesser, Neil (March 1983). "The Life and Times of Q*bert & Joust". *Video Games* (Pumpkin Press) (Volume 1, Number 8): 26–30.

[3] The Coca-Cola Company acquired Columbia Pictures, Gottlieb's owner, in 1982, and renamed the company to Mylstar Electronics, in 1983.

4.43.7   References

[1] Perron, Bernard; Wolf, Mark J.P. (eds.). *The Video Game Theory Reader 2*. Routledge. p. 158.

[2] "Cursing Q*Bert: @!#?@! you, Coily!". *Electronic Fun with Computers & Games* (Fun & Games Publishing) (Volume 1, Number 5): 92. March 1983.

[3] Sharpe, Roger C. (May 1983). "Is This the Next Arkie Winner?". *Electronic Games* (Reese Publishing Company) (Volume 1, Number 15): 78–79.

[4] "Arcade Action Close-Up: Crazy For Q&bert's Cube". *Vidiot* (Creem Publications): 30–31. April–May 1983.

[5] "'Q' Up for this One". *Computer and Video Games* (EMAP) (18): 31. April 1983.

[6] Kent, Steven (2001). "The Fall". *Ultimate History of Video Games*. Three Rivers Press. pp. 222–224. ISBN 0-7615-3643-4.

[7] Tesser, Neil (March 1983). "The Life and Times of Q*bert & Joust". *Video Games* (Pumpkin Press) (Volume 1, Number 8): 26–30.

[8] Wild, Kim (September 2008). "The Making of Q*bert". *Retro Gamer* (Imagine Publishing) (54) 70–73.

[9] Edge Staff (January 2003). "The Making of Q*bert". *Edge* (132): 114–117. Retrieved 2010-01-07.

[10] Davis, Warren. "The Creation of Q*Bert". *Coinop.org*. Retrieved 26 September 2011.

[11] "Q*bert Videogame by Gottlieb (1982)". Killer List of Videogames. Retrieved 2009-05-31.

[12] Greenebaum, Ken; Barzel, Ronen, eds. (2004). "Retro Game Sound: What We Can Learn from 1980s Era Synthesis". *Audio Anecdotes: Tools, Tips, and Techniques for Digital Audio, Volume 1*. A K Peters, Ltd. pp. 164–165. ISBN 1-56881-104-7.

[13] Sellers, John (August 2001). *Arcade Fever: The Fan's Guide to The Golden Age of Video Games*. Running Press. pp. 108–109. ISBN 0-7624-0937-1.

[14] *Q-bert (Registration Number PA0000164088)*, The Library of Congress, 1983-02-10, retrieved 2014-04-22

[15] "Top Ten Hits". *Video Games* (Pumpkin Press) (Volume 1, Number 7): 66. March 1983.

[16] *Q*bert: Your Best New Videogame Buy for 1983* (advertisement), Gottlieb, 1983, retrieved 2014-07-15

[17] Brohaugh, William (Fall 1983). "Q*bert: A Player's Guide". *Creative Computing Video & Arcade Games* 1 (2): 28.

[18] Ellis, David (2004). "Arcade Classics". *Official Price Guide to Classic Video Games*. Random House. p. 402. ISBN 0-375-72038-3.

[19] Pugliese, Mike (January 1983). "The Amoa Show". *The Coin Slot* (Rosanna B. Harris) (Volume 8, Number 4): 27–29.

[20] "1984 Arcade Awards". *Electronic Games* (Reese Communications) (Volume 2, Number 11): 68–81. January 1984.

[21] "1983 Golden Joystick Awards". *Electronic Games* (Reese Communications) (Volume 2, Number 11): 49–51. August–September 1983.

[22] Kent, Steven (2001). "The Golden Age (Part 2: 1981–1983)". *Ultimate History of Video Games*. Three Rivers Press. p. 177. ISBN 0-7615-3643-4.

[23] Cassidy, William (2002-06-23). "Hall of Fame: Q*bert". GameSpy. Archived from the original on 2012-05-05. Retrieved 2014-05-01.

[24] Ellis, David (2004). "A Brief History of Video Games". *Official Price Guide to Classic Video Games*. Random House. p. 7. ISBN 0-375-72038-3.

[25] Parish, Jeremy (2007-02-26). "Retro Roundup 2/26: Ocarina of Time, Q*Bert, Chew Man Fu". 1UP.com. Retrieved 2009-06-02.

[26] Dunham, Jeremy (2007-02-23). "Q*Bert Review". IGN. Retrieved 2009-06-02.

[27] Rusel DeMaria, Johhny L. Wilson (2004). "Q*Bert". *High Score! the illustrated history of electronic games* (second ed.). McGraw-Hill/Osborne. p. 84. ISBN 0-07-223172-6.

[28] "Parker Grabs Two Hot Licenses". *Electronic Games* (Reese Publishing Company) (Volume 1, Number 14): 8 April 1983.

[29] Brownstein, Mark (March 1984). "Follow the Leader: Spin-offs Jump To The Q*bert Challenge". *Video Games* (Pumpkin Press) (Volume 2, Number 6): 28–31.

[30] "How to Get Q*bert Out of your System". *Electronic Games* (Reese Communications) (Volume 2, Number 10): 101. December 1983.

[31] "Q*Bert". *Computer Games* (Carnegie Publications) (Volume 3, Number 2): 60. June 1984.

[32] Worley, Joyce (January 1984). "The Block Bouncer Busts Loose!". *Electronic Games* (Reese Communications) (Volume 2, Number 11): 122–125.

[33] Ellis, David (2004). "Classics Handheld and Tabletop Games". *Official Price Guide to Classic Video Games*. Random House. p. 237. ISBN 0-375-72038-3.

[34] "Parker Video Game Cartridge: Q*bert". RetroMO. Retrieved 2014-04-30.

[35] "Q*bert". SMS Power!. Retrieved 2014-04-30.

[36] "Q*bert for NES". MobyGames. Retrieved 2014-04-30.

[37] P., D. (December 1983). "Q*Bert". *Video Games* (Pumpkin Press) (Volume 2, Number 3): 65–66.

[38] Richardson, Will (January 1984). "Get Hopping with Q*bert!". *Electronic Games* (Reese Communications) (Volume 2, Number 11): 102.

[39] Hacker, Randi (November 1983). "Q*Bert". *Video Games* (Fun & Games Publishing): 58.

[40] Buchanan, Levi (2008-03-17). "Top 10 Worst Atari 2600 Arcade Ports". IGN. Retrieved 2009-06-01.

[41] Berman, Marc (December 1983). "Q*Bert". *Video Games* (Fun & Games Publishing): 62.

[42] B., M. (April 1984). "Q*Bert" . *Video Games* (Pumpkin Press) (Volume 2, Number 7): 60.

[43] Duberman, David (December 1983). "Product Reviews: Two from Parker Brothers" . *Antic* **2** (9): 124.

[44] Leyenberger, Arthur (January 1984). "The 1983 Outpost: Atari Computer Game Awards". *Creative Computing* **10** (1): 242–247.

[45] Bang, Derrick (Jan–Feb 1984). "Q*Bert" . *Softline*. pp. 56–57. Retrieved 29 July 2014.

[46] "The Best and the Rest" . *St.Game*. Mar–Apr 1984. p. 49. Retrieved 28 July 2014.

[47] "Vodafone calls Sony Pictures Mobile for new games and entertainment services". Vodafone. 2003-09-05. Retrieved 2014-05-01.

[48] Falcon, Robert (2006). "Q*Bert Mobile Review". Modojo. Retrieved 2014-05-01.

[49] French, Michael (2006-02-12). "Q*Bert: An arcade classic hops to mobile" . Pocket Gamer. Retrieved 2014-05-01.

[50] "Q*Bert". Airgamer. 2007-04-18. Retrieved 2014-05-01.

[51] Avery Score, Wireless Gaming Review Staff (2004-04-28). "An Introduction to Mobile Gaming". Gamespot. Retrieved 2014-05-01.

[52] Sinclair, Brendan (2007-02-16). "Q*Bert hops to PS3" . Gamespot. Retrieved 2014-05-01.

[53] Gerstmann, Jeff (2007-02-27). "Q*bert Review". GameSpot. Retrieved 2009-06-02.

[54] Leadbetter, Richard (2007-04-14). "Q*Bert" . Eurogamer.net. Retrieved 2014-05-01.

[55] Craig Glenday, ed. (2008-03-11). "Top 100 Arcade Games: Top 20–6" . *Guinness World Records Gamer's Edition 2008*. Guinness World Records. Guinness. p. 234. ISBN 978-1-904994-21-3.

[56] Gutman, Dan (April 1984). "The Clones of Q*Bert" . *Computer Games* (Carnegie Publications) (Volume 3, Number 1): 48–51.

[57] "The Thrill is Gone" . *PC Magazine* (Ziff-Davis) **3** (10): 286. May 29, 1984.

[58] Murphy, Brian J. (May 1984). "Cubit" . *InCider* (Ziff-Davis): 127–128.

[59] "Pogo" . *Crash* (Newsfield) (4): 84. May 1984.

[60] "Spellbound" . *Crash* (Newsfield) (6): 51–52. July 1984.

[61] "Hubert" . *Crash* (Newsfield) (10): 139. November 1984.

[62] "Ruby-Spears Productions – About Us". Ruby-Spears Productions. Retrieved 2009-05-31.

[63] Sharkey, Scott. "Top 5 Classic Videogame Cartoons". 1UP.com. Retrieved 2009-05-31.

[64] "Q*bert @ The Cartoon Scrapbook". Retrieved 2014-05-02.

[65] Zeitchik, Steven (2012-11-03). "*Wreck-It Ralph* Cheat Code: Which Video Games Get Shout-Outs?". *The Los Angeles Times*. Retrieved 2012-11-05.

[66] Johnson, Phil; Lee, Jennifer, *Wreck-It Ralph (screenplay)* (PDF), Walt Disney Studios, retrieved 2014-07-11

[67] "Classic video game characters unite via film 'Pixels'". *Philstar*. July 23, 2014. Retrieved July 23, 2014.

[68] Seppala, Timothy (2015-07-24). "'Pixels' is somehow even worse than I thought it could be" . *engadget*. AOL Inc. Retrieved 2015-08-13.

[69] "All of this to play Q-Bert?!". *Ultima Adventures*. 2010-10-06. Retrieved 2014-07-11.

[70] Chester, Nick (2009-05-20). "Games Ghostbusters play: Q-Bert" . *Destructoid*. Retrieved 2014-07-15.

[71] "Chick Cancer". *Family Guy*. Season 5. Episode 7. 2006-11-26. Fox Broadcasting Company.

[72] "Anthology of Interest II". *Futurama*. Season 3. Episode 18. 2002-01-06. Fox Broadcasting Company.

[73] Canning, Robert (2009-03-23). "The Simpsons: "In the Name of the Grandfather" Review". IGN. Retrieved 2009-05-30.

[74] "In the Name of the Grandfather". *The Simpsons*. Season 20. Episode 14. 2009-03-22. Fox Broadcasting Company.

[75] Stillman, Josh (2012-10-10). "'Robot Chicken' tackles 'Street Fighter'". Entertainment Weekly. Retrieved 2014-04-30.

[76] "James Bond: Reply All/Randy Savage: 9th Grade Wrestler (2013): Connections" . IMDb. Retrieved 2014-04-30.

[77] "Imaginationland: Episode III (2007): Connections" . IMDb. Retrieved 2014-04-30.

[78] "Q*bert High Score Marathon Rankings" . Twin Galaxies. Archived from the original on 2009-06-23. Retrieved 2009-11-13.

[79] Epstein, Rick (2013-02-18). "Man claims world record by playing *Q*bert* for 84 hours in Hunterdon arcade" . nj.com. Retrieved 2014-04-30.

[80] Morris, Chris (2013-02-19). "Man plays Q*bert for more than 80 hours, breaks 30-year-old record" . Yahoo! Games. Retrieved 2014-04-30.

[81] Craig Glenday, ed. (2008-03-11). "About Twin Galaxies". *Guinness World Records Gamer's Edition 2008*. Guinness World Records. Guinness. p. 9. ISBN 978-1-904994-21-3.

[82] "Q*bert High Score Tournament Rankings". Twin Galaxies. Archived from the original on 2008-10-04. Retrieved 2009-11-13.

[83] "George Leutz Q*Bert world records (Marathon and Tournament)". Twin Galaxies. Retrieved February 7, 2015.

[84] *Recording of Q*Bert Tournament Track world record*. Retrieved February 7, 2015.

[85] "Q*bert Interview". Tomorrow's Heroes. Retrieved 2014-04-30.

[86] Campbell, Stuart (January 2008). "A Whole Different Ball Game". *Retro Gamer* (Imagine Publishing) (45): 49.

[87] *Q*bert's Qubes Instruction Manual*, Mystar, 1984, p. 36

[88] "Q*bert's Qubes Videogame by Mylstar (1983)". Killer List of Videogames. Retrieved 2009-06-01

[89] Ahl, David H. (April 1985). "1985 Winter Consumer Electronics Show". *Creative Computing* (Ziff-Davis) 11 (4): 51.

[90] Weiss, Brett A. "Q*bert's Qubes – Overview". Allgame. Retrieved 2009-06-01.

[91] "Q*bert's Qubes for Colecovision – Technical Information". GameSpot. Retrieved 2009-06-01.

[92] "Q*bert's Qubes for Atari 2600 – Technical Information". GameSpot. Retrieved 2009-06-01.

[93] "Q*Bert (Konami)". *AM Life* (in Japanese) (Kabushiki Kaisha Amusement) (3): 10. March 1983.

[94] "Qbert: De toutes les couleurs!". *MSX News* (in French) (Sandyx S.A.) (5): 12. September–October 1987

[95] "Q*Bert: The Arcade Hit Leaps to Your Game Boy". *Game Informer* (Funco) (Spring 1992): 46–47. Spring 1992.

[96] "Q*bert 3 for SNES". MobyGames. Retrieved 2014-04-30.

[97] Weiss, Brett A. "Q*bert 3 – Overview". Allgame. Retrieved 2009-06-02.

[98] "IGN: Q*bert 3". IGN. Retrieved 2009-06-02.

[99] Weiss, Brett A. "Q*bert – Review". Allgame. Retrieved 2009-06-02.

[100] Rice, Kevin (May 2001). "Q*bert Review". *Next Generation Magazine* (Imagine Media): 82.

[101] "Archive 1995 - 1999". Artech Studios. Archived from the original on 2012-05-05. Retrieved 2014-07-15.

[102] "Q*Bert". Sony Pictures. Archived from the original on 2004-12-10. Retrieved 2014-04-30.

[103] "iTunes Preview: Q*Bert Deluxe". Apple. Archived from the original on 2010-03-11. Retrieved 2014-04-30.

[104] "Q*bert 2005". *Download.com powered by Cnet*. CBS Interactive Inc. 2005-05-02. Retrieved 2014-06-10.

[105] "Q*bert Rebooted brings the franchise back to Steam, mobile and tablets". *Polygon*. Vox Media. 2014-07-02. Retrieved 2014-07-03.

[106] Campbell, Evan (2015-02-06). "Q*Bert Rebooted Coming to PlayStation Systems". *IGN*. Ziff Davis. Retrieved 2015-08-12.

[107] Desat, Marla (2014-07-04). "Arcade Hit Q*bert Coming to Steam with Classic, Rebooted Modes". *The Escapist*. Defy Media. Retrieved 2014-07-14.

[108] Woolsey, Cameron (2015-03-04). "Q*Bert Rebooted Review". *Gamespot*. CBS Interactive Inc. Retrieved 2015-08-12.

[109] Thurmond, Joey (2015-02-24). "Q*Bert Rebooted". *Push Square*. Retrieved 2015-08-12.

4.44 Radar Rat Race

Radar Rat Race (レーダーラットレース *Rēdā Ratto Rēsu*) is a 1981 game made by HAL Laboratory for the Commodore VIC-20, later converted to the Commodore MAX Machine and Commodore 64. A clone of Namco's *Rally-X* arcade game, it was among thirty game titles marketed by Commodore on cartridges. It was originally released in Japan as *Rally-X* (ラリーX) from Commodore Japan K.K.

Radar Rat Race is cartridge number VIC-1910 for the Commodore VIC-20

4.44.1 Overview

The player guides a mouse through a large maze. The camera follows the mouse and shows only a small portion of the maze at any given time. The player is pursued by at least three rats. The goal is to eat all of the pieces of cheese, shown for the entire maze on a radar screen, without getting caught by a rat or bumping into a stationary cat. By pressing the joystick button, the mouse can disperse a limited amount of magical dust (called "star screen") which confuses the rats for about five seconds.

Once the round is complete, the game starts again, with more rats and faster play.

The gameplay is accompanied by a frenetic, rhythmically altered version of a phrase from *Three Blind Mice* which cycles endlessly.

4.45   Rescue at Rigel

Rescue at Rigel is a 1980 science fiction computer role-playing game written and published by Automated Simulations (later known as Epyx), and later branded as part of the *Starquest* series. The game was released for the Apple II, DOS (PC Booter), TRS-80, VIC-20, and Atari 8-bit. *Rescue at Rigel* was soon followed by *Star Warrior* in the "Starquest" series, although *Star Warrior* used a more heavily modified game engine than *Rigel*.

4.45.1   Gameplay

Players take on the role of adventurer Sudden Smith. Smith must try to rescue captives from the interior of an asteroid orbiting the star Rigel. Players have 60 minutes to rescue 10 human captives from the alien moon base. They must first find the captives before delivering them to the rescue ship (via a transport beam). Players must defeat or avoid the enemies wandering the base: the alien Tollahs, two types of armed robots, a six-legged "cerbanth", and a huge amoebic slug. As players forge deeper into the alien stronghold, they have the opportunity to acquire better weapons.

The playfield is presented as a top-down view of the current location of the hero. The game is turn-based, with the player given a certain number of "points" to spend on various actions, completing their turn when the points ran out. *Rescue at Rigel* is very similar to *Temple of Apshai*, a popular dungeon crawl by Epyx, part of their "Dunjonquest" series. *Rescue at Rigel* had a timer similar to *The Datestones of Ryn*, an earlier Dunjonquest game.

Rescue at Rigel used the concept of providing room descriptions similar to those used in some Dunjonquest games, but instead of unique descriptions for numbered rooms, the game had multiple rooms labeled "Sanctum", for example, and a detailed description of what typical Sanctums contained was provided in the manual along with about a dozen other room types.

4.45.2   Plot

Although nominally a science fiction setting, the plot of rescuing hostages was perhaps derived from the Iran Hostage Crisis, which was headline news when the game was written. Additionally, one type of enemy which the player must defeat is the High Tollah, a name that resembles the title of Ayatollah Ruhollah Khomeini, the Iranian religious leader whose supporters took the American hostages.

4.45.3   See also

- *2400 A.D.*

4.45.4   External links

- *Rescue at Rigel* at MobyGames

4.46   River Rescue

River Rescue is a 1982 game for the Commodore VIC-20, Commodore 64,[*][1] ZX Spectrum, and the Atari 400 and 800 computers.[*][2] It is one of a number of game titles produced by Thorn EMI Computer Software (later known as *Creative Sparks*) during 1982 and 1983.

4.46.1   Description

The aim is to rescue explorers lost in the jungle.

The player guides a boat along a river, which scrolls from right to left across the screen. The river and boat are viewed from above. The boat must avoid colliding with various floating hazards in the river - these objects include crocodiles and canoes. After a while, an SOS call in morse code is received and soon a jetty comes into view. The boat must dock with this jetty and a lost explorer will walk aboard. There are three explorers to rescue.

Once the round is complete, the game starts again, faster.

4.46.2   Gameplay

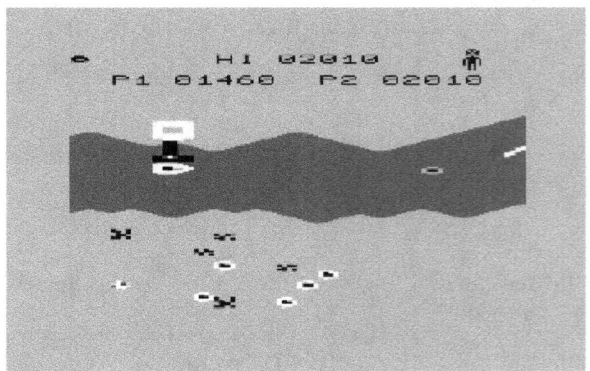

The Vic 20 version of River Rescue

When a game starts, and in between goes, action pauses until the player moves the joystick up or down (or presses the up or down key).

Before and between games, the game plays in randomised demonstration mode.

4.46.3 References

[1] Creative Sparks, Gamebase 64 forum. Article retrieved 2007-04-22.

[2] River Rescue, Atarimania. Article retrieved 2010-08-28.

- Thorn EMI Video Programmes: (1982) River Rescue Game Manual

4.46.4 External links

- Archive of VIC-20 game images

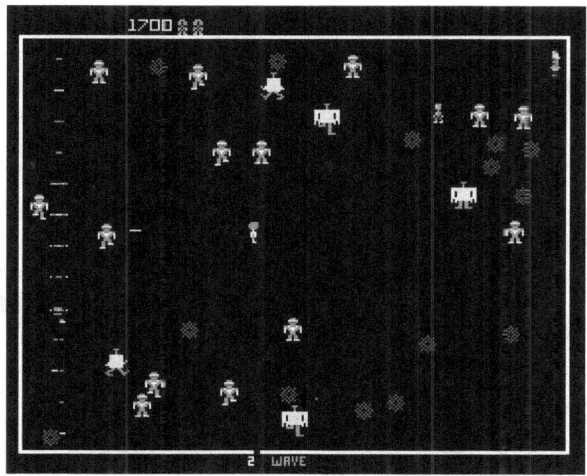

The protagonist (center) shoots the robots while dodging their attacks and attempting to rescue the human (top right).

4.47 Robotron: 2084

This article is about the video game. For other uses, see Robotron (disambiguation).

Robotron: 2084 (also referred to as ***Robotron***) is an arcade video game developed by Vid Kidz and released by Williams Electronics (part of WMS Industries) in 1982. It is a shoot 'em up with two-dimensional graphics. The game is set in the year 2084 in a fictional world where robots have turned against humans in a cybernetic revolt. The aim is to defeat endless waves of robots, rescue surviving humans, and earn as many points as possible.

Robotron popularized the twin joystick control scheme, one that had previously been used in Taito's Space Dungeon.

Robotron: 2084 was critically and commercially successful. Praise among critics focused on the game's intense action and control scheme. The game is frequently listed as one of Jarvis's best contributions to the video game industry. *Robotron: 2084* arcade cabinets have since become a sought-after collector's item. It was ported to numerous platforms.

4.47.1 Gameplay

Robotron is a 2D multi-directional shooter game in which the player controls the on-screen protagonist from a topdown perspective. The game is set in the year 2084 in a fictional world where robots ("Robotrons") have taken control of the world and eradicated most of the human race. The main protagonist is a nameless superhuman attempting to save the last human family.[*][1][*][2][*][3]

The game uses a two-joystick control scheme; the left joystick controls the on-screen character's movement, while the right controls the direction the character's weapon fires. Both joysticks allow for an input direction in one of eight ways. Each level, referred to as a "wave" , is a single screen populated with a large number of various enemy robots; types include invincible giants to robots that continually manufacture other robots that shoot the protagonist. Coming into contact with an enemy or enemy projectile results in the character dying. Waves also include human family members which can be rescued to score additional points. Defeating all the onscreen-robots allows the player to progress to a more difficult wave; the cycle continues until the player depletes extra attempts to continue the game.[*][1][*][2][*][3]

4.47.2 Development

Robotron: 2084 features monaural sound and raster graphics on a 19-inch CRT monitor.[*][2] It uses a Motorola 6809 central processing unit that operates at 1MHz.[*][4] To produce multiple sounds on a single audio channel, the game uses a priority scheme to generate sounds in order of importance.[*][5] A custom graphics coprocessor—which operates as a blitter chip—generates the on-screen objects and visual effects. The coprocessor increases the transfer speed of memory, which allows the game to simultaneously animate a large number of objects.[*][4][*][6]

The game was developed in six months by Eugene Jarvis and Larry DeMar, founders of Vid Kidz.[*][4] Vid Kidz served as a consulting firm that designed games for Williams Electronics (part of WMS Industries), whom Jarvis and DeMar had previously worked for.[*][7] The game was designed to provide excitement for players; Jarvis described the game as an "athletic experience" derived from a "physical element" in the two joystick design. *Robotron: 2084*'s gameplay is based on presenting the player with conflicting goals: avoid enemy attacks to survive, defeat enemies to progress, and save the family to earn points.[*][8] It was first inspired by Stern Electronics' 1980 arcade game *Berzerk* and the Commodore PET computer game *Chase*. *Berzerk* is a shooting game in which a character traverses a maze to shoot robots, and *Chase* is a text-based game in which players move text characters into others.[*][4][*][9] The initial concept involved a passive main character; the object was to get robots that chased the protagonist to collide with stationary, lethal obstacles.[*][4][*][5] The game was deemed too boring compared to other action titles on the market and shooting was added to provide more excitement.[*][4][*][10]

The dual joystick design was developed by Eugene Jarvis, and resulted from two experiences in Jarvis's life: an automobile accident and playing *Berzerk*. Prior to beginning development, Jarvis injured his right hand in an accident—his hand was still in a cast when he returned to work, which prevented him from using a traditional joystick with a button. While in rehabilitation, he thought of *Berzerk*.[*][7][*][9] Though Jarvis enjoyed the game and similar titles, he was dissatisfied with the control scheme; *Berzerk* used a single joystick to move the on-screen character and a button to fire the weapon, which would shoot the same direction the character was facing.[*][7][*][10] Jarvis noticed that if the button was held down, the character would remain stationary and the joystick could be used to fire in any direction.[*][4][*][10] This method of play inspired Jarvis to add a second joystick dedicated to aiming the direction projectiles were shot.[*][10] Jarvis and DeMar created a prototype using a *Stargate* arcade system board and two Atari 2600 controllers attached to a control panel.[*][4][*][7] In retrospect, Jarvis considers the design a contradiction that blends "incredible freedom of movement" with ease of use.[*][5]

The developers felt a rescue theme similar to *Defender*—one of their previous games—was needed to complete the game, and added a human family as a method to motivate players to earn a high score.[*][9][*][10] The rescue aspect also created a situation where players had to constantly reevaluate their situation to choose the optimal action: run from enemies, shoot enemies, or rescue humans.[*][5][*][8] Inspired by George Orwell's *Nineteen Eighty-Four*, Jarvis and DeMar worked the concept of an Orwellian world developed into the plot. The two noticed, however, that 1984 was approaching, but the state of the real world did not match that of the book. They decided to set the game further in the future, the year 2084, to provide a more realistic timeframe for their version of "Big Brother". Jarvis, a science fiction fan, based the Robotrons on the idea that computers would eventually become advanced entities that helped humans in everyday life. He believed the robots would eventually realize that humans are the cause of the world's problems and revolt against them.[*][7]

Jarvis and DeMar playtested the game themselves, and continually tweaked the designs as the project progressed.[*][5] Though games at the time began to use scrolling to have larger levels, the developers chose a single screen to confine the action.[*][10] To instill panic in the player, the character was initially placed in the center of the game's action, and had to deal with projectiles coming from multiple directions, as opposed previous shooting games such as *Space Invaders* and *Galaxian*, where the enemies attacked from a single direction. This made for more challenging gameplay, an aspect Jarvis took pride in.[*][7] Enemies were assigned to stages in different groups to create themes.[*][4] Early stages were designed to be relatively simple compared to later ones. The level of difficulty was designed to increase quickly so players would struggle to complete later stages. In retrospect, Jarvis attributes his and DeMar's average player skills to the game's balanced design. Though they made the game as difficult as they could, the high end of their skills ended up being a good challenge for expert players.[*][5] The graphics were given a simple appearance to avoid a cluttered game screen, and object designs were made distinct from each other to avoid confusion. Black was chosen as the background color to help characters stand out and reduce clutter.[*][10]

Of special note is that Robotron had a major defect where the game would reset (Carpet pattern reset/watchdog reset) if a specific scenario was experienced while shooting an Enforcer in a corner of the screen. In 1987, Christian Gingras evaluated the code to find the problem. A visit to Williams headquarters to consult with the VidKidz resulted in code fixes that eventually made it into all later ports of the game.

Enemy designs

Each enemy was designed to exhibit a unique behavior toward the character; random elements were programmed into the enemies' behaviors to make the game more interesting.[*][4][*][5] The first two designed were the simplest: Electrodes and Grunts. Electrodes are stationary objects that are lethal to the in-game characters, and Grunts are simple robots that chase the protagonist by plotting the shortest path to him.[*][4][*][10] Grunts were designed to overwhelm the player with large groups.[*][5] While testing the game with the new control system and the two enemies, Jarvis and DeMar were impressed by the gameplay's excitement and fun. As a result, they began steadily increasing the number of on-screen enemies to over a hundred to see if more enemies would generate more enjoyment.[*][4][*][10]

Other enemies were created to add more variety. Large, indestructible Hulks, inspired by an enemy in *Berzerk*, were added to kill the humans on the stage. Though they cannot be destroyed, the developers decided to have the protagonist's projectiles slow the Hulk's movement as a way to help the player. Levitating Enforcers were added as enemies that could shoot back at the main character; Jarvis and DeMar liked the idea of a floating robot and felt it would be easier to animate. A projectile algorithm was devised for Enforcers to simulate enemy artificial intelligence. The developers felt a simple algorithm of shooting directly at the protagonist would be ineffective because the character's constant motion would always result in a miss. Random elements were added to make the projectile more unpredictable; the Enforcer aims at a random location in a ten pixel radius around the character, and random acceleration curves the trajectory. To further differentiate Enforcers, Jarvis devised the Spheroid enemy as a robot that continually generated Enforcers, rather than have them already on the screen like other enemies. Brains were conceived as robots that could capture humans and brainwash them into enemies called Progs. DeMar devised the final enemies as a way to further increase the game's difficulty; Tanks that fire projectiles which bounce around the screen, and Quarks as a tank-producing robot.*[10]

In the summer of 2012, Eugene Jarvis wrote a comprehensive evaluation of the Robotron Enemy Dynamics: The game is hard-coded with 40 waves, whereupon the game repeats wave 21 to 40 over and over until the game restarts back to the original wave 1, once the player completes wave 255. In the summer of 2012, Larry DeMar also provided details about how to trigger the secret room in Robotron.*[11]

4.47.3 Reception

Robotron: 2084 was commercially successful; Williams sold approximately 19,000 arcade cabinets, and mini cabinets and cocktail versions were later produced.*[7]*[9] The different arcade versions have since become varying levels of rarity; the cocktail and cabaret versions are very rare, while the upright cabinets are more available.*[13] It is one of the most collected arcade games, and is consistently higher priced than other titles among collectors.*[7]*[14]

The game has been positively received by critics. Author David Ellis called *Robotron: 2084* a "classic favorite" of its time, and stated that, despite the game's difficulty, it is among the most popular video games in the industry.*[14]*[15] *Retro Gamer* rated the game number two on their list of "Top 25 Arcade Games", citing its simple and addictive design.*[16] In 2008, Guinness World Records listed it as the number eleven arcade game in technical, creative and cultural impact.*[17] Brett Alan Weiss of Allgame called it one of the industry's "most exciting and intense" games. He complimented the gameplay, graphics and audio, calling them addictive, colorful and energized respectively.*[18] The game has garnered praise from industry professionals as well. Midway Games's Tony Dormanesh and Electronic Arts' Stephen Riesenberger called *Robotron: 2084* their favorite arcade game.*[19] David Thiel, a former Gottlieb audio engineer, referred to the game as the "pinnacle of interactive game design".*[20] Xot's John Leffingwell described *Robotron: 2084* as "the perfect blend of mayhem and simplicity", and commented that its plot was an interesting commentary. Jeff Peters from GearWorks Games praised the playing field as "crisp and clear", and described the strategy and dexterity required to play as a challenge to the senses. He summarized the game as "one of the best examples of game play design and execution."*[19]

Critics lauded *Robotron: 2084*'s gameplay. Authors Rusel DeMaria and Johnny Wilson enjoyed the excitement created by the constant waves of robots and fear of the character dying. They considered it one of the more impressive games produced from the 80s and 90s.*[21]*[22] Author John Vince considered the reward system (saving humans) and strategic elements as positive components.*[23] *ACE* magazine's David Upchurch commented that despite the poor graphics and basic design, the gameplay's simplicity was a strong point.*[24] The control scheme in particular was well received. DeMaria and Wilson considered it a highlight which provided the player a tactical advantage.*[21] Owen Linzmayer of *Creative Computing Video & Arcade Games* praised the freedom of movement afforded by the controls.*[25] Ellis commented that the unique control scheme was a factor in the game's success.*[14] *Retro Gamer* lauded the controls, describing them as "one of the greatest control systems of all time."*[16] In retrospect, DeMar felt players continued to play the game because the control scheme offered a high level of precision.*[26]

4.47.4 Legacy

Jarvis's contributions to the game's development are often cited among his accolades.*[27]*[28] Vince considered him as one of the originators of "high-action" and "reflex-based" arcade games, citing *Robotron: 2084*'s gameplay among other games designed by Jarvis.*[23] In 2007, IGN listed Eugene Jarvis as a top game designer whose titles (*Defender*, *Robotron 2084*, and *Smash TV*) have influenced the video game industry.*[28] *GamesTM* referred to the game as the pinnacle of his career.*[5] Shane R. Monroe of RetroGaming Radio called *Robotron* "...the greatest twitch and greed game of all time."*[29]

Bill Loguidice and Matt Barton of Gamasutra commented that *Robotron* 's success, along with *Defender*, illustrated that video game enthusiasts were ready for more difficult games with complex controls.[*][30] Though not the first to implement it, *Robotron: 2084* 's use of dual joysticks popularized the design among 2D shooting games, and has since been copied by other arcade-style games.[*][10][*][16][*][19][*][31] The control scheme has appeared in several other titles produced by Midway Games:[*][Note 1] *Inferno*, *Smash TV*, and *Total Carnage*.[*][31] Many shooting games on Xbox Live Arcade and PlayStation Network use this dual control design.[*][32][*][33][*][34] The 2003 title *Geometry Wars* and its sequels also use a similar control scheme.[*][31][*][35] The input design was most prominent in arcade games until video games with three-dimensional (3D) graphics became popular in the late 1990s. Jarvis attributes the lack of proliferation in the home market to the absence of hardware that offered two side-by-side joysticks. Most 3D games, however, use the dual joystick scheme to control the movement of a character and a camera. Few console games, like the 2004 title *Jet Li: Rise to Honor*, use two joysticks for movement and attacking.[*][30]

Remakes and sequels

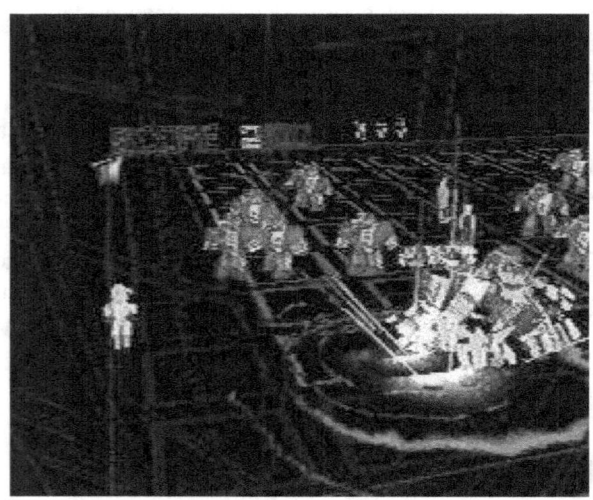

The sequel Robotron X *features gameplay similar to the original, but with three-dimensional graphics.*

Jarvis planned to produce sequels, but the North American video game crash of 1983 halted most video game production for a few years.[*][10] Prior to the full effects of the crash, Vid Kidz developed an unofficial sequel (*Blaster*) in 1983. The game is set in the same universe and takes place in 2085 in a world overrun by Robotrons.[*][10][*][36][*][37] Williams considered creating a proper sequel in the mid-1980s as well as a movie adaptation.[*][38][*][39] The com-

pany released a sequel with 3D graphics titled *Robotron X* in 1996 for the Sony PlayStation and personal computers. It was ported two years later to the Nintendo 64 as *Robotron 64*. In addition to the graphical update, the game includes new audio and multiple camera angles.[*][40][*][41] Though the game features similar gameplay as the original, it was not as well received.[*][5][*][40] Authors Andrew Rollings and Ernest Adams considered the moving camera in the 3D environment a negative update. They felt the original format—an overhead perspective of a single screen—presented the player with all the necessary information and relied on the player's skill. The moving camera angle, however, obscured areas of the playing field and could result in the player being shot by an enemy that suddenly appeared.[*][40] Vince echoed similar statements, stating that the gameplay suffered from the loss of important aspects from the original.[*][42] Rollings and Adams, however, attribute the fad of classic video game remakes in the late 1990s in part to *Robotron X* 's release.[*][40]

Robotron: 2084 has been remade on different platforms. Beginning in 1983, the game was ported to several platforms including the Atari 5200, Atari 7800, Apple IIe, Commodore 64, and TI-99/4A.[*][43] Most conversions did not have a dual joystick and were received less favorably by critics.[*][10][*][30][*][44] In 2000, a web-based version of *Robotron: 2084*, along with nine other classic arcade games, were published on Shockwave.com (a website related to Adobe Shockwave).[*][45] Four years later, Midway Games also launched a website featuring the Shockwave versions.[*][46] The game has been included in several multi-platform compilations: the 1996 *Williams Arcade's Greatest Hits*, the 2000 *Midway's Greatest Arcade Hits*, the 2003 *Midway Arcade Treasures*, and the 2012 *Midway Arcade Origins*.[*][47][*][48][*][49][*][50] In 2004, Midway Games planned to release a plug and play version of *Robotron: 2084* as part of a line of TV Games, however, it was never released.[*][51][*][52] *Robotron: 2084* became available for download via Microsoft's Xbox Live Arcade in November 2005.[*][53] In February 2010, however, Microsoft removed it from the service citing permission issues.[*][54] The Xbox Live version included high-definition graphics and two-player cooperative multi-player with one player controlling the movement and another the shooting. Scores were tracked via an online ranking system.[*][53] The game has also inspired other titles. The 1990 arcade game *Smash TV*, also designed by Jarvis, features a similar design—two joysticks used to shoot numerous enemies on a single screen—as well as ideas he intended to include in sequels.[*][10][*][40] In 1991, Jeff Minter released a shareware game titled *Llamatron* based on *Robotron: 2084* 's design.[*][55] Twenty years later, Minter released an upgraded version titled *Minotron: 2112* on the iPhone.[*][56]

4.47.5 Robotron II

The video game crash of the early 80's saw the VidKidz disband before creating Robotron's sequel. DeMar went back to pinball development and Jarvis went back to college. Later, both went on to entertainment industry successes with hits in pinball, racing games, light gun games, casino games, and social media games Robotron II remained a forgotten project, never to be implemented.

4.47.6 In other media

The robots attacking the player show up in the movie "Pixels", near the end where the arcade characters begin a full-on attack on Earth.

4.47.7 Notes

[1] Williams Electronics purchased Midway in 1988. and later transferred its games to the Midway Games subsidiary.

4.47.8 References

[1] Sellers, John (August 2001). "Robotron: 2084". *Arcade Fever: The Fan's Guide to The Golden Age of Video Games.* Running Press. pp. 110–111. ISBN 0-7624-0937-1.

[2] "Robotron: 2084 Videogame by Williams (1982)". Killer List of Videogames. Retrieved 2009-02-10.

[3] Cook, Brad. "Robotron: 2084 - Overview - allgame". Allgame. Retrieved 2009-02-10.

[4] James Hague, ed. (1997). "Eugene Jarvis". *Halcyon Days: Interviews with Classic Computer and Video Games Programmers.* Dadgum Games.

[5] GamesTM Staff (October 2005). "Robotron: 2084 Behind the Scenes". *GamesTM* (36): 146–149.

[6] Vince, John (2002). *Handbook of Computer Animation.* Springer Science+Business Media. p. 4. ISBN 1-85233-564-5.

[7] Kent, Steven (2001). "The Fall". *Ultimate History of Video Games.* Three Rivers Press. pp. 220–222. ISBN 0-7615-3643-4.

[8] Sellers, John (August 2001). "The Creator". *Arcade Fever: The Fan's Guide to The Golden Age of Video Games.* Running Press. pp. 52–53. ISBN 0-7624-0937-1.

[9] Digital Eclipse (2003-11-18). *Midway Arcade Treasures.* PlayStation 2. Midway Games. Level/area: The Inside Story On Robotron 2084.

[10] Grannell, Craig (March 2009). "The Making of Robotron: 2084". *Retro Gamer* (Imagine Publishing) (60): 44–47.

[11] "Larry DeMar shows Robotron secret". Retrieved 2015-08-18.

[12] Weiss, Brett Alan. "Robotron: 2084 - Review". Allgame. Retrieved June 4, 2014.

[13] Ellis, David (2004). "Arcade Classics". *Official Price Guide to Classic Video Games.* Random House. p. 405. ISBN 0-375-72038-3.

[14] Ellis, David (2004). "Arcade Classics". *Official Price Guide to Classic Video Games.* Random House. pp. 337–340. ISBN 0-375-72038-3.

[15] Ellis, David (2004). "A Brief History of Video Games". *Official Price Guide to Classic Video Games.* Random House. p. 7. ISBN 0-375-72038-3.

[16] Retro Gamer Staff (September 2008). "Top 25 Arcade Games". *Retro Gamer* (Imagine Publishing) (54): 68.

[17] Craig Glenday, ed. (2008-03-11). "Top 100 Arcade Games: Top 20–6". *Guinness World Records Gamer's Edition 2008.* Guinness World Records. Guinness. p. 235. ISBN 978-1-904994-21-3.

[18] Weiss, Brett Alan. "Robotron: 2084 - Review - allgame". Allgame. Retrieved 2009-02-10.

[19] Hong, Quang (2005-08-05). "Question of the Week Responses: Coin-Op Favorites?". Gamasutra. Retrieved 2009-05-12.

[20] Kent, Steven (2001). "The Fall". *Ultimate History of Video Games.* Three Rivers Press. p. 219. ISBN 0-7615-3643-4.

[21] DeMaria, Rusel; Wilson, Johnny L. (2003). *High Score!: The Illustrated History of Electronic Games* (2 ed.). McGraw-Hill Professional. p. 86. ISBN 0-07-223172-6.

[22] DeMaria, Rusel; Wilson, Johnny L. (2003). *High Score!: The Illustrated History of Electronic Games* (2 ed.). McGraw-Hill Professional. p. 339. ISBN 0-07-223172-6.

[23] Vince, John (2002). *Handbook of Computer Animation.* Springer Science+Business Media. pp. 1–2. ISBN 1-85233-564-5.

[24] Upchurch, David (February 1992). "Robotron 2084". *Advanced Computer Entertainment* (53): 77.

[25] Linzmayer, Owen (Spring 1983). "Mastering Robotron: 2084". *Creative Computing Video & Arcade Games* **1** (1): 21.

[26] Digital Eclipse (2003-11-18). *Midway Arcade Treasures.* PlayStation 2. Midway Games. Level/area: Interview Clip 1 – Robotron's Controls.

[27] Maragos, Nich (2005-02-17). "Eugene Jarvis To Receive IGDA Lifetime Achievement Award". Gamasutra. Retrieved 2009-05-12.

[28] IGN Staff (2007-07-24). "Top 10 Tuesday: Game Designers". IGN. Retrieved 2009-03-16.

[29] *Passenger Seat Radio*, Episode 2008-08-07 3:24

[30] Loguidice, Bill; Matt Barton (2009-08-04). "The History of Robotron: 2084 - Running Away While Defending Humanoids". Gamasutra. Retrieved 2009-10-15.

[31] Harris, John (2007-12-06). "Game Design Essentials: 20 Unusual Control Schemes". Gamasutra. Retrieved 2009-05-12.

[32] Donahoe, Michael (November 2007). "Online Scene: Robocopied". *Electronic Gaming Monthly* (Ziff Davis) (221): 50.

[33] Retro Gamer Staff (April 2008). "Retro Rated: Omega Five". *Retro Gamer* (Imagine Publishing) (49): 88.

[34] Retro Gamer Staff (October 2008). "Retro Rated: Commando 3". *Retro Gamer* (Imagine Publishing) (55): 89.

[35] Gamasutra Staff (2008-12-23). "Gamasutra's Best Of 2008: Top 10 Games Of The Year". Gamasutra. Retrieved 2009-05-12.

[36] "Blaster Videogame by Williams (1983)". Killer List of Videogames. Retrieved 2009-03-17.

[37] Green, Earl. "Blaster - Overview - allgame". Allgame. Retrieved 2009-03-17.

[38] Digital Eclipse (2003-11-18). *Midway Arcade Treasures*. PlayStation 2. Midway Games. Level/area: Joust 2 Interview Clip #2.

[39] Digital Eclipse (2003-11-18). *Midway Arcade Treasures*. PlayStation 2. Midway Games. Level/area: The Inside Story On Joust.

[40] Rollings, Andrew; Adams, Ernest (2003). *Andrew Rollings and Ernest Adams on Game Design*. New Riders. p. 283. ISBN 1-59273-001-9.

[41] Weiss, Brett Alan. "Robotron X - Overview - allgame". Allgame. Retrieved 2009-03-17.

[42] Vince, John (2002). *Handbook of Computer Animation*. Springer Science+Business Media. pp. 19–20. ISBN 1-85233-564-5.

[43] "MobyGames Quick Search: Robotron: 2084". MobyGames. Retrieved 2009-03-17.

[44] Buchanan, Levi (2008-02-27). "The Atari 5200 Buyer's Guide". IGN. Retrieved 2009-03-19.

[45] Parker, Sam (2000-05-05). "Midway Coming Back At You". GameSpot. Retrieved 2009-03-20.

[46] Kohler, Chris (2004-09-24). "Midway Arcade Treasures Web site goes live". GameSpot. Retrieved 2009-03-17.

[47] Weiss, Brett Alan. "Williams Arcade's Greatest Hits - Overview - allgame". Allgame. Retrieved 2009-03-18.

[48] All Game Staff. "Midway's Greatest Arcade Hits: Vol. 1 - Overview - allgame". Allgame. Retrieved 2009-03-18.

[49] Harris, Craig (2003-08-11). "Midway Arcade Treasures". IGN. Retrieved 2009-02-10.

[50] http://www.ign.com/articles/2012/11/14/midway-arcade-origins-review

[51] Harris, Craig (2004-02-17). "Midway's TV Games". IGN. Retrieved 2009-02-10.

[52] Vavasour, Jeff (2009-02-16). "Jeff Vavasour's Video And Computer Game Page". Retrieved 2009-05-01.

[53] Gerstmann, Jeff (2005-12-20). "Robotron: 2084 Review". GameSpot. Retrieved 2009-03-16.

[54] Sinclair, Brendan (2010-02-17). "Midway XBLA games pulled". GameSpot. Retrieved 2011-02-15.

[55] "Llamasoft – 16 Bit". Llamasoft. Retrieved 2009-03-16.

[56] Minter, Jeff (2011-02-24). "I have been a busy ox. Again." Llamasoft. Retrieved 2011-03-04.

4.47.9 External links

- *Robotron: 2084* guide at StrategyWiki

- The arcade version of *Robotron: 2084* can be played for free in the browser at the Internet Archive

- Robotron2084Guidebook Comprehensive training including detailed Enemy Analysis by Eugene Jarvis, commentary from Larry DeMar, and walk-through's.

- Robotron: 2084 on Coinop.org

- Time Extend: Robotron 2084 at Edge-Online

- Eurogamer Retrospective: Robotron: 2084

- Robotron: 2084 information on Arcade-History

4.48 Scramble (video game)

This article is about the 1981 Konami game. For the 1982 Grandstand game, see Scramble (tabletop electronic game).

Scramble (スクランブル *Sukuranburu*) is a 1981 side-scrolling shoot 'em up arcade game. It was developed by Konami, and manufactured and distributed by Leijac in Japan and Stern in North America. It was the first side-scrolling shooter with forced scrolling and multiple distinct levels.*[3] The Konami Scramble arcade system board hardware uses two Zilog Z80 microprocessors for the central processing unit, two AY-3-8910 sound chips for the sound,*[4] and Namco Galaxian video hardware for the graphics.*[5]*[6]

The game was a success, selling 15,136 video game arcade cabinets in the United States within five months, by August 4, 1981, becoming Stern's second best-selling arcade classic after *Berzerk*. Its sequel *Super Cobra* sold 12,337 cabinets in the US in four months that same year, adding up to 27,473 US cabinet sales for both, by October 1981.*[2]

4.48.1 Gameplay

The player controls an aircraft, referred to in the game as a "Jet," and has to guide it across a scrolling terrain, battling obstacles along the way. The ship is armed with a forward-firing weapon and bombs; each weapon has its own button. The player must avoid colliding with the terrain and other enemies, while simultaneously maintaining its limited fuel supply which diminishes over time. More fuel can be acquired by destroying fuel tanks in the game.

The game is divided into six sections, each with a different style of terrain and different obstacles. There is no intermission between each section; the game simply scrolls into the new terrain. Points are awarded based upon the number of seconds of being alive, and on destroying enemies and fuel tanks. In the final section, the player must destroy a "base" . Once this has been accomplished, a flag denoting a completed mission is posted at the bottom right of the screen. The game then continues by returning to the first section once more, with a slight increase in difficulty.

4.48.2 Reception and legacy

Scramble was commercially successful and critically acclaimed in its time. In its February 1982 issue, *Computer and Video Games* magazine said it "was the first arcade game to sent you on a mission and quickly earned a big following." *[7] In 1982, *Arcade Express* gave the dedicated Tomytronic version of the game a score of 9 out of 10, describing it as an "engrossing" game that "rates as one of the year's best so far." *[8] The Vectrex version of the game was also praised in a review by *Video* magazine where reviewers praised its fidelity to the original arcade game and described it as their favorite among the Vetrex titles they had reviewed.*[9]*:120 The game's overlays were singled out for praise, with reviewers commenting that "when you're really involved with a Vectrex game like *Scramble*, it's almost possible to forget that the program is in black-and-white." *[9]*:32

The direct sequel to *Scramble* was the helicopter arcade game *Super Cobra*. Unlike *Scramble*, *Super Cobra* was widely ported to video game systems and home computers of the time.

An updated version of *Scramble* is available in *Konami Collector's Series: Arcade Advanced* by inputting the Konami Code in the game's title screen. This version allows three different ships to be chosen: the Renegade, the Shori and the Gunslinger. The only difference between the ships besides their appearance are the shots they fire. The Renegade's shots are the same as in the original Scramble, the Shori has rapid-fire capabilities triggered by holding down the fire button, and the Gunslinger's shots can pierce through enemies, meaning they can be used for multiple hits with a single shot.

According to the Nintendo Game Boy Advance Gradius Advance intro and the *Gradius Breakdown* DVD included with *Gradius V*, *Scramble* is considered the first in the "Gradius" series. However, the Gradius Collection guidebook issued a few years after by Konami, lists *Scramble* as part of their shooting history, and the Gradius games are now listed separately.

Scramble was included on *Konami Arcade Classics* in 1999.

Scramble joined the Xbox Live Arcade library for the Xbox 360 on September 13, 2006, its release having been delayed from September 6, 2006 due to bugs.

Scramble made the list of Top 100 arcade games in the Guinness World Records Gamer's Edition

Scramble was made available on Microsoft's *Game Room* service for its Xbox 360 console and for Windows-based PCs on March 24, 2010.

Its emulated version was re-released in 2005 for PlayStation 2 in Japan as part of the *Oretachi Geasen Zoku Sono*-series.

4.48.3 Legal history

In Stern Electronics, Inc. v. Kaufman, 669 F.2d 852, the Second Circuit held that Stern could copyright the images and sounds in the game, not just the source code that produced them.*[10]

4.48.4 Notes and references

[1] http://www.arcade-history.com/?n= scramble-model-gx387&page=detail&id=2328

[2] "Stern Production Numbers and More CCI Photos". 1 May 2012. Retrieved 21 July 2013.

[3] Game Genres: Shmups, Professor Jim Whitehead, January 29, 2007, Accessed June 17, 2008

[4] *Scramble* at the Killer List of Videogames

[5] https://github.com/mamedev/mame/tree/master/src/
mame/drivers/scramble.c

[6] https://github.com/mamedev/mame/tree/master/src/
mame/includes/galaxold.h

[7] http://www.solvalou.com/subpage/arcade_reviews/161/
432/scramble_tips.html

[8] "The Hotseat: Reviews of New Products" (PDF). *Arcade
Express* **1** (1): 6–7 [6]. August 15, 1982. Retrieved 3 February 2012.

[9] Kunkel, Bill; Katz, Arnie (October 1982). "Arcade Alley:
The First Portable Video Game System" . *Video* (Reese
Communications) **6** (7): 32, 118–120. ISSN 0147-8907.

[10] Brandon Rash. "Case: Stern Elec. v. Kaufman (2nd Cir.
1982)". Patent Arcade. Retrieved 2006-09-16.

- *Gradius Portable Official Guide*. Konami. 2006. ISBN
4-86155-111-0.

4.48.5   See also

- *Caverns of Mars*
- *Cosmic Avenger*
- *Vanguard*

4.48.6   External links

- Official Arcade Archives website
- Official PlayStation Minisite (Japanese)
- *Scramble* at the Killer List of Videogames
- *Scramble* at the Arcade History database
- *Scramble* at MobyGames

4.49   Sea Wolf (video game)

Sea Wolf is an arcade game by Midway, originally released in 1976.*[1] It was a video game update of
an earlier coin-operated electro-mechanical (em) Midway game, *Sea Devil*,*[2] itself based on Sega's 1966
coin-op electro-mechanical arcade submarine simulator
Periscope.*[3] Midway's video game version was designed
by Dave Nutting and eventually sold 10,000 video game arcade cabinets. A color sequel, *Sea Wolf II*, was released in
1978 that sold another 4,000 units.*[4]

4.49.1   Gameplay

The player looks through a large periscope to aim at ships
moving across the virtual sea line at the top of the screen,
using a thumb button on the right handle of the scope
to fire torpedoes. The periscope swivels to the right and
left, providing horizontal motion of a targeting cross-hair.
The cabinet features a mixture of video game and older
electro-mechanical technology for player feedback. Using back-lit transparencies reflected inside the scope, the
number of torpedoes remaining are displayed, as well as a
red "RELOAD" light which lights up momentarily when
the player has launched five torpedoes. Additionally, when
ships are hit on the screen, an explosion "light" is reflected
inside the scope. A blue overlay is affixed to the screen to
provide a "water color" to the sea. Sounds include a sonar
ping and the sound of the PT Boat racing across the screen.

Sea Wolf is time-limited, with the player having an opportunity to win bonus time by reaching an operator-set score.
The player's score is shown on the bottom half of the screen
as well as the high score, one of the first known instances of
a high score in a video game. Targets include destroyers, a
fast moving PT Boat, and mines floating across the screen
can that serve as obstructions.

4.49.2   Legacy

Sea Wolf was followed by Sea Wolf II in 1978.

In 2008, Coastal Amusements released a "retro video"*[5]
redemption game based on the original Sea Wolf, released
by Midway in 1976.*[6] It is a 3D remake.*[7]

4.49.3   Ports

In 1982 Commodore International produced ports of
Sea Wolf for the Commodore VIC-20 and then-new
Commodore 64 computers, released in cartridge form.*[8]

In 1983 Epyx ported *Sea Wolf II* and another Midway game,
Gun Fight, to the Atari 8-bit family, and released them in
an "Arcade Classics" compilation.*[9]

4.49.4   Highest score

The current world record holder for *Sea Wolf* is Alan Radue
with a score of 11,300 points. The record was set on October 2, 2011 at the Tranquility Base Arcade and verified by
Twin Galaxies International on October 9, 2011.

4.49.5 External links

- *Sea Wolf* at the Killer List of Videogames

4.49.6 References

[1] "Sea Wolf Killer List of Video Games Entry". Retrieved 2007-05-25.

[2] Marvin's Marvelous Mechanical Museum. "1976 Midway Sea Wolf". Retrieved 2007-05-25.

[3] Steve L. Kent (2001), *The ultimate history of video games: from Pong to Pokémon and beyond: the story behind the craze that touched our lives and changed the world*, p. 102, Prima, ISBN 0-7615-3643-4

[4] Steven L. Kent (2000), *The first quarter: a 25-year history of video games*, BWD Press, p. 83, ISBN 0-9704755-0-0, retrieved 2011-04-09, Sea Wolf, which was another creation of Dave Nutting, did solid business, selling more than 10,000 machines. (A later color version sold an additional 4000 units.)

[5]

[6] "Sea Wolf Redemption". Highwaygames.com. Retrieved 2011-06-28.

[7] Shaggy. "Shaggy's Review – Sea Wolf by Coastal Amusements". Arcade Heroes. Retrieved 2011-06-28

[8] "Sea Wolf for Commodore 64 (1982) - MobyGames". Retrieved 2012-06-07.

[9] "Atarimania - Arcade Classics: Sea Wolf II / Gun Fight". Retrieved 2011-02-01.

4.50 The Secret of Bastow Manor

The Secret of Bastow Manor is a 1983 graphical adventure game for the Commodore VIC-20 and Commodore 64 home computers. The Commodore 64 version is formally titled **The Secret of Bastow Manor 64** as per the tradition at the time to distinguish C64 titles by addition of the **64** moniker.

The Secret of Bastow Manor is notable for being an early video game title on these platforms and for being a pioneer in 8-bit adventure games.

4.50.1 External links

- Secret of Bastow Manor at Lemon

- Walkthrough solution by Andrew Williams

Opening puzzle, entering the manor.

4.51 Serpentine (video game)

Serpentine is a 1982 action computer game developed by David Snider and published by Brøderbund. The gameplay and visuals are similar to that of the Konami arcade game Jungler, released the previous year. Serpentine was originally written for the Apple II [*][1] and ported to the VIC-20, Commodore 64 and Atari 8-bit computers.

4.51.1 Description

The player controls (rides, by game description [*][2]) a multi-segmented blue 'good' serpent in a maze with the objective of eating all computer-controlled 'evil' (red or orange or green) serpents. Eating the tail segments of serpents makes them shorter, and a red or orange serpent turns green when shorter than the player. Hitting a green serpent headfirst eliminates it, and causes the player's serpent to grow an additional segment. Hitting a red or orange serpent headfirst causes the player's serpent to die. A frog appears at random intervals and gives any serpent eating it an additional segment. Once all opponents have been eliminated, the player's serpent automatically returns to a protected area.

As the game progresses, opposing serpents are faster and longer, increasing the difficulty, and each advancing level the existing players serpent gets slower. If the player's serpent dies, the replacement regains its original speed, but loses any additional segments gained by the previous incarnation.

One unique aspect of the game is how extra lives are gained. The playing serpent will lay an egg (losing a segment in the process) and, if given enough time, the egg hatches into another serpent, which hurries to the protected area. Enemy serpents will also lay eggs; if one hatches, a new two-segment opponent appears. It is possible to lose the last segment to an egg, resulting in the death of that serpent, but this can only happen to the player' serpent. If a frog happens to appear while an egg is on the map, it will head towards the egg and eat it as well. This will occur even at the end of a level when the player's serpent is operating on autopilot, making the choice of position where the last enemy serpent is killed tactically important.

Most versions of the game include 20 different mazes, but the Atari version only has 5.

4.51.2 Reception

Softline in 1982 called *Serpentine* "devilishly addicting, being endowed with the qualities that make arcade games worth the bother".*[3]

Serpentine ranked #13 for most popular game of 1982 according to *Softalk* magazine.*[4]*[5]

4.51.3 References

[1] Alan Hewston (Feb 2007), "The Many Faces of Serpentine", *Retrogaming Times*: 1

[2] Chris Vogeli (Fall 1983), "Five Great Games For The Apple Computer", *Creative Computing Video & Arcade Games*: 100

[3] Ferris, Michael; Albert, Dave (September 1982). "Serpentine". *Softline*. p. 22. Retrieved 27 July 2014.

[4] "It's Choplifter in '82", *Softalk*, April 1983: 76–82

[5] Broderbund Software, Inc (August 1983). "Games, Games, Games". *Compute! Gazette*. p. 102. Retrieved 4 December 2013.

4.52 Shamus (video game)

Shamus is a computer game written by William Mataga (now Cathryn Mataga) and published by Synapse Software. Originally released for the Atari 8-bit computers in 1982, it was ported to the VIC-20, Commodore 64, TRS-80 Color Computer, TI-99/4A, and IBM PC. Several of these ports were made by Atarisoft. It was followed by a sequel *Shamus: Case II*, available on the Atari and C64.

William Mataga's original version was 16K in size and released on disk, tape, and cartridge for the Atari 8-bit family. The VIC-20 port was 8K and contained only 32 levels (unlike the 128 in every other version). Other releases were either on disk or tape.

In 1999, Mataga developed a remake for the Game Boy Color.

Official ports of both *Shamus* and *Shamus: Case II* are available for iOS devices.

4.52.1 Gameplay

Inspired by the arcade game *Berzerk*, the objective of the game is to navigate the eponymous robotic detective through a 4-skill level, 128-room maze of electrified walls. The ultimate goal at the end of this journey is "The Shadow's Lair". *Shamus* differs from *Berzerk* in that there's a persistent world instead of rooms that are randomly generated each time they are entered. There are also items to collect: bottles containing extra lives, mystery question marks, and keys which open exits.

Opposing the player are a number of robotic adversaries, including spiral drones, robo droids and snap jumpers. Shamus is armed with "Ion SHIVs", SHIV being an acronym for Short High Intensity Vaporizer, and is able to hurl up to two at a time at his enemies. Like many other games in this genre, touching an electrified wall results in instantaneous death. Upon the completion of each level, the gameplay speeds up, increasing the chances of running into a wall.

The main gameplay involves clearing the room of all enemies, picking up special items on the way and then leaving through an exit. Upon returning to the room, the enemies are regenerated and returned to their original positions. In exactly the same way as *Berzerk*, the player is attacked if he or she spends too much time in a room. In this case, the Shadow himself emerges from off-screen and hops directly at Shamus, unhindered by the walls. If shot, the Shadow briefly freezes in place.

The game was unique in that its combination of locks and keys required the player to complete each of its four levels in a particular order. To complete the game in its entirety would take several hours, which combined with the lack of a pause function (except on the IBM version), the necessity of remembering the location of dozens of rooms and keys, and the frenetic gameplay meant that this was extremely difficult to accomplish.

Funeral March of a Marionette, the theme song from *Alfred Hitchcock Presents*, played on the title screen.

4.52.2 Trivia

The various maze layouts are all named after famous fictional detectives or agents, such as "Clouseau", "Marlowe", "Holmes" or "Bond".

4.52.3 Reception

Softline in 1983 stated that "*Shamus* is the best cross between arcade and adventure games currently on the Atari market ... To know it is to love it, play it constantly, and not get enough of it".[1] That year its readers named the game seventh on the magazine's Top Thirty list of Atari 8-bit programs by popularity,[2] and in 1984 they named *Shamus* in tenth place for 1983.[3] *Ahoy!* wrote in 1984 that *Shamus* for the Commodore 64 "is a thoroughly enjoyable game with all the action and suspense that both novices and sophisticated gamers will demand".[4]

Softline in 1983 called *Shamus: Case II* "another masterpiece of compressed programming", and advised readers to "run out and buy it so Mataga will be encouraged to create Case 3!"[5]

4.52.4 References

[1] Bang, Derrick (January 1983). "Shamus". *Softline*. p. 42. Retrieved 27 July 2014.

[2] "The Most Popular Atari Program Ever". *Softline*. March 1983. p. 44. Retrieved 28 July 2014.

[3] "The Best and the Rest". *St.Game*. Mar–Apr 1984. p. 49. Retrieved 28 July 2014.

[4] Sodaro, Robert J. (February 1984). "Shamus". *Ahoy!*. p. 51. Retrieved 27 June 2014.

[5] Bang, Derrick (July–August 1983). 'Shamus: Case 2". *Softline*. pp. 25–26. Retrieved 28 July 2014.

4.52.5 External links

- Mazes of Shamus (Feature about Atari 8-bit version)

4.53 Sky Hop

Sky Hop is a 1983 horitontally scrolling platform game for the Commodore VIC-20 programmed by Nick Sharp and published by Your Computer Magazine (Volume 3, Number 11, November 1983).[1] Sky Hop was inspired by the arcade game Jump Bug.

4.53.1 References

[1] http://www.computinghistory.org.uk/det/5168/Your-Computer-November-1983/

4.54 Snooker (video game)

Snooker is a 1983 sports simulation video game published by Visions Software Factory.

4.54.1 Summary

This video game simulates the game of snooker on the major home computers of that era, including Commodore 64, Commodore VIC-20, ZX Spectrum, Acorn Electron, and BBC Micro.

The players take turns to hit a white cue ball against the reds or colors following the rules of snooker. The strength of the shot and the spin can be selected using the space bar and cursor keys respectively.

The limited color selections of the home computers of the time (often limited to 8 colors) along with the memory sizes (the VIC-20 version ran in less than 6K of RAM) meant the user experience was limited compared to more modern implementations.

4.54.2 References

4.54.3 External links

- Video of the game play.

4.55 Squish 'em

Squish'em, also known as *Squish'em Sam*, is a video game published in 1983 for a number of home computers as well as the Colecovision game console. It was designed by Tony Ngo and published by Sirius Software. The Colecovision version features digitised speech without additional hardware and was published as *Squish'em Featuring Sam*. The game is the sequel to *Sewer Sam*.[1] Unofficial homebrew remakes of the game were released for Windows in 2003 and the Atari 2600 in 2007.[2][3]

4.55.1 Gameplay

Squish'em is a vertically scrolling game likened to the 1980 arcade game *Crazy Climber*. Players guide Sam to the top of uncompleted 48 storey buildings in order to collect suitcases of money.[4] Each time Sam reaches the top of a building he grabs the money and parachutes down in order to attempt the next building. Each building is represented by a grid of girders connected vertically and horizontally in different patterns; sometimes there is only a single way up to the next storey of the building. The style of buildings does not change, but the grid layout and colors change as Sam progresses through the game. Sam moves by shimmying horizontally across the grid and climbing upwards; once he has moved up a storey he cannot climb back down.[*][5][*][6]

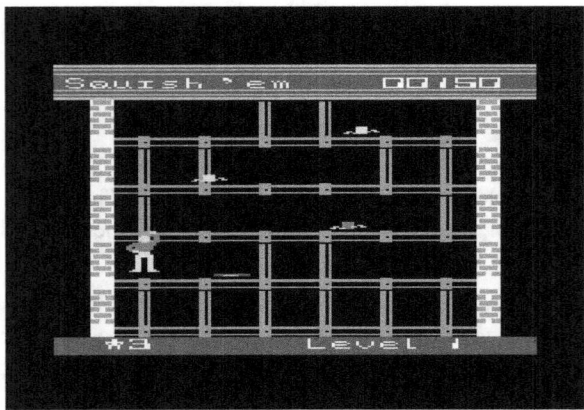

Screenshot of the C64 version.

The buildings are patrolled by dangerous enemies which move horizontally across the grid. Debris is dropped from above which Sam must avoid. Making contact with an enemy without jumping on them or being hit by falling debris results in Sam being knocked from the building and losing a life. The player starts the game with four lives; extra lives can be collected during play and up to 128 can be held by the player. Enemies can be 'squished' by being jumped on, they can also be jumped over or avoided by ascending to the next storey. Once an enemy has been squished it is rendered harmless for a short period, before turning white and becoming invulnerable. Once an enemy has turned white it must be avoided or jumped over; jumping over enemies is more difficult than landing on and squishing them. As play progresses enemies become taller and faster, making them harder to squish or avoid. The falling debris prevents players from climbing too many storeys at once, so enemies must be avoided skillfully in order to prevent Sam being cornered by a now invulnerable enemy which is too tall to be easily jumped over.[*][5][*][6]

In the Colecovision version of the game digitised speech is employed when Sam performs certain actions, for instance he exclaims "squish 'em" after successfully attacking an enemy and "money, money, money" after collecting the suitcase at the top of each building. It is one of the few Colecovision games to contain speech, as unlike the rival Intellivision console the Colecovision lacks a speech module.[*][5][*][6]

4.55.2 Reception

Craig Holyoak of *Deseret News* rated the Colecovision version 3 out of 5, praising the game's voice effects but criticizing it for being unoriginal, despite the boundaries of technology being pushed. He stated that the game contains "plenty of cute", enabling it to appeal to all ages and both genders. He also stated that "there is little new here that will keep an experienced gamer long at the screen." Holyoak played the game intensely for a number of days, but lost interest after mastering the ability to squish the more dangerous enemies. He finished by stating, "if you are less jaded and are looking for a climbing game with a new twist, Squish 'em may well be it." [*][4]

Allgame's Brett Weiss praised the Colecovision version of the game's smooth learning curve in a retrospective review. He stated, "Unlike many home videogames in which you can recall a specific level that the action started getting hard in, the transition from easy to challenging in this game is almost invisible." [*][6] He praised the game's animation, but added, "the graphics are clean and solid, but are lacking in an abundance of detail." [*][6] Weiss criticized the game's repetitiveness, stating that it is "fun, but it borders on redundant. Building after building after building, the objective is the same." [*][6] He stated that a bonus level or different building types would have improved the game. "Overall, the game is a solid addition to any gamer's ColecoVision library, especially those who are crazy about climbing." [*][6]

4.55.3 References

[1] The cover of Atari version shows "Squish'em" as game's title (see http://www.atarimania.com/game-atari-400-800-xl-xe-squish-em_4967.html), while Colecovision's one says "Squish'em Featuring Sam" (see http://spong.com/asset/187212/3/11016877/entity)

[2] http://www.mksztsz.hu/games.html

[3] http://www.atariage.com/software_page.html?SoftwareLabelID=2777

[4] Holyoak, Craig (June 13, 1984). "A Game to Spur Shrieking, Sneaking and Talking Back". *Deseret News* (Deseret News Publishing Company): 75.

[5] Weiss, Brett. "Overview - Squish'em Featuring Sam". allgame. Retrieved 2011-10-15.

[6] Weiss, Brett. "Review - Squish'em Featuring Sam". allgame. Retrieved 2011-10-15.

4.55.4 External links

- *Squish 'em* at Lemon 64

4.56 Star Trek (arcade game)

Star Trek: Strategic Operations Simulator is a space combat simulation arcade game based on the original *Star Trek* television program, and released by Sega in 1983.*[2]*[3]

It is a vector game, with both a two-dimensional display and a three-dimensional first-person perspective.*[4] The player controls the Starship Enterprise, and must defend sectors from invading Klingon ships.

The game was presented in two styles of cabinets: an upright standup, and a sit-down/semi-enclosed deluxe cabinet with the player's chair modeled after the *Star Trek Motion Picture's* bridge chairs with controls integrated into the chair's arms.

4.56.1 Gameplay

The game makes use of painstakingly synthesized speech, since memory costs at the time made the use of sampled audio almost prohibitive.

Unlike most arcade games of the time, the player is presented with multiple views of the playfield. Throughout the game, survival depends on the player's ability to accumulate shields. These are rewarded by docking with starbases, which sometimes must be saved from destruction at the hands of the Klingons.

The control system for *Star Trek* employed the use of a weighted spinner for ship heading control, while a series of buttons allowed the player to activate the impulse engines, warp engines, phasers, and photon torpedoes. The warp button was deliberately placed farther away from the rest of the buttons, in order to force the player to reach for them in heated battle. The sit-down (AKA: environmental) version of the game had convenient location of the warp button at the right hand thumb.

Downloadable link: A_Comprehensive_Guide_To_Winning_Sega_Star_Trek.pdf

4.56.2 Ports

Star Trek - Strategic Operations Simulator was ported to most of the contemporary computers and consoles of the era; namely Commodore 64, TI-99/4A, the Atari 8-bit family, and Atari 5200 in 1983, Tandy Color Computer in 1984 (as *Space Wrek*), the Atari 2600, Commodore VIC-20, ColecoVision and the Apple II.

4.56.3 Reception

The review in the August 1983 issue of *Electronic Games* said that "*Star Trek* is sure to be a top-grosser in the arcades this year. If you can squeeze through the crowd around the machine, you may never want to leave." *[5]

4.56.4 References

[1] System16.com. *Game hardware page*. Retrieved August 5, 2006.

[2] "Star Trek" . *The Arcade Flyer Archive*. Killer List of Videogames. Retrieved 7 June 2012.

[3] US Copyright Database listed date of publication 1983-01-21

[4] *Star Trek* at the Killer List of Videogames

[5] Forman, Tracie (August 1983). "Insert Coin Here" . *Electronic Games* **2** (6): 100.

4.56.5 External links

- *Star Trek: Strategic Operations Simulator* at MobyGames

- *Star Trek: Strategic Operations Simulator* at GameFAQs

- *Star Trek: Strategic Operations Simulator* at the Internet Archive

- *A comprehensive tactics guide to winning at Sega Star Trek* at [Afternight.com]

Software Collection

4.57 Strange Odyssey

Strange Odyssey was a text-based adventure program written by Scott Adams and Neil Broome .

4.57.1 Description

Published by Adventure International, this text-based adventure game was one of many from Scott Adams.

Gameplay involved moving from location to location. picking up any objects found there, and using them somewhere else to unlock puzzles. Commands took the form of verb and noun, e.g. "Take Shovel" . Movement from location to location was limited to North, South, East, West, Up and Down.

The game begins with the player stranded on a tiny asteroid in a damaged spaceship. The player must use an alien teleportation device to travel to distant worlds, collect treasure, and find the materials to repair the spacecraft.

4.57.2 Reception

Kilobaud Microcomputing stated that *Strange Odyssey* was inferior in quality to *Adventureland* despite being released later, stating that the older game had "many more treasures and situations to figure out" and criticizing *Strange Odyssey* 's lack of help for novice players.[1] The game was reviewed in issue #42 of *The Dragon* magazine. The reviewer, Mark Herro, stated that "My present situation in this game is opposite that of *Pirate Adventure*. I' ve found treasures but I don't know where to take them! ... The game starts in the control room of a disabled spaceship. It took me a good half hour just to find my way out of the spaceship! To compound problems, a space suit must be worn when outside the spaceship. When the air is gone, that' s it, my friend." [2]

4.57.3 References

[1] Colsher, William L. (September 1980). "Role-Playing Games Reviewed" . *Kilobaud*. pp. 106–108. Retrieved 23 June 2014.

[2] Herro, Mark (October 1980). "The Electric Eye" . *The Dragon* (42): 42–43.

4.57.4 External links

- Strange Odyssey at MobyGames

4.58   Sword of Fargoal

Sword of Fargoal is a 1982 video game by Jeff McCord, published by Epyx. The November 1996 anniversary issue of *Computer Gaming World* listed *Sword of Fargoal* as #147 on the "Top 150 Best Video Games of All Time." [1]

4.58.1 Development

Sword of Fargoal was created by author and programmer Jeff McCord and based on his original dungeon adventure, *Gammaquest II,* which was programmed in BASIC for the Commodore PET computer and written in 1979-1981

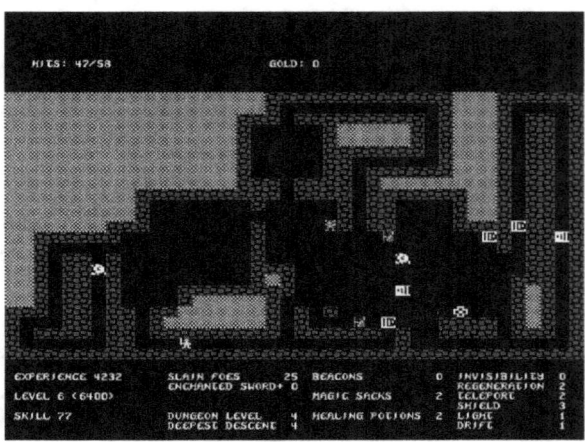

Screenshot of a modern PC remake of Sword of Fargoal, *using a graphics set similar to the original C64 tiles.*

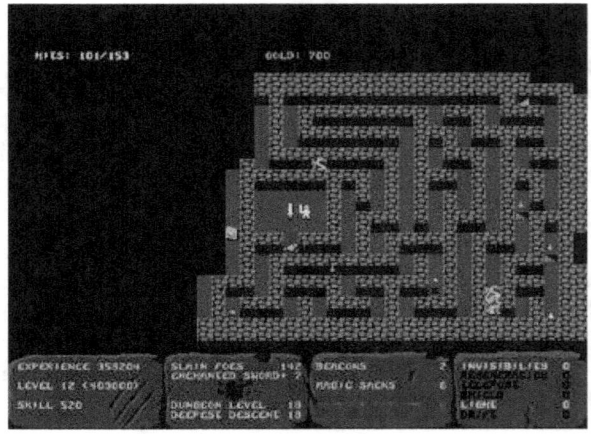

Screenshot from Sword of Fargoal *ported to the PC. Here the Sword has been found by the player.*

while he was still in high school in Lexington, Kentucky. *Gammaquest II* created randomly generated dungeons that were revealed piece-by-piece as the character explored the map, and stayed "lit" behind the character as it moved, emulating the "mapping" of a dungeon level. The game graphics, however, were limited to the character set of the computer.[2]

McCord accepted an offer to publish the game from the video game developer and publisher Epyx in 1982 on the Commodore VIC-20. His original name for the new version was *Sword of Fargaol*, deriving the name from the Old English spelling of jail (*gaol*), but his producer at Epyx, Susan Lee-Merrow, convinced him to change it to its present form.[2]

The following year, with the release of the Commodore 64 (C64), McCord was asked to release a version of *Sword of Fargoal* for that machine as well. McCord was unable to implement the conversion as it was written in BASIC, and the sprite-based graphics required machine language programming. McCord's friend, Scott Corsaire (then Carter) and Steve Lepisto wrote all the machine language code that was needed so that game would perform fast enough for the C64 version of the game (including the main redrawing of the dungeon levels, clearing of the screen in a spiral pattern effect, monster AI, collision detection, and joystick control).

Sword of Fargoal is a roguelike game, with the player controlling an adventurous warrior attempting to reclaim the "Sword of Fargoal" from the depths of a monster-infested, treasure-stocked, randomly generated dungeon. The Sword is placed randomly somewhere between the fifteenth and twentieth dungeon level. This so-called 'Sword Level" also has the unique characteristic of being a randomly generated, twisty maze of single tile-width passages. rather than a conventional dungeon level like the others. This helped make reaching the "Sword Level" an exciting event in the gameplay; once the player sees the maze design, they know the Sword is nearby.

Sword of Fargoal is noteworthy for being one of the first microcomputer games to introduce elements later used by so-called roguelike games, such as dungeons that are randomly generated for each session of play. and gave a nod to earlier games such as *Colossal Cave Adventure,* which was played without graphics on mainframe computers of the day using Unix terminals.

Sword of Fargoal has remained somewhat notorious within C64 fandom as being extremely difficult to win. Due to the random design of the "Sword Level," it is possible that the player may enter it with no way of actually reaching the Sword room, and he or she must exit and return to that level for another chance. Further, once the Sword was claimed by the player, they have exactly 2,000 seconds (33 minutes and 20 seconds) to escape the dungeon by going back through each level, or the Sword would be destroyed by a curse. Of course, since all levels are newly generated when the player returns to them, they must be fully explored to find the correct staircases leading upward, of which there is usually only one per level on this return trip. Complicating matters further was the fact that if the Sword was lost for any reason (such as being stolen by a wandering foe), the player must return to the level he or she originally found the Sword to reclaim it, and the clock did not stop or reset when this occurred.

The game was originally released on computer cassette tape and 5¼" floppy disk formats. An open source remake exists in both PC and Macintosh versions. An iPhone version was also released in December 2009.

4.58.2 Gameplay

In the game, the player controls a warrior who explores numerous dungeon levels in search of the legendary "Sword of Fargoal" artifact. The levels become progressively harder to survive as the player descends deeper and deeper into the dungeon. Each dungeon is covered in complete darkness that illuminates as the dungeon rooms and corridors are explored. When the Sword of Fargoal is successfully found, a clock countdown begins where the player must successfully escape the dungeon without it being stolen before the time expires, or the Sword is lost.

The warrior gains character levels by gaining experience points, which increase the character's fighting ability and hit points, (called Hits), as they progress through the dungeon. There are several items in the dungeon that help the character, which can be found in treasure chests or on slain adversaries.

Combat in the game is controlled by the computer, and the player has no control over how well or how bad their warrior fights. A warrior can flee an attack at anytime, unless they fall victim to a sneak attack (which is when a monster engages in combat before the warrior has a chance to move). The warrior can move freely about the dungeon, whereas monsters take intermittently timed steps.

Each dungeon has a number of staircases that go up or down. In the iPhone version of the game, there are even slippery staircases that will move a character down two levels. Because each map is randomly generated, a level the player returns to will not be the same as when they left it. Stairs also provide an entry for wandering monsters that, over time, replace slain ones on a level. In the iPhone version of the game. floors stay the same on a single game. but change when the player dies or starts a new game.

Characters can find bags of gold scattered around the dungeon. The bags can be taken by enemies if they step over them. Gold can also be stolen from the character by humanoid enemies. If those thieves are killed, the gold is returned to the warrior. A warrior can only carry 100 pieces of gold, and Magic sacks must be located that allow the warrior to carry more.

Each dungeon level contains a temple. Every time the warrior steps on a temple, their gold is sacrificed to their deity, which earns additional experience. In the iPhone version, shields the player is carrying will be renewed when they are blessed. If a warrior remains standing on a temple, it acts as a sanctuary where they become invisible to enemies around them.

Chests in the game are both a bane and a boon to the player. Some contain something useful, or contain a deadly trap. Some chests explode, causing damage, and others release crumbling ceiling or pit traps. A player doesn't know, however, if a chest contains a trap or a useful item, and must take a chance of encountering either. Chests can be picked up by enemies if they step over them. There are six spells that can be found in the dungeon.

There are several enemies in the dungeon. In general, "human" type enemies are more dangerous than creatures. Some new monsters appear in the iPhone version of the game, and "human" type enemies also carry (and use) treasure, such as potions. Often the game will describe whether the monster one encounters is strong or weak.

The latest version of the game ported to modern computers allows the player to adjust the settings and difficulty of the game. The player can choose such things as graphics themes and monster behavior. The player can also trade increased skill in combat over hit points, or vice versa.

4.58.3   Reception

Computer Gaming World noted some bugs and inconsistencies with the documentation, but called *Sword of Fargoal* "an exciting and intriguing adventure game. The graphics are beautifully crafted".[3] *Ahoy!* called the VIC-20 version "an engrossing adventure-type maze game".[4] The magazine stated that the Commodore 64 version was "nearly addictive", but criticized the lack of a savegame feature. More seriously, it stated that the randomized dungeons removed mapping and solving mysteries, important aspects of adventure gaming, and concluded "This game is so close to its goal, and yet so far".[5]

The iOS port has a Metacritic score of 85% based on 6 critic reviews.[6]

4.58.4   References

[1] *Computer Gaming World*: 150 Best Games of All Time from CDAccess.com

[2] Official *Sword of Fargoal* website

[3] Wilson, Dr. Johnny L. (July–August 1983). "The Commodore Key". *Computer Gaming World*. p. 42.

[4] Salm, Walter (March 1984). "VIC Game Buyer's Guide". *Ahoy!*. p. 49. Retrieved 27 June 2014.

[5] Herring, Richard (November 1984). "Sword of Fargoal". *Ahoy!*. p. 40. Retrieved 27 June 2014.

[6] http://www.metacritic.com/game/ios/sword-of-fargoal

4.58.5   External links

- Official website

- *Sword of Fargoal* at MobyGames

4.59   The Perils of Willy

Miner Willy is the protagonist in a series of platform games for the ZX Spectrum, MSX, Amstrad CPC and the Commodore 64 home computers. The first two games - *Manic Miner* and *Jet Set Willy* - were written by computer-game programmer Matthew Smith during the early 1980s.

A third game in the series was planned, *Miner Willy Meets The Taxman*.[1]

The series started in 1983 with the release of *Manic Miner*, and was followed up a year later with *Jet Set Willy* (one of the best selling video games of all time) and *Jet Set Willy II*. Another game in the series, *The Perils of Willy*, was released solely for the Commodore VIC-20.[2] *"Andre's night off"* was published as a type-in listing in the June 1984 issue of Computer & Video Games.[3] In addition quite a few unofficial sequels, remakes, homages and updates have been released.[4]

4.59.1   Games in the series

- *Manic Miner*, (1983), Bug-Byte / Software Projects

- *Jet Set Willy*, (1984), Software Projects

- *The Perils of Willy*, (1984), Software Projects

- *Andre's Night Off*, 1984, Matthew Smith

- *Jet Set Willy II*, (1985), Software Projects

- *Jet Set Racing*,[5] (2005), Numfum

4.59.2   References

[1] "Matthew Smith interview with Iain Lee". Thumb Candy. Retrieved 2009-08-10.

[2] The Perils of Willy at VIC 20 Gamer Archived January 6, 2015 at the Wayback Machine

[3] Andre's night off instructions

[4] World of Spectrum listing, showing the large number of fan-made *Jet Set Willy* sequels.

[5] Notingham Screenplay Festival interview, February 2005 Referred to as the *Vodafone version*

4.60 Traxx (video game)

Traxx is a computer game released in 1983 by Quicksilva for the ZX Spectrum and Commodore VIC-20. It was inspired by the arcade game *Amidar*.[1]

4.60.1 Overview

The player moves along a rectangular grid painting all of its sections. Various enemies also inhabit the grid and will try to kill the player. Unlike Amidar, the sections of the grid are not captured when surrounded; the goal is purely to color all of the lines.

4.60.2 References

[1] http://www.llamasoft.co.uk/lc-8bit.php

4.60.3 External links

- *Traxx* at World of Spectrum

4.61 Ultima: Escape from Mt. Drash

Ultima: Escape from Mt. Drash is a video game for Commodore VIC-20 home computer.

4.61.1 Description and plot

In the game, creatures called "garrintrots" have imprisoned the player in Mt. Drash, and the player's task is to escape the dungeons.

The game itself is a very simple series of three-dimensional randomly generated dungeons, and the idea is to destroy all monsters that stand in way and exit to the next level. There is a time limit as well. The game doesn't employ custom graphics, but rather uses the VIC-20's standard set of graphical characters to draw the game scene.

The game was also notable for its soundtrack; using the 4-voice sound capabilities of the MOS Technology VIC video chip in the VIC-20 to provide background music. Very few games written for the VIC-20 featured such a background soundtrack that would become commonplace in games for the Commodore 64, Nintendo NES, etc.

The source code was written in VIC-BASIC, with assembly language routines for faster animation of graphics and to handle the background music. Unlike most of the more popular VIC-20 games, it wasn't published as a ROM cartridge but rather on cassette tape. Due to the complexity of the source code and the unusual (by VIC-20 standards) length of the game, as well as the fact that unlike a cartridge game all of it had to fit in RAM, an 8k or 16k RAM memory expansion cartridge was required to be installed in the VIC-20 before running the game, further limiting the target audience.

Copy protection consisted of the RUN-STOP and RESTORE keys on the VIC-20 keyboard being disabled (to prevent "breaking in" to the BASIC code), as well as the original cassette being recorded and mastered in a way which made duplicating on a dual-cassette deck troublesome.

The game itself doesn't tie in to the *Ultima* series in too many ways. Both have a fantasy setting, Mt. Drash is the name of a dungeon in *Ultima I*, and the name "garrintrots" is an obvious pun on Richard Garriot's surname; but there the similarity ends.

4.61.2 History

The game was written by Keith Zabalaoui for Sierra On-Line in 1983. Sierra, which had published *Ultima II*, named the game an *Ultima* to improve its sales. Richard Garriott gave permission to the company and Zabalaoui—a friend who had worked on previous games for him—to do so. Sierra was skeptical of the game's appeal given the declining VIC-20 market and need for memory expansion, and only manufactured the few thousand copies needed to meet contractual requirements, with one advertisement in the July 1983 *Compute!* describing it as "A real-time, fantastic adventure" and part of the SierraVenture series.[1][2] Sierra even denied the game ever existed, until Zabalaoui confirmed it actually was finished and was actually shipped to retailers. Approximately 3000 units were made, though exact numbers are not available.

For a long time, a lot of the details surrounding the game were very vague. For example it was believed the game was a cartridge, while in fact it was released on cassette tape. One of the rumors about the game was that Sierra sold a minimal number of the games, barely enough to break even, then buried the remaining stock at a foot of a mountain somewhere. (This parallels the ultimate fate of Atari's *E.T.* game - the remainder of unsold *E.T.* cartridges were allegedly buried in a landfill.) In fact, some retailer near Vancouver, B.C. had dumped unsold software over a cliff, and this is where one of the only known complete copies were eventually found.[3] Many falsely believed that Sierra named the game *Ultima* without Garriott's knowledge or authorization.[1]

In recent years, the game has been extremely sought after by collectors. First copies of the game were discovered and announced in 2000. The first online auction of a copy was in September 2003. Since then, there have been some very rare sightings, but due to high demand, there have been quite a few counterfeit games on the market. The first complete copy of the game sold on eBay in March 2004 for US$3,605 to collector Peter Olafson; the next complete copy to emerge a couple of months later went for a thousand less.

On 20 June 2009, another boxed copy (without manual) of the game was listed on eBay. The seller from Tucson, Arizona, had bought the game about a decade earlier from a Commodore enthusiast, along with other games, and had it in his closet for ten years. Not knowing the extreme rarity of the game, he listed it on eBay to make a quick buck, listed at a starting bid of US$4.99. The extremely rare auction closed on 25 June 2009 with a final bid of US$1,875. There were a total of 31 bids in all.

On 11 June 2003, the game was ported to PC by Kasper Fauerby. The original VIC-20 cassette is also available in a format suitable for VIC-20 emulators, although to detract from the greater likelihood of counterfeiting tapes, the TAP file (a recreation of the entire tape itself) has never been made available.

4.61.3   References

[1] Maher, Jimmy (2013-05-16). "The Legend of Escape from Mt. Drash". *The Digital Antiquarian*. Retrieved 10 July 2014.

[2] "VIC-20 Owners". *Compute!* (advertisement). July 1983. p. 79. Retrieved 10 July 2014.

[3] Mt. Drash at The Computer and Book RPG/Adventure Museum

4.61.4   External links

- *Ultima: Escape from Mt. Drash* at MobyGames

4.62   Voodoo Castle

Voodoo Castle is a text adventure program written by Scott Adams and his wife Alexis.

4.62.1   Description

Published by Adventure International, this text-based adventure game is one of many from Scott Adams.

Gameplay involves moving from location to location, picking up any objects found there, and using them somewhere else to unlock puzzles. Commands take the form of verb and noun, e.g. "Climb Tree". Movement from location to location is limited to North, South, East, West, Up and Down.

The aim of the game is to wake up Count Cristo, who is lying in a coffin at the starting location in the game. In order to do so, the player needs to obtain certain items, which requires overcoming certain obstacles, such as an exploding test tubes and a doorway that's too small to pass through normally.

4.62.2   External links

- A Memorial to AI Games including solution

4.63   Wacky Waiters

Wacky Waiters is a 1982 two-dimensional platform game for the Commodore VIC-20 home computer.

4.63.1   Description

Published by Imagine Software, *Wacky Waiters* was one of Imagine's more difficult games. Gameplay involved controlling a waiter from one side of the screen to the other and back again hopping in and out of lifts trying not to spill the drinks tray he was carrying to the beckoning customer on the opposite side of the screen.

4.63.2   References

- Gordon Laing: The golden years of gaming, *Personal Computer World*, 23 June 2001.

Chapter 5

Text and image sources, contributors, and licenses

5.1 Text

- **Commodore VIC-20** *Source:* https://en.wikipedia.org/wiki/Commodore_VIC-20?oldid=688194568 *Contributors:* Damian Yerrick, Mav, Valhalla, MrH, Edward, Mahjongg, Nixdorf, Dave Farquhar, Alfio, Adam Conover, Emperorbma, Kaare, Wernher, Morn, SD6-Agent, Ke4roh, Boffy b, Psychonaut, Paul G, Dehumanizer, Tom harrison, Mikko Paananen, Apalsola, Rich Farmbrough, Pixel8, Smyth, Solkoll~enwiki, Xezbeth, Bender235, Ylee, CanisRufus, Richard W.M. Jones, Femto, Thu, Spalding, Deathawk, Polluks, Courtarro, Jumbuck, Borisborf, Enigmar, Wtshymanski, Uucp, Michaelm, Mathmo, Dah31, Graham87, Taestell, ScottAdams, Jquarry, Fred Bradstadt, Pkorossy, Mirror Vax, JeremyMcCracken, Geimas5~enwiki, YurikBot, RobotE, Crotalus horridus, StuffOfInterest, Gaius Cornelius, Wgungfu, Rbarreira, Zwobot, Crisco 1492, Closedmouth, Petri Krohn, KJBracey, SmackBot, Video99, Eskimbot, KelleyCook, Hmains, KD5TVI, Chris the speller, Letdorf, Nixeagle, RolandR, Gang65, Bilby, Omnedon, SQGibbon, Scarlet Lioness, CmdrObot, Scirocco6, Nczempin, Safalra, URORIN, Cbmeeks, Landroo, Thijs!bot, Al Lemos, Electron9, Phooto, InnovationsUnlimited, Salavat, Gioto, Luna Santin, L0b0t, Bona Fides, Kung Fu Man, Cyclonius, MegX, Magioladitis, Robotman1974, Frotz, Desktop1341, Borisser, Shawn in Montreal, Mrceleb2007, Hitekmastr, Siteobserver, VolkovBot, Mike Yaloski, TXiKiBoT, Bdb484, Sarenne, SPL68, Gona.eu, Andy Dingley, Phantomdj, Fnagaton, Spikey Meister, Triwbe, Lightmouse, ImageRemovalBot, Martarius, ClueBot, CiudadanoGlobal, Hippo99, MikeVitale, Mild Bill Hiccup, Michael2695, Mikaey, A plague of rainbows, DumZiBoT, InternetMeme, Spitfire, Petchboo, Ost316, Addbot, Mortense, Arxiloxos, Luckas-bot, Yobot, Jason Recliner, Esq., AnomieBOT, Ogliast, Danno uk, Xqbot, Ubcule, Brutalben, Doulos Christos, Chaheel Riens, Retropaul, Plbyrd, Tduk, EmausBot, WikitanvirBot, Unrulyevil5, Dewritech, Мирослав Ћика, ZéroBot, NormalTheBunny, RolaPL, KenBowd, JosJuice, Sbmeirow, Evan-Amos, RedBeard7, Pantergraph, Helpful Pixie Bot, Compilation finished successfully, Wbm1058, SuperX9, NukeofEarl, Josvebot, Aelthorn01, Commodore82, The1337gamer, BattyBot, ChrisGualtieri, Oldeowl, SFK2, Ianpurton, RickMelick, Retroisler, Monkbot, OMPIRE, Sagitus1, Jack Martinelli and Anonymous: 163

- **CARDCO** *Source:* https://en.wikipedia.org/wiki/CARDCO?oldid=645810089 *Contributors:* Rich Farmbrough, Fastily, Dsimic, Sbmeirow, Mhiji and FSII

- **Commodore 1540** *Source:* https://en.wikipedia.org/wiki/Commodore_1540?oldid=649712499 *Contributors:* Dave Farquhar, Wernher, Coldacid, Klemen Kocjancic, Pixel8, CanisRufus, Jeodesic, Mirror Vax, YurikBot, Crotalus horridus, P.o.h, Black and White, Electron9, Jenigmat429, Addbot, Dawynn, Yobot, ChrisGualtieri, OMPIRE and Anonymous: 4

- **Commodore Datasette** *Source:* https://en.wikipedia.org/wiki/Commodore_Datasette?oldid=659201071 *Contributors:* Liftarn, Smack, Wik, Wernher, Boffy b, Psychonaut, Guanabot, Pixel8, Ylee, Drhex, Jef-Infojef, Dah31, GregorB, Kbdank71, Mirror Vax, Todd Vierling, Where next Columbus?, Rwwww, SmackBot, Arny, Thumperward, Rolypolyman, Pipatron, Kubanczyk, Luminifer, Electron9, Mk*, John a s. Austin-murphy, √2, Destynova, Badn3wz~enwiki, AlleborgoBot, Michael Frind, Snideology, Holothurion, Badmachine, Petchboo, Addbot, LaaknorBot, Verbal, Luckas-bot, Yobot, AnomieBOT, Chaheel Riens, Surv1v4l1st, Elpiades, MaGa, Pantergraph, Wbm1058, BattyBot, Dexbot, Sofia Koutsouveli, OMPIRE, SPRDEF and Anonymous: 38

- **MOS Technology VIC** *Source:* https://en.wikipedia.org/wiki/MOS_Technology_VIC?oldid=679795910 *Contributors:* Mahjongg, Nixdorf, Wernher, Robbot, RedWolf, Mboverload, Pixel8, Indrian, Thu, Polluks, Mirror Vax, YurikBot, Crotalus horridus, DGJM, SmackBot, Ronaldvd, Thijs!bot, Electron9, Xeno, Daibot~enwiki, VVVBot, ImageRemovalBot, Shape84, Addbot, LaaknorBot, Lightbot, Luckas-bot, AnomieBOT, CoolingGibbon, Updatehelper, BG19bot and Anonymous: 10

- **PETSCII** *Source:* https://en.wikipedia.org/wiki/PETSCII?oldid=666286871 *Contributors:* Mjb, Mahjongg, Nixdorf, Wernher, Psychonaut, DavidCary, Mboverload, Bumm13, Abdull, PhennPhawcks, Rich Farmbrough, Lovelac7, Jaberwocky6669, Shlomital, Wtshymanski, Apckrif, Gudeldar, FlaBot, Mirror Vax, Random user 39849958, Zwobot, Calcwatch, Krótki, SmackBot, Radak, Loadmaster, TastyPoutine, Seek100, DevinCook, Nczempin, Thijs!bot, Gioto, Magioladitis, R'n'B, Kernal 7.1, Austriacus, Ogre lawless, DumZiBoT, Addbot, Yobot, Coroboy, Uhmgawa, LawBot, Nevin.williams, Angrytoast, Akazik, Pantergraph, Matthiaspaul, BG19bot, Digital Brains, Synthetoonz, MartUK2012 and Anonymous: 33

- **PWP** *Source:* https://en.wikipedia.org/wiki/PWP?oldid=668042647 *Contributors:* Pakaran, Gwaur~enwiki, Qutezuce, Gargaj, Vossanova, Tim!, Mirror Vax, Viznut, Wavelength, BazookaJoe, Gilliam, Aeternus, Cydebot, Lambdacore, Mstuomel, Addbot, Lightbot, Armbrust, Mikko J. Putkonen, KasparBot and Anonymous: 11

141

- **Stack Light Rifle** *Source:* https://en.wikipedia.org/wiki/Stack_Light_Rifle?oldid=644677219 *Contributors:* Rich Farmbrough, Pak21, Polluks, Jtalledo, Welsh, Larsinio, Whobot, JLaTondre, SmackBot, Bluebot, OrphanBot, Remember the dot, WOSlinker, Yobot and Anonymous: 5

- **Super Expander** *Source:* https://en.wikipedia.org/wiki/Super_Expander?oldid=607515758 *Contributors:* Wernher, Tobias Bergemann, CanisRufus, Shenme, Polluks, Ae-a, GregorB, Jquarry, Mirror Vax, Eubot, Crotalus horridus, Jameboy, Electron9, YK Times, LS650, Petchboo, Lightbot, AnomieBOT, OMPIRE and Anonymous: 7

- **VICE** *Source:* https://en.wikipedia.org/wiki/VICE?oldid=678106894 *Contributors:* Nixdorf, Wwwwolf, Jll, Nikai, Wernher, Psychonaut, Sysy, Rich Farmbrough, Perfecto, Yono, Devil Master, Polluks, Trevj, TheParanoidOne, CyberSkull, Mikenolte, Woohookitty, FlaBot, Mirror Vax, Eubot, NotInventedHere, SmackBot, Strobe, Bluebot, Dinjiin, Vorburger, Tomraider, Jac16888, Cydebot, .anacondabot, Marko75, Inclusivedisjunction, AlleborgoBot, Badger Drink, Str4nd, XLinkBot, SilvonenBot, Addbot, Roadstaa, Pcap, Mark Schierbecker, Full-date unlinking bot, Dewritech, ZéroBot, Helpful Pixie Bot, WikiTryHardDieHard, Aisteco, Khazar2, Dexbot, So-retro-it-hurts and Anonymous: 34

- **The Automatic Proofreader** *Source:* https://en.wikipedia.org/wiki/The_Automatic_Proofreader?oldid=515055870 *Contributors:* Wernher, Rich Farmbrough, Crotalus horridus, Chris the speller, Remember the dot, GrahamHardy, ImageRemovalBot, FrescoBot and Anonymous: 2

- **Commodore BASIC** *Source:* https://en.wikipedia.org/wiki/Commodore_BASIC?oldid=684548954 *Contributors:* The Anome, Cyrek, Maury Markowitz, Edward, Nixdorf, Alfio, Andrewman327, Greenrd, Zoicon5, Wernher, Robbot, Fredrik, RedWolf, Gtrmp, DNewhall, Gazpacho, Jkl, Pixel8, Paul August, Jaberwocky6669, Closeapple, CanisRufus, Devil Master, JIP, Mirror Vax, Margosbot~enwiki, Crotalus horridus, Rhmoore, Musicomputer, Ed de Jonge, Cojoco, Jaysbro, Maxamegalon2000, SmackBot, LocalH, Imzadi1979, Jeffro77, Thumperward, MaxxFordham, Chris Thornett, Harryboyles, RCX, Gang65, TastyPoutine, DevinCook, Nczempin, Muzer, Cydebot, Lofote, After Midnight, CharlotteWebb, TheGiantHogweed, Gioto, Guy Macon, East718, Magioladitis, SineWave, Dozen, Jcea, Roland Hermans, BuickCenturyDriver, NYCDA, Lightmouse, Sfan00 IMG, The Thing That Should Not Be, Rpawson, Daudzoss, Niceguyedc, Trivialist, Petchboo, Addbot, PJonDevelopment, Wikipedian314, Locobot, ClickRick, LinDrug, Dexter Nextnumber, Wirepath, Dewritech, GoingBatty, Δ, RichardOSmith, Helpful Pixie Bot, Wbm1058, Philippe97, OMPIRE, Helios crucible, SPRDEF and Anonymous: 44

- **Contiki** *Source:* https://en.wikipedia.org/wiki/Contiki?oldid=684814515 *Contributors:* Olivier, Edward, Liftarn, Wernher, Samsara, Northgrove, Chris 73, RedWolf, Tim Ivorson, Stewartadcock, Tobias Bergemann, Brouhaha, DavidCary, Alexander.stohr, AlistairMcMillan, OverlordQ, Stuart hc, Djj, Night Gyr, WegianWarrior, CanisRufus, Polluks, Jason One, TheParanoidOne, Sligocki, Wgw2024, Suruena, RJFJR, Allen Moore, Mirror Vax, Margosbot~enwiki, Bgwhite, Raggha, Dangph, SmackBot, Thumperward, Rrelf, Pegua, Frap, Matgraham, Moala, TastyPoutine, Flibble, TJ Spyke, Mblumber, ChristTrekker, CieloEstrellado, Thijs!bot, Gioto, Isilanes, Dougher, Addw, Sykemyke, JohnLongSilver, Beyond the classroom, SieBot, Jerryobject, Badger Drink, Kl4m-AWB, Alexbot, Stlman, DumZiBoT, Addbot, Mortense, Ghettoblaster, Pmod, Nikita Bukhvostov, Luckas-bot, Yobot, Sissssou, Uusijani, ZéroBot, BlueEntity, Kendall-K1, BattyBot, Dexbot, Kephir, Adnk, Lgfcd, Comp.arch, Monkbot and Anonymous: 58

- **Delta Drawing** *Source:* https://en.wikipedia.org/wiki/Delta_Drawing?oldid=667136030 *Contributors:* Ringbang, Gioto and Anonymous: 1

- **Commodore DOS** *Source:* https://en.wikipedia.org/wiki/Commodore_DOS?oldid=667496355 *Contributors:* Wwwwolf, Furrykef, Wernher, Owen, Boffy b, Psychonaut, Uzume, Bumm13, Pixel8, Ylee, Thu, Guy Harris, Wtshymanski, Firsfron, Mandarax, Kbdank71, JIP, Mirror Vax, YurikBot, Crotalus horridus, Blueyoshi321, SmackBot, Chris the speller, Droll, GVnayR, WhosAsking, Sassan Sanei, Washi, Editor at Large, PhilKnight, Ekotkie, Bigdumbdinosaur, Kernal 7.1, JayLevitt, SoxBot, Addbot, Ghettoblaster, Lightbot, Luckas-bot, Boxstaa, Dexter Nextnumber, John of Reading, Angrytoast, Tijfo098, ClueBot NG, Pantergraph, Matthiaspaul, Frietjes, DieSwartzPunkt, Wbm1058, Gary ml0gmwuuc, BattyBot, Hydradix, Monkbot, OMPIRE, So-retro-it-hurts and Anonymous: 25

- **SpeedScript** *Source:* https://en.wikipedia.org/wiki/SpeedScript?oldid=661299023 *Contributors:* Maury Markowitz, Tedernst, Psychonaut, Alexander.stohr, Bobblewik, Coldacid, Murple, Ylee, Polluks, TheParanoidOne, Rajanala83, Mirror Vax, Crotalus horridus, Alynna Kasmira, Hmcnally, CmdrObot, Salavat, Magioladitis, GermanX, R'n'B, ImageRemovalBot, FrescoBot, ChronoKinetic, Helpful Pixie Bot, Pantser, ArmbrustBot, OMPIRE, SPRDEF and Anonymous: 5

- **List of Commodore VIC-20 games** *Source:* https://en.wikipedia.org/wiki/List_of_Commodore_VIC-20_games?oldid=679739230 *Contributors:* Frecklefoot, Bearcat, Alan Liefting, Timvasquez, Adashiel, Ylee, Frodet, CyberSkull, Metron4, Marasmusine, Woohookitty, Hailey C. Shannon, Taestell, Pkorossy, D.brodale, Elfguy, Wiki alf, Welsh, 21stCenturyDigitalBoy, JuJube, 2fort5r, Tuli, Cus07, Cryptoboy, Egsan Bacon, JeffW, Noebse, ShakespeareFan00, Cydebot, Libro0, PV250X, Odie5533, Avicennasis, WhatamIdoing, Duffaboy, Charliechuck, Chriswiki, Retro junkie, Izno, Andyajoflaherty, WOSlinker, Maxtremus, - tSR - Nth Man, Gelliant, SchreiberBike, Ost316, Addbot, Yobot, Deltasim, SporkBot, BG19bot, Northamerica1000, Bushnrvn, AntonTchekhov, MatthewHoobin, Communal t, SafcforeverFTM and Anonymous: 71

- **3D Silicon Fish** *Source:* https://en.wikipedia.org/wiki/3D_Silicon_Fish?oldid=636447548 *Contributors:* MatthewHoobin and Stamptrader

- **Adventureland (video game)** *Source:* https://en.wikipedia.org/wiki/Adventureland_(video_game)?oldid=680653585 *Contributors:* Maury Markowitz, Frecklefoot, EALacey, Alan De Smet, Jonabbey, Ben Standeven, 23skidoo, Frodet, Czolgolz, Marasmusine, Jeff3000, KevinOKeeffe, David Levy, ScottAdams, Remurmur, Nihiltres, Metropolitan90, TnS, Morgan Leigh, Nimbex, SmackBot, Clampton, Oatmeal batman, BOZ, Hu12, Mika1h, Drinibot, Mathsgeek, SHOlafsson, Ntsimp, Alaibot, BetacommandBot, X201, AgentPeppermint, Jay Firestorm, Geniac, Diggernet, Make, Duffaboy, Mr WR, Charliechuck, R'n'B, Retro junkie, SJP, Varnent, Rjm at sleepers, Wjl2, MrKIA11, Bobmurder, Addbot, Tanhabot, Yobot, Playclever, Molotron, Surv1v4l1st, Citation bot 1, RjwilmsiBot, NeoGenPT, Eekerz, GoingBatty, Raven-14, The1337gamer and Anonymous: 17

- **Alpha Blaster** *Source:* https://en.wikipedia.org/wiki/Alpha_Blaster?oldid=673934471 *Contributors:* Ringbang, X201 and AdrianAssist

- **Apple Panic** *Source:* https://en.wikipedia.org/wiki/Apple_Panic?oldid=676962082 *Contributors:* Frecklefoot, Mahjongg, Furrykef, Haeleth, Bumm13, Rich Farmbrough, Ylee, Dgpop, Marasmusine, ADeveria, NeonMerlin, The wub, Bgwhite, Xihr, Javeryt, N. Harmonik, Huwmanbeing, GVnayR, Shane Lawrence, Rodrigo.siqueira, Hans Bauer, TJ Spyke, Amalas, Mika1h, Drinibot, Btonyb, Cydebot, BetacommandBot, Zron, X201, TigerK 69, East718, Magioladitis, Mr WR, Sam Blacketer, WOSlinker, AlleborgoBot, Xot, MikeVitale, Bfried2, Addbot, Cavarrone, Deltasim, Trappist the monk, Eekerz, The1337gamer, ChrisGualtieri and Anonymous: 6

- **Arcadia (video game)** *Source:* https://en.wikipedia.org/wiki/Arcadia_(video_game)?oldid=673932355 *Contributors:* Frodet, Remurmur, N. Harmonik, Jagged 85, Phoenixrod, Drinibot, Tectar, Alaibot, BetacommandBot, X201, East718, Duffaboy, Thibbs, Mr WR, Retro junkie, Signalhead, WOSlinker, Martarius, Dawynn, Chaheel Riens, ActiveExpression, AdrianAssist and Anonymous: 7

- **Artillery Duel** *Source:* https://en.wikipedia.org/wiki/Artillery_Duel?oldid=676962631 *Contributors:* Thunderbrand, Interiot, Jreid, KevinOKeeffe, Sadangel, Mirror Vax, D.brodale, Bgwhite, Sus scrofa, GeeJo, Flipkin, SmackBot, GVnayR, Mika1h, Alaibot, BetacommandBot, Stonic, X201, Magioladitis, SharkD, Captain99, WOSlinker, JohnnyMrNinja, Hippo99, Plastikspork, Randomran, Dawynn, Lightbot, Mcjakeqcool, Greg Tyler, Lorson, DASHBot, RenamedUser01302013, 23W and Anonymous: 5

- **Atlantis (video game)** *Source:* https://en.wikipedia.org/wiki/Atlantis_(video_game)?oldid=673933910 *Contributors:* Kuralyov, Ylee, Interiot, Yamla, ReyBrujo, KevinOKeeffe, Sadangel, Dubkiller, Mirror Vax, D.brodale, GeeJo, Ali Karbassi, Larsinio, Noidner, Flipkin, SmackBot, Kisholi, Chubz123, CardinalFangZERO, JoeBot, CmdrObot, Drinibot, Alaibot, BetacommandBot, Dwight Schrute, Stonic, X201, Thedrij, Thibbs, 28bytes, Floppydog66, WOSlinker, Zuzuu, JohnnyMrNinja, Cyfal, ImageRemovalBot, Trivialist, Addbot, Dawynn, Lightbot, Materialscientist, Neurolysis, Xqbot, Balph Eubank, Zzzjim, The1337gamer, The Interocitor, 5ives, AdrianAssist and Anonymous: 15

- **Avenger (1981 video game)** *Source:* https://en.wikipedia.org/wiki/Avenger_(1981_video_game)?oldid=673942972 *Contributors:* Gtrmp, Dgpop, Brainy J, CyberSkull, Eubot, Metropolitan90, TrS, Cryptoboy, Woodshed, MrBoo, Drinibot, BetacommandBot, X201, PrimroseGuy, Avicennasis, Duffaboy, Mr WR, JasonAQuest, Dawynn, Lightbot, Black Squirrel 2, Lorson, Moritz37, AntonTchekhov, AdrianAssist and Anonymous: 1

- **Battlezone (1980 video game)** *Source:* https://en.wikipedia.org/wiki/Battlezone_(1980_video_game)?oldid=686306975 *Contributors:* Modemac, Frecklefoot, Kchishol1970, Liftarn, Wik, Furrykef, Chris Rodgers, Kizor, Xanzzibar, Dbenbenn, Gtrmp, FriedMilk, Jrdioko, Wmahan, ChicXulub, D4, Bumm13, Krupo, Sam Hocevar, Spottedowl, Asbestos, Tjansen, BACbKA, CanisRufus, Kross, AshSert, DaveGorman, Tyan23, Frodet, Ahruman, ProhibitOnions, 2mcm, Drat, Sonicrazy, Tsuba~enwiki, New Age Retro Hippie, Skyraider, ^demon, Combination, JIP, Rjwilmsi, Mirror Vax, Master Thief Garrett, RexNL, Czar, D.brodale, Windharp, Hibana, DVdm, Subwayguy, YurikBot, Borgx, Rtkat3, Wgungfu, InformationalAnarchist, TakingUpSpace, Larsinio, Rmky87, Sandman1142, Pelladon, N. Harmonik, Wizou~enwiki, AtariKee, Clayhalliwell, Romanista, Jams Watton, Curpsbot-unicodify, Krótki, Flipkin, SmackBot, Jagged 85, Doc Strange, Srnec, Betacommand, Hraefen, Parrothead1983, Kurt S Koller, Emurphy42, KieferSkunk, Mr.Z-man, GVnayR, Jmlk17, Thegraham, Bdushaw, Ndrly, Jacquismo, That CS Guy, Violncello, TJ Spyke, Hu12, Walter Day, Asmpgmr, HDCase, Nczempin, Mika1h, Drinibot, Equendil, Cydebot, Farine, Dumb3OT, Jay32183, Chris Henniker, BetacommandBot, Thijs!bot, JAF1970, Qwyrxian, X201, Pogogunner, RJLong, Rees11, RobotG, MER-C, Msell, Magioladitis, Diego bf109, Thibbs, MartinBot, Sm8900, TechnoFaye, Maurice Carbonaro, Retro junkie, Harani66, Xgmx, STBotD, Tagus, Varnent, Terrell.bailey, WOSlinker, RaddishIoW, JayC, Ashlandchemist, Dogah, Luigihann, Miremare, Wageslave, WikiTracker, Galactic Explorer, ImageRemovalBot, Martarius, Sfan00 IMG, 31stCenturyMatt, EoGuy, Spark240, Bbb2007, Randomran, Thingg, Yun-Yuuzhan (lost password), Eik Corell, Hotcrocodile, OpusAtrum, Ost316, Rockgardenmagic, Addbot, Next-Genn-Gamer, Dwbern, Tomtheedito-, Lightbot, Yobot, AnomieBOT, VanishedUser sdu9aya9fasdsopa, Fabzyboy, Danno uk, Citation bot, Sasikiran 10, Chaheel Riens, FrescoBot, Benzol-Bot, Concernedresident's butler, NoKk0m3riT, Miracle Fen, Lorson, DASHBot, ZéroBot, H3llBot, SlimGoodbody, Htimreimer, Leicester Pete, Kendall-K1, RichardMills65, RPall BZ, Cyberbot II, Khazar2, Crivvee, Nsdfnsdf, Maplestrip and Anonymous: 111

- **Blitz (video game)** *Source:* https://en.wikipedia.org/wiki/Blitz_(video_game)?oldid=673943365 *Contributors:* Jareha, Thunderbrand, Pearle, RJFJR, JoshuacUK, Jasonglchu, Fozi999, N Harmonik, SmackBot, Tzu7, BetacommandBot, Phooto, InnovationsUnlimited, X201, TigerK 69, Avicennasis, CommonsDelinker, Mankind 2k, Fratrep, Mild Bill Hiccup, Ost316, Addbot, Xqbot, Erik9bot, Dougofborg, AdrianAssist and Anonymous: 10

- **Cannonball Blitz** *Source:* https://en.wikipedia.org/wiki/Cannonball_Blitz?oldid=684942701 *Contributors:* Greenrd, Gerrit, Ylee, Dgpop, Clarkbhm, Ian Pitchford, D.brodale, Bgwhite, Toquinha, N. Harmonik, SmackBot, Klytos, Mika1h, Cydebot, X201, X96lee15, RobJ 981, Magioladitis, Just H, FasterPussycatWooHoo, Eaowens, JohnnyMrNinja, Black Squirrel 2, П8, The1337gamer, BattyBot, OccultZone and Anonymous: 1

- **Chariot Race** *Source:* https://en.wikipedia.org/wiki/Chariot_Race?oldid=622152592 *Contributors:* GregorB, Combination, BD2412, SmackBot, GVnayR, Kuru, Mika1h, Drinibot, Alaibot, BetacommandBot, RobJ1981, Duffaboy, Thibbs, Mr WR, STBot, Varnent, Dawynn, Blue Square Thing, Lightbot, Mmmmwii, Lava476, Erik9bot, FrescoBot, DivineAlpha, Tellier and Anonymous: 5

- **Choplifter** *Source:* https://en.wikipedia.org/wiki/Choplifter?oldid=680088286 *Contributors:* Deb, Frecklefoot, Kchishol1970, Haukurth, Tempshill, Fredrik, Xanzzibar, Nifboy, FriedMilk, Horst F JENS, Bumm13, Kevin Rector, Tjansen, N328KF, Pak21, Ylee, Fuxx, Polluks, DaveGorman, Tyan23, Vslashg, Jtalledo, Katana, Drat, Duke33, Lkseitz, Woohookitty, Combination, Rjwilmsi, Mirror Vax, Xmoogle, D brodale, Hibana, Bgwhite, YurikBot, Splash, MacMog, Pelladon, N. Harmonik, Krótki, SmackBot, Chief of Naval Operations, Doktor Wilhelm, Piotor, Chris the speller, Bluebot, Thumperward, Huwmanbeing, GVnayR, NickPenguin, Dickclarkfan1, CmdrObot, Drinibot, Equendil, Cydebot, BetacommandBot, MrMarmite, Magioladitis, Thibbs, RoseTech, VolkovBot, WOSlinker, Suriel1981, Ridabewa, Android Mouse Bot 3, Cashfoley, AutocracyBot, Hippo99, Chessage, Trivialist, MrKIA11, Arimis~enwiki, Addbot, Lightbot, Ptbotgourou, Brightgalrs, SirFilipoAswaldoII, Lackett, Hellknowz, Black Squirrel 2, Mono, Religious Burp, Hydao, Pweasel, Helpful Pixie Bot, PoorlyNamedWikiUser, The1337gamer, BattyBot, DrDevilFX, Ilstrtr75 and Anonymous: 46

- **Chuck Norris Superkicks** *Source:* https://en.wikipedia.org/wiki/Chuck_Norris_Superkicks?oldid=686209591 *Contributors:* Paul A, Tregoweth, Wtmitchell, Hibana, RussBot, GVnayR, George100, Cydebot, X201, Scepbot, Arx Fortis, Bongwarrior, AuburnPilot, Harry~enwiki, El Pantera, WikiLaurent, Plastikspork, Trivialist, Jax 0677, Addbot, Luckas-bot, AnomieBOT, George2001hi, Suffusion of Yellow, CueBot NG, Cgbuff, Calabe1992, Magicman69, Asdasdasd1234321, 23W, Piercehovey15, Tentinator, Edwin Belcher, Mario Weaboo and Anonymous: 7

- **Clowns (video game)** *Source:* https://en.wikipedia.org/wiki/Clowns_(video_game)?oldid=676968308 *Contributors:* Alansohn, Woohookitty, Rjwilmsi, D.brodale, Bgwhite, Dream out loud, ABurness, CmdrObot, Mika1h, JettaMann, Libro0, Carlwev, Station1, Antonio Lopez, JL-Bot, Martarius, ClueBot, La Pianista, Qwerty1234567890123456789O, Spittlespat, MuZemike, Yobot, 爆笑連合, Jackiechanster, Gelatart Calmer Waters, Black Squirrel 2, Samwalton9 and Anonymous: 6

- **Cops 'n' Robbers** *Source:* https://en.wikipedia.org/wiki/Cops_'n'_Robbers?oldid=676969953 *Contributors:* Nihiltres, Bgwhite, DMS, Retro junkie, Zach425, TubularWorld, MrKIA11, Ost316, LilHelpa, Ubcule, Deltasim, Samwalton9 and Anonymous: 2

- **The Count (video game)** *Source:* https://en.wikipedia.org/wiki/The_Count_(video_game)?oldid=676970294 *Contributors:* EALacey, Alan De Smet, 23skidoo, CyberSkull, Czolgolz, Bgwhite, Wgungfu, SmackBot, Hu12, Mika1h, Drinibot, Alaibot, BetacommandBot, RobJ1981, Cartoon Boy, Diggernet, Torchiest, Duffaboy, Charliechuck, Retro junkie, Misterz99, JohnnyMrNinja, Martarius, Someone another, Bearsona, Lightbot, The Bushranger, FrescoBot, Samwalton9, Spiderjerky and Anonymous: 10

- **Deadly Duck** *Source:* https://en.wikipedia.org/wiki/Deadly_Duck?oldid=667617955 *Contributors:* Derek R Bullamore, Noian, MrKIA11, Dawynn, Gamesforall129o, BG19bot, Meatsgains, The1337gamer, Mogism, AntonTchekhov and Anonymous: 1

- **Demon Attack** *Source:* https://en.wikipedia.org/wiki/Demon_Attack?oldid=677393307 *Contributors:* DrBob, Gnossie, Bumm13, Indrian, Ylee, Dgpop, CyberSkull, BD2412, Remurmur, FlaBot, Czar, Stormwatch, D.brodale, Bgwhite, Larsinio, N. Harmonik, 2fort5r, Flipkin, Smack-Bot, ERobson, Hu12, Ultrogonic, Tawkerbot2, Mika1h, Drinibot, Cydebot, BetacommandBot, Andrzejbanas, Magioladitis, Thibbs, Varnent, 28bytes, WOSlinker, W5WMW, - tSR - Nth Man, SheepNotGoats, Samquinn, Wrassedragon, Martarius, Zekeeper, Trivialist, Sspmig, Addbot, Dawynn, Lightbot, Yobot, Wurzel91, Black Squirrel 2, Lorson, RenamedUser01302013, The1337gamer and Anonymous: 16

- **Dig Dug** *Source:* https://en.wikipedia.org/wiki/Dig_Dug?oldid=684907966 *Contributors:* Bryan Derksen, Frecklefoot, Kwertii, Liftarn, Andrevan, WhisperToMe, Paul Richter, TOttenville8, Kapow, Brian Kendig, Niteowlneils, Wmahan, Kusunose, SAMAS, Bumm13, Kevin B12, Spottedowl, TiMike, Asbestos, Ultrarob, Jh51681, Ollinaie, Tjansen, Monkeyman, Pak21, Ylee, Nelson339, Aaronbrick, Pikawil, DaveGorman, Tyan23, LostLeviathan, DCEdwards1966, CyberSkull, WTGDMan1986, SidP, ProhibitOnions, ReyBrujo, Sonicrazy, Czolgolz, Stemonitis, Kelly Martin, Woohookitty, Hbdragon88, Ppk01, Combination, Eluchil, Ashmoo, BD2412, JamesHenstridge, Staecker, Jazy510, Nandesuka, Dar-Ape, Yamamoto Ichiro, Mirror Vax, Master Thief Garrett, Brianreading, D.brodale, Windharp, Liontamer, Kjammer, Eric B, Banaticus, Rtkat3, Herzog, Lar, DaKing, Gaius Cornelius, Wgungfu, Lavenderbunny, Anetode, Larsinio, PhilipO, Tony1, Falcon9x5, Joestump, Pelladon, N. Harmonik, Nikkimaria, Th1rt3en, Neilka, Jeff Silvers, Krótki, Flipkin, SmackBot, Moez, Jagged 85, Bwithh, John9276, Underwater, Plotor, Parrothead1983, Bluebot, Raymondluxuryacht, Tree Biting Conspiracy, Leoni2, Nbarth, Emurphy42, Can't sleep, clown will eat me, OrphanBot, Rwatson73, Folksong, Gingerfield rocks, Khoikhoi, Knuckles Dawson, Salamurai, Zaphraud, Nighthawknz, E. Megas, TPIRFanSteve, TJ Spyke, Walter Day, Iridescent, Asmpgmr, Mathfan, RaviC, SkyWalker, Tanthalas39, FunPika, Waffle88, Mika1h, Drinibot, Cydebot, Farine, Dancter, Lkermel, Quetzalcoatl45, BetacommandBot, Thijs!bot, RFerreira, Salavat, Why My Fleece?, Farosdaughter, Reallybored999, .anacondabot, VoABot II, Ecksemmess, Rswindell, Chivista~enwiki, DerHexer, Squirrelburrito, Ariel., Dphower, Lonelybirthday, Tgeairn, Itzcuauhlti, Sem-Dem, Sethnessatwikipedia, Warut, Masky, C. Foultz, Casper10, Signalhead, DSRH, WarddrBOT, WOSlinker, Sombody, Superjustinbros., SonicBoom95, Mr. DigDug, Ashum Besher, Mr dubois, Entirelybs, DSFanatic, AlleborgoBot, SuperNESPlayer, MuzikJunky, Joel clingerman, Vanished user 82345ijgeke4tg, Namcorules's Temporary Account, Kailey elise, Celainre, Asher196, Digidude220, Martarius, Sfan00 IMG, Extremus2, ClueBot, Immblueversion, EoGuy, Iuhkjhk87y678, Froman1, Boneyard90, Gtstricky, RTaptap, Thingg, Nakomaru, Super-marine9, XLinkBot, Blackwatch21, Ost316, Wikigonish, SilvonenBot, Addbot, CL, Jhuhn, MuZemike, JeanHavoc, Yoda3539, Luckas-bot, Yobot, AnomieBOT, SnickySnacks, Xqbot, Bellegarde, FrescoBot, PrBeacon, Jm2222, Serf3469, GerbilSoft, Pjs239, Mjs1991, Black Squirrel 2, Glorioussandwich, Lorson, Canyq, Tbhotch, Kwpolska, John of Reading, Hydao, H3llBot, Qatder, ChainsawKilla66, ClueBot NG, MrJoshbumstead, Primergrey, Atxbryan, Gabriel Yuji, Vanished user ivweij23ijwefk4, Himan1238569, SuperSherbet, Achowat, BattyBot, Khazar2, Dobie80, Makecat-bot, Joey108, Magicperson6969, Funshine97, SJ Defender, Cwisnie1, Radiotron, DuppyDutch, Landingdude13, Maplestrip and Anonymous: 215

- **Donkey Kong (video game)** *Source:* https://en.wikipedia.org/wiki/Donkey_Kong_(video_game)?oldid=688048202 *Contributors:* Damian Yerrick, AxelBoldt, The Anome, Frecklefoot, Liftarn, Chinju, CatherineMunro, Darkwind, Amcaja, Arteitle, Andrevan, Slark, Doradus, StAkAr Karnak, WhisperToMe, Furrykef, Whcernan, Tempshill, Miterdale, Postdlf, Dehumanizer, Nifboy, Brian Kendig, Taviso, FriedMilk, Pascal666, Timbatron, ChicXulub, Utcursch, OverlordQ, MisfitToys, DNewhall, Bumm13, Gscshoyru, Asbestos, Tjansen, Abdull, Freakofnurture, Monkeyman, Poccil, Slady, Brutannica, Pak21, Qutezuce, Solkoll~enwiki, Shadow Hog, Martpol, Indrian, DcoetzeeBot~enwiki, Rubicon, Swid, Ylee, Matteh, AshSert, Dgpop, Longhair, Xevious, DaveGorman, Tyan23, Apostrophe, DCEdwards1966, Maebmij, Jason One, Frodet, Alansohn, Richard Harvey, MrItty, CyberSkull, InShaneee, Mysdaao, Wtmitchell, SidP, Drat, Sciurinæ, Kaibabsquirrel, Nintendo Maximus, H2g2bob, Ringbang, Tsuba~enwiki, Kouban, New Age Retro Hippie, Sir Slush, Duke33, Y0u, Natalya, Stephen, Arnim, Ian Moody, Smoke, LOL, Daniel Case, Robert K S, Trogga, Hbdragon88, Bbatsell, Ppk01, KevinOKeeffe, Combination, Rad Racer, Robfergusonjr, Rjwilmsi, OneWeirdDude, Vary, Staecker, Sdornan, Bruce1ee, DennyJay, Ligulem, NeonMerlin, Brighterorange, Remurmur, Platypus222, A Man In Black, SNIyer12, Mirror Vax, RobertG, Master Thief Garrett, SuperDude115, Gurch, Briguy52748, Czar, Jonny2x4, D.brodale, DrippingInk, Tofergregg, Hibana, Imnotminkus, Chobot, Igordebraga, Eric B, Hahnchen, Mr.Do!, Eclipsed Moon, Borgx, Jcam, Sceptre, Rtkat3, RussBot, Jobie, Hede2000, Crumbsucker, Chaser, WikidSmaht, IanManka, Hydrargyrum, Wgungfu, Rhindle The Red, NawlinWiki, Smash, Pagrashtak, RattleMan, Nutiketaiel, Retired username, Larsinio, Rmky87, Vivaldi, Tony1, Occono, Xompanthy, Zythe, Falcon9x5, Mr Fist, EmiOfBrie, Gadget850, Pelladon, AnnaKucsma, Druff, Zzuuzz, Conan-san, Theda, Mike Selinker, Alakazam, Clayhalliwell, Th1rt3en, Careax, Jedidrunkenllama, Bly1993, Moomoomoo, Paul Erik, Rehevkor, Victor falk, Ryūkotsusei, Noidner, Locke Cole, Neier, Flipkin, SmackBot, 87th, KnowledgeOfSelf, C.Fred, Prototime, AndyZ, Jagged 85, Pennywisdom2099, AnOddName, AKismet, Born2killx, Master Deusoma, SmartGuy Old, Oscarthecat, Tennekis, Plotor, Parrothead1983, Bluebot, AndrewRT, Sirex98, Tree Biting Conspiracy, AtmanDave, Zachkudrna18@yahoo.com, JONJONAUG, Ctbolt, ACupOfCoffee, Nintendude, KieferSkunk, Dethme0w, Silent Tom, TKD, Rsm99833, Treygdor, Oscara, Oanabay04, Gamgee, Artie p, Trieste, Hogtied, NES Boy, Kenan&Kel 97, Sljaxon, Wizardman, Fireswordfight, Ceoil, Ohconfucius, Trickster 1!, Axem Titanium, Kuru, 041744, Berenlazarus, Fryguy64, SQGibbon, Geologyguy, AdultSwim, Ryulong, TPIRFanSteve, Ryanjunk, TJ Spyke, Walter Day, Ace of Sevens, Ricardoread, Mvent2, Judgesurreal777, Wjejskenewr, Twas Now, Cpro, Thewebdruid, Courcelles, TheHorseCollector, Bentendo24, The Prince of Darkness, Smiloid, DevinCook, Mika1h, Drinibot, Dgw, Porterhse, MrFish, HalJor, Cydebot, Gogo Dodo, Thaddius, DumbBOT, DiScOrD tHe LuNaTiC, DBaba, Kozuch, Guyinblack25, Quetzalcoatl45, Brad101, Rlk89, Satori Son, BetacommandBot, Epbr123, Dbarnes99, Bytebear, TK421, FangzofBlood, Oliver202, Marek69, Arkracer, Dk7991, Roelzzz, Silver Edge, Scottandrewhutchins, Salavat, Escarbot, SnoopingAsUsual, Demonofthefall, RobotG, Loves Nintendo!, DarkAudit, RobJ1981, Chill doubt, Jhsounds, Terrabull, Rvn5gt6rafejd, Andrzejbanas, ThomasO1989, MER-C, Yuut, IanOsgood, Power Slave, Deltopia, Justin The Claw, Jthorsen3315, Y2kcrazyjoker4, Magioladitis, Connormah, Frankyboy5, Bongwarrior, VoABot II, AuburnPilot, Brandt Luke Zorn, Careless hx, Diego bf109, Ecksemmess, Avicennasis, Shocking Blue, 28421u2232nfenfcenc, Boffob, Make, Mr. Corncob, Thibbs, Jpallanza, DerHexer, Seansinc, Baristarim, Mykas0, Pikazilla, MartinBot, Rsegrest, Kateshortforbob, CommonsDelinker, PrestonH, Tgeairn, Trick-the-Peak Guy, J.delanoy, Pharaoh of the Wizards, Bogey97, Richiekim, Uncle Dick, Malcolmo, SemDem, Nigholith, Dispenser, DarkFalls, Samtheboy, The Glory Boy, Jcldude, Tomvec, AntiSpamBot, Gamenac, Retro junkie, Mrceleb2007, NewEnglandYankee, Cometstyles, Stvoyager, WJBscribe, Jamesofur, Casper10, Donmike10, Marioman12, Varnent, Martial75, HamatoKameko, VolkovBot, Nburden, DreamingLady, Tomer T, WOSlinker, RevRocks92, Philip Trueman, Fran Rogers, Oshwah, JulianB14, Anna Lincoln, Melsaran, Cremepuff222, BotKung, DCico, Ianjones50, Ashnard, Y, Wolfrock, Enviroboy, RaseaC, Sesshomaru, Victory93, Brianga, HiDrNick, Slash2x, Legoktm, W4chris, Pickles27, Trey, Telemachus Claudius Rhade, Unused000702, Peter Fleet, MuzikJunky, Addit, Parhamr, Caltas, RJaguar3, Triwbe, Vanished user 82345ijgeke4tg, Nullo247, Dolnk-jp, Erushford, Radon210, Angry Sun, JSpung, Karawane 71, Xe7al, AnonGuy, Lightmouse, BrightRoundCircle, FarichFenand, Meleemaster2001, Xot, JohnnyMrNinja, Cyfal,

Shooke, Wonchop, Escape Orbit, Asher196, Wikipedian Marlith, Martarius, Elassint, Phyte ClueBot, Strongsauce, Suiseiseki23, The Thing That Should Not Be, Dave fsm, SuperHamster, CounterVandalismBot, Blanchardb, Harland1, Mmark089, Wwefan981, Jusdafax, Jpeg 15, Resoru, Thales of Miletus, NG 4 90, Crazzy yetti, Shadow959. Dkpwnerer, Jamezbond275, Xblaznone, Solsticedhiver, Cbullen, Cheststory, Chunky-monkey22, Snake7man, Charlieclegg1995, Canuker, Robindarjeeling, Gigglelaugh225, SAT instructor, Amsexy, Winkidude7001, Zombie007, Rafff18, Homie joe, Trmooney, Vivio Testarossa, Sun Creator, The Dork Knight, Xxxdogscoolxx, Okiefromokla, Holothurion, Dexisugi, Trud the Borg, La Pianista, Aitias, Versus22, Tezero, SoxBot III, Notaknowitall, Apparition11, Recharge330, Eik Corell, Dulcem, XLinkBot, Jovian-eye, Rror, Ost316, Dogbreath123456, Fixedgerald, Wordo, FightingStreet, Colliric, Nintendo1379, RyanCross, Thatguyflint, Addbot, Xp54321, Cheesefajita, Willking1979, Some jerk on the Internet, Guoguo12, Megata Sanshiro, Ironholds, Scientus, Leszek Jańczuk, Gufferdexter, Chemicalblink1445, Davezap, Cst17, AndersBot, Favonian, LinkFA-Bot, DigDug231, RanmaSuper, Tassedethe, Wattson-jr, Surfthetsu, Tide rolls, MuZemike, Slyguy1025, Luckas-bot, Yobot, GamerPro64, Jewido, Kg4wrm, Green Ambush, AnomieBOT, Kingpin13, Flewis, Basilisk4u, Kwanzi, Snorlax Monster, LilHelpa, Belasted, Es138, Abram151, Brownsam, Ptrf, Anonymous from the 21st century, Doctorx0079, Jongo211, Michael10100, Sergecross73, JulianDelphiki, Dmnmnk1234, Shadowjams, Geierkrächz, Sesu Prime, Saudijp, Benbenben987654321, Kiccorydude, Citation bot 1, Javert, Schmeater, TheRustyBanana, Smuckola, Calmer Waters, A8UDI, Toonmore, Nikedude18, Trioculus1, Salvidrim!, Talk of the Toast, ArtistScientist, Tsepelcory, Mariacar Cervantes, Theteddyt, Aslanio, Deerwood2009, SinisterTonic, Dudefly. Crazy333, Jhenderson777, Lorson, Canyq, Tbhotch, Maryamirana, Jamilah10, NeoGenPT, ITzMe73, Skamecrazy123, EmausBot, Acather96 Eekerz, Ajraddatz, Niwi3, RenamedUser01302013, Ischa1, Wikipelli, Bleakgadfly, Sirius 128, Chya47, John Cline, Liquidmetalrob, Francism2000, Qwerty&scrabble, Nintendocan, Lolololololollolo, Shigoroku, Caspertheghost, Dallin800, Wayne Slam, Expertbombermanloged, White x Tee, Coconut bro, Smartie2thaMaxXx, Zelshafeoo, AndyTheGrump, Evan-Amos, Targaryen. Thejfh1999, ClueBot NG, Kingc95, MrJoshbumstead, Darknessthecurse, Pillow2011, P10922572, Austin98981, AtlasBurden, Mtking, Popedcap, Helpful Pixie Bot, Dinner4theking, Swiftwinter, BG19bot, Zzyxzaa26, HtotheJtotheOPA, Silverhedgehog0, TheLoverofLove, Bloodbandit, PTJoshua, Shakeandbake15, Monkeybuts11, Legiteditordontban, WeatherBoyWheeler, DawzDaBozz, Legend1000, Arkhandar, C3F2k, Dannybean614, Uglymew, Patdarat78, Pasq243, SlyStallone123, Minsbot, BattyBot, Jonathan Tennyson. Iyiyiyiy, Khazar2, Lildude55, Lildude56, Dexbot, Mogism, BDE1982, 93, Epicgenius, G33k Enterprises, Magicperson6969, EvergreenFir, HamishMorgan, Richolmes14, Pressstarttoplay, BartSmith85, Lildude69, Monkbot, Bria nix, Aftertastebutthole, DSCrowned, ClassicOnAStick, Qhoeger, Zacharyalejandro, PenguinoManGamez, Kethrus, TheTMOBGaming2, Colton V-burg, Markman999, Pittsburghpaige and Anonymous: 714

- **Dragonfire (video game)** *Source:* https://en.wikipedia.org/wiki/Dragonfire_(video_game)?oldid=681314394 *Contributors:* Ylee, Dgpop, CyberSkull, D.brodale, TnS, Bgwhite, Pagrashtak, N. Harmonik, GVnayR, Mika1h, Drinibot, Mathsgeek, BetacommandBot, RobJ1931, Duffaboy, Thibbs, Mr WR, SolotaireDeaton, Monkeys rule 99, Flyer22 Reborn, Trivialist, Someone another, JasonAQuest, DumZiBoT, Addbot, MuZemike, Luckas-bot, Deltasim, Figmentcry, Lorson, ClueBot NG, Shaddim, LoneWolf1992, Dsprc and Anonymous: 6

- **Frantic (video game)** *Source:* https://en.wikipedia.org/wiki/Frantic_(video_game)?oldid=677691823 *Contributors:* D.brodale, Bgwhite, Dweller, X201, Davewho2, Duffaboy, Slysplace, Triwbe, Cyfal, Gene93k, Dawynn, Lightbot and Anonymous: 1

- **Frogger** *Source:* https://en.wikipedia.org/wiki/Frogger?oldid=686516392 *Contributors:* Bryan Derksen, William Avery, Frecklefoot, Infrogmation, DopefishJustin, Liftarn, Error, Zoicon5, Furrykef, K1Bond007, Morn, Branddobbe, Boffy b, Noplasma, Mirv, Academic Challenger, Meelar, LGagnon, Paul Richter, Gtrmp, Brian Kendig. Jrquinlisk, Average Earthman, ChicXulub, CryptoDerk, Abu badali, Mike Storm, Bumm13, Kevin B12, Spottedowl, Tyler McHenry, Luigi, Ehudshapira, Asbestos, Neutrality, Tjansen, Discospinster, Rhobite, Kevin Jones, Sc147, Ylee, Dgpop, Bobo192, DaveGorman, Tyan23, DCEdwards1966, Payal, Jason One, Frodet, Alansohn, Gargaj, Hackwrench, Jtalledo, Fritzpoll, Greenghoulie, Wtmitchell, ProhibitOnions, Tsuba~enwiki, New Age Retro Hippie, Dismas, Feezo, Marasmusine, Mindmatrix, Robnjr, ThomasHarte, Combination, Rad Racer, MrLeo, Rjwilmsi, Koavf, Wikibofh, The wub. Nandesuka, Yamamoto Ichiro, FlaBot, Mirror Vax, Master Thief Garrett, RAMChYLD, Brianreading, Nivix, SuperDude115, Gurch, D.brodale, Liontamer, Psantora, DVdm, Kummi, LittleSmall~enwiki, Rob T Firefly, Jimp, Aaron Walden, Cpc464, Gaius Cornelius, Wgungfu, Rsrikanth05, RadioKirk, Knyght27, NawlinWiki, DragonHawk, Larsinio, PhilipO, Rmky87, Mooveeguy, Palpalpalpal, Rwalker, N. Harmonik, FF2010, Mike Selinker, Clayhalliwell, Reyk Neilka, Americanidiot, DaProx, DVD R W, Krótki, SmackBot, Wildqat, Reedy, Dr. B, Jagged 85, Gilliam, Ohnoitsjamie, Doktor Wilhelm, Cscarthecat, Hillsc, Plotor, GoneAwayNowAndRetired, Mdwh, Kevin Ryde, Apple2gs, SquarePeg, Emurphy42, KieferSkunk, Can't sleep, clown will eat me, Big Cowboy Kev, Rsm99833, Kcordina, GVnayR, Ianmacm, Nakon, Sbluen, Crv1, BryanEkers, Wturrell, KLLvr283, Klytos, Khazar, MilborneOne, Rpachico, 16@r, Andypandy.UK, Special-T, TJ Spyke, Walter Day, Iridescent, Asmpgmr, JoeBot, FancyPants, Dlohcierekm, CmdrObot, Wafulz, Unionhawk, Mika1h, Drinibot, Neave, ShelfSkewed, Equendil, Cydebot, Farine, Adolphus79, Tectar, Dancter, UberMan5000, Nyk78raider, Quetzalcoatl45, After Midnight, Mtijn, BetacommandBot, JAF1970, Epbr123, Darlenemm, Marek69, X201, RFerreira, Pcapitola, Silver Edge, Scottandrewhutchins, Jay Firestorm, Cbiwankenobi, MirrorKnight, Aruffo, Darklilac, MECU, Student Driver, Kariteh. Andrzejbanas, Ozba, Thefinerminer, MER-C, Stardotboy, Acroterion, Magioladitis, Bongwarrior, VoABot II, Antmusic, Pan Dan, FredMSloriker, Cocytus, Gwern, ClubOranje, MartinBot, Theclaw1, Miraculousrandomness, Rettetast, J.delanoy, DandyDan2007, Lg16spears, USN 977, SemDem, Captain Infinity, McSly, Mdumas43073, Retro junkie, Mrceleb2007, Cue the Strings, C. Foultz, SixteenBitJorge, Bryko614, Mr Wesker, X!, ABF, Jeff G., WOSlinker, Philip Trueman, Tavix, SeanMooney, BlueLint, Jibbajaba, Martin451, Sb2007, Phantomdj, Bitbut, VeolenBot, HansHermans, IndulgentReader, Austriacus, Sunookitsune, BlueDragonLegends, Luigihann, Ardicius Greenknight, Realdeals228, Taco pie2, Yintan, Merotoker1, Radon210, Samquinn, Jc47906, Oxymoron83, Aspects, Ryan Holloway, AutocracyBot, JohnnyMrNinja, Denisarona, Troy 07, ImageRemovalBot, Sfan00 IMG, ClueBot, Corkwatchr, Executioner0, Guthrierocksa, Trivialist, Excirial, Thecovertpanda, Bob101101, Kazoozoo, Holothurion, Mixerof98800, Froggerr, Apparition11, Classicrockfan42, DumZiBoT, Rizz768, Spitfire, JosephGoh, Blackwatch21, Ost316, Sergay, Cmungall, Khanttzarate, Stunteltje, Addbot, Cst17, Download, LaaknorBot, Cameronbell76, Simeon24601, AndersBot, Angelczek, R3ap3R, Tide rolls, Lightbot, Luckas-bot, Yobot, Eric-Wester, A More Perfect Onion, Jim1138, Ismashed, Bluerasberry, Zum Zamim, Materialscientist, Arctix, Zachhassinger, ArthurBot, LilHelpa, Jeremysteele, Averis1011, Ekwos, RedKiteUK, Usc213abc, Sergecross73, Defil3d, Locobot, Shadowjams, Spongefrog, FrescoBot, Mit771, Oldlaptop321, Recognizance, Emags41, Agbwiki, BenzolBot, Piscowie, Pegasos2, Hamtechperson, Wikitanvir, Secret Saturdays, Banej, Caryq, Hajatvrc, Slon02, Chuck369, EmausBot, John of Reading, Eekerz, Flavio.mprado, Mlkrogman, MikeyMouse10, ZéroBot, Fæ, Nater666, Akerans, H3llBot, AndrewN, Nybcy2, L Kensington, CrestwoodRocks, BornonJune8, ClueBot NG, MrWii000, Satellizer, Easy4me, WPSamson, The guy who edits stuff, Neptune's Trident, Gandolf the black, Nednerb1999, BinaryLust, Gabriel Yuji, Atomician, DirkDigglerz, Snow Blizzard, NaBoKill, LoneWolf1992, The1337gamer, H8kct, BattyBot, LinuxPower, Cyberbot II, Mogism, Joey108, AntonTchekhov, Mar6638, Alex schullo, Everymorning, Taikoguide, Planeflyer121, Mr. Lama, Zenibus, J158n, Manul, Noyster, Konveyor Belt, JaconaFrere, VinylTavi, Monkbot, Landingdude13, DangerousJXD, Ninendo, Yomaamaa, Maplestrip, LMParadis, Ashcastle, Anthony Sull0ven, 123fudgesticks, 忍者くん and Anonymous: 358

- **Galaxian** *Source:* https://en.wikipedia.org/wiki/Galaxian?oldid=685887712 *Contributors:* Frecklefoot, Liftarn, Eurleif, Pjedicke, FriedMilk, Wmahan, Zeimusu, DNewhall, Bumm13, Spottedowl, Hellisp, Tjansen, Canterbury Tail, The stuart, Monkeyman, Paul August, Indrian, Ylee, Cmdrjameson, DaveGorman, Frodet, Interiot, CyberSkull, WTGDMan1986, Ashley Pomeroy, ProhibitOnions, Jef-Infojef, Hbdragon88, Combination, Cuchullain, Rjwilmsi, Doppelgangland, ABot, Nandesuka, FlaBot, Mirror Vax, Master Thief Garrett, Felixdakat, D.brodale, Windharp, Hibana, Yoshi348, Crotalus horridus, Michael Slone, Wgungfu, TakingUpSpace, Larsinio, Zwobot, Pelladon, N. Harmonik, LepricahnsGold, Krótki, Crystallina, Flipkin, SmackBot, DuoDeathscyther 02, McGeddon, Pschelden, Jagged 85, Oscarthecat, Chris the speller, KieferSkunk, Gingerfield rocks, GVnayR, ShadowUltra, Ohconfucius, Khazar, Jacquismo, Legal Tender, TJ Spyke, Walter Day, Asmpgmr, Robot Chicken, Waffle88, Mika1h, Drinibot, Madsci, Ken Gallager, Cydebot, Farine, PembrokeWKorgi, BetacommandBot, Thijs!bot, JAF1970, Stonic, X201, Jhsounds, DOSGuy, Pejorative.majeure, Celithemis, Ecksemmess, Thibbs, Pennyman, Brittany Ka, Macmelvino, Blueerica, The Glory Boy, Skier Dude, Namcorules, SixteenBitJorge, Varnent, Jeff G., Bolt, WOSlinker, PNG crusade bot, Martin451, Superjustinbros., Mr. DigDug, Truthanado, SieBot, Namcorules's Temporary Account, Neogura, Anyeverybody, DeviantMan, Asher196, Martarius, ClueBot, Badger Drink, Meisterkoch, LizardJr8, Froman1, Grebenkov, Alexbot, ChrisHamburg, Thingg, GloryQuestor, Londonclanger, XLinkBot, Ost316, Addbot, Jhuhn, Tide rolls, Lightbot, Heebeegeebee, Yobot, Tyrathect, AnomieBOT, Jordiferrer, Doctorx0079, Chaheel Riens, FrescoBot, Black Squirrel 2, Lorson, Canyq, Reach Out to the Truth, Eekerz, Mk5384, ZéroBot, Stoo bloke, Rescuehero942, EdoBot, ClueBot NG, Despatche, Helpful Pixie Bot, Cameron1999, Vanished user ivweij23ijwefk4, Suzukaze-c, BattyBot, Sumguy1994, Hmainsbot1, Magicperson6969, DallTX314, Mondo Films, Inc., Monkbot, Markman999 and Anonymous: 121

- **Ghost Manor** *Source:* https://en.wikipedia.org/wiki/Ghost_Manor?oldid=672989982 *Contributors:* Cyde, Rjwilmsi, Nihiltres, Czar, D.brodale, Hibana, TnS, N. Harmonik, Jagged 85, GVnayR, Drinibot, Alaibot, TonyTheTiger, Stoshmaster, Onyxraven1979, T@nn, Ding Chavez, Shadow460, Jordoniscool, Caltas, Android Mouse Bot, Plastikspork, Bruplex, DeltonaCharger, Lightbot, AnomieBOT, Pinched, Rickraptor707, 23W, BattyBot, Hmainsbot1, Helsabot and Anonymous: 14

- **Gorf** *Source:* https://en.wikipedia.org/wiki/Gorf?oldid=683074654 *Contributors:* Damian Yerrick, Frecklefoot, Ken Arromdee, Ixfd64, Scott, Furrykef, FriedMilk, Gracefool, Sonjaaa, Spottedowl, Biot, Rcv, Fade~enwiki, Indrian, Ylee, DreamGuy, ProhibitOnions, Kouban, Hbdragon88, KevinOKeeffe, Combination, Angusmclellan, JHMM13, Nandesuka, FlaBot, Mirror Vax, Xmoogle, D.brodale, Gaius Cornelius, Wgungfu, Lavenderbunny, Pagrashtak, Larsinio, Moe Epsilon, Gmaletic, Pelladon, N. Harmonik, Josh3580, Red Jay, Nothings, JDspeeder1, Smack-Bot, DuoDeathscyther 02, Frymaster, Plotor, Thumperward, Emurphy42, KieferSkunk, Zachdms, Walter Day, Asmpgmr, Clarityfiend, Haus, Mika1h, Drinibot, ShelfSkewed, Equendil, Cydebot, PV250X, BetacommandBot, Qwyrxian, Arthur Ellis, Credema, Kdokeeffe, Jimrselleck, Bendavidson1950, Thibbs, Brittany Ka, Gwern, Macmelvino, CommonsDelinker, Retro junkie, WOSlinker, Pengwiin, DeviantMan, Year 2144, ClueBot, Fyyer, Badmachine, XLinkBot, Addbot, LaaknorBot, Lightbot, OlEnglish, Pextris, Bt8257, KLBot2, Krswans, The1337gamer, Samwalton9, BattyBot, ChrisGualtieri, BDE1982 and Anonymous: 71

- **Hareraiser** *Source:* https://en.wikipedia.org/wiki/Hareraiser?oldid=677855904 *Contributors:* Robbot, Alansohn, Rjwilmsi, Bgwhite, N. Harmonik, SmackBot, McGeddon, Raphie, BetacommandBot, Cain Mosni, X201, Magioladitis, Duffaboy, Retro junkie, Sfan00 IMG, WikHead, Addbot, Ettrig, Yobot, Materialscientist, Deltasim, Black Squirrel 2, ChrisGualtieri, Monkbot and Anonymous: 6

- **Hunchback (video game)** *Source:* https://en.wikipedia.org/wiki/Hunchback_(video_game)?oldid=671798158 *Contributors:* Woohookitty, Ck lostsword, Cydebot, BetacommandBot, Thijs!bot, X201, QuiteUnusual, Jllm06, Boston, Retro junkie, Austriacus, Moonriddengirl, El Pantera, JohnnyMrNinja, ClueBot, Malcjennings, Addbot, Gangcai, Lightbot, Patreides, Philsan, Neurolysis, Chaheel Riens, Ididit4thelulz, Tim1357, Shaddim and Anonymous: 11

- **Jetpac** *Source:* https://en.wikipedia.org/wiki/Jetpac?oldid=681624500 *Contributors:* Mrwojo, Liftarn, PiaCarrot, Honzik, Jh51681, Slady, Pak21, DcoetzeeBot~enwiki, Evice, Richard W.M. Jones, Thunderbrand, BFunk, Frodet, Anthony Appleyard, Plumbago, InShaneee, The-DotGamer, Sir Slush, Marasmusine, Mestesso, Exxolon, ThomasHarte, Remurmur, StuartBrady, RobyWayne, Czar, Wgungfu, Insouciance, N. Harmonik, Ledow, GrinBot~enwiki, SmackBot, Jagged 85, Oscarthecat, Nintendude, Sculpher, Zagrebo, GVnayR, Ritchie333, Valenciano, Ravimakkar, Hope(N Forever), Jetman, Mika1h, Drinibot, SHOlafsson, Dogman15, Cydebot, Danrok, BetacommandBot, SNS, X201, Salavat, RobJ1981, PresN, Xeno, DietLimeCola, Alsee, The Glory Boy, WOSlinker, Lots42, SeanMooney, WinTakeAll, Miremare, ImageRemovalBot, Martarius, Hippo99, Niceguyedc, Alexbot, Holothurion, Dynesclan, Sikor soft~enwiki, Basilicofresco, Avengah, Legobot, Luckas-bot, Subzerosmokerain, AnomieBOT, Chaheel Riens, Jaguar, Full-date unlinking bot, Lorson, NeoGenPT, Eekerz, Niwi3, H3llBot, SporkBot, Shaddim, Frietjes, Neji56565onyoutube, DrRockso87, SolarStarSpire, Serpinium, Game4brains, Monkbot and Anonymous: 25

- **Jungle Hunt** *Source:* https://en.wikipedia.org/wiki/Jungle_Hunt?oldid=675131756 *Contributors:* Maury Markowitz, Frecklefoot, Haukurth, Bumm13, Luvcraft, Dgpop, Longhair, SYS21604, Tyan23, Jason One, CyberSkull, Cburnett, RainbowOfLight, Electrode, Jeff3000, ThomasHarte, Combination, Feydey, FlaBot, Mirror Vax, D.brodale, Melodia, Wgungfu, Larsinio, Pelladon, N. Harmonik, Lando242, Smack-Bot, Jagged 85, ZS, Poeman~enwiki, Plotor, Slo-mo, Onorem, MeekSaffron, Zagrebo, Victor Lopes, Jetman, Drinibot, Equendil, Cydebot, BetacommandBot, X201, Aruffo, Fayenatic london, Jfarajr, Dream Focus, Thibbs, USN1977, WOSlinker, Melsaran, JohnnyMrNinja, ImageRemovalBot, Trivialist, Addbot, Eauhomme, Luckas-bot, Locobot, Martin IIIa, ChoHyeri, NickyChooChoo, 忍者くん and Anonymous: 28

- **Log Run** *Source:* https://en.wikipedia.org/wiki/Log_Run?oldid=687425175 *Contributors:* Dgpop, Salavat, Dawynn, MuZemike, Markbrennand, Masssly and Anonymous: 1

- **Lunar Leepers** *Source:* https://en.wikipedia.org/wiki/Lunar_Leepers?oldid=678893777 *Contributors:* Psychonaut, Ylee, Bgwhite, RussBot, Addbot, LaaknorBot, Deltasim and The1337gamer

- **Mole Attack** *Source:* https://en.wikipedia.org/wiki/Mole_Attack?oldid=641893740 *Contributors:* BOZ and WereSpielChequers

- **Mountain King** *Source:* https://en.wikipedia.org/wiki/Mountain_King?oldid=622628522 *Contributors:* Bumm13, Ylee, KevinOKeeffe, Wgungfu, ShelfSkewed, Captpackrat, OberRanks, Martarius, Danielcg, MrKIA11, Urbanchampion, Ray and jub, Ndboy, Darthpineapple401, MoldyFred08, BattyBot and Anonymous: 4

- **Mutant Herd** *Source:* https://en.wikipedia.org/wiki/Mutant_Herd?oldid=679000546 *Contributors:* Klemen Kocjancic, Ylee, Bgwhite, WOSlinker, Gorobay, The1337gamer, BattyBot, AntonTchekhov and Anonymous: 1

- **Mystery Fun House (video game)** *Source:* https://en.wikipedia.org/wiki/Mystery_Fun_House_(video_game)?oldid=679000953 *Contributors:* Ylee, Elipongo, Czolgolz, D.brodale, Bgwhite, SmackBot, BetacommandBot, Duffaboy, Retro junkie, SimonTrew, Rockfang, Lightbot, Frostie Jack, Vajayjaythejetplane, Citation bot, Samwalton9, BattyBot, Greefan443 and Anonymous: 5

- **Omega Race** *Source:* https://en.wikipedia.org/wiki/Omega_Race?oldid=679063611 *Contributors:* Dimadick, Ylee, CyberSkull, Marasmusine, Firsfron, GregorB, D.brodale, Bgwhite, Epolk, N. Harmonik, Closedmouth, SmackBot, Plotor, Bluebot, Evenly, GVnayR, Radagast83, Ryan Roos, J 1982, CmdrObot, Drinibot, JettaMann, Cydebot, Farine, JAF1970, ColourBurst, Bunnytron, Thibbs, Macmelvino, Meow07, VolkovBot, Phantomdj, Trivialist, XLinkBot, Anticipation of a New Lover's Arrival, The, Addbot, Cavarrone, Mlpearc, FrescoBot, Helpful Pixie Bot, The1337gamer, Samwalton9, BattyBot, Hmainsbot1 and Anonymous: 15

- **Pac-Man** *Source:* https://en.wikipedia.org/wiki/Pac-Man?oldid=686773988 *Contributors:* Damian Yerrick, AxelBoldt, Mav, Bryan Derksen, Robert Merkel, Koyaanis Qatsi, Rmhermen, PierreAbbat, Bkellihan, Montrealais, Modemac, Atlan, Ericd, Jim McKeeth, Frecklefoot, Ken Arromdee, Dante Alighieri, Dominus, Nixdorf, Liftarn, Ixfd64, Goatasaur, Tregoweth, Card~enwiki, Mdebets, Ahoerstemeier, ZoeB, NicoNet, Ijon, DropDeadGorgias, Xamian, Julesd, Lupinoid, Vzbs34, Scott, Big iron, Sethmahoney, Schneelocke, Dyss, Jengod, Timwi, Andrevan, WhisperToMe, Zoicon5, Foodman, Furrykef, Jgrn, Taxman, K1Bond007, Djungelurban, Lensi, Morn, Topbanana, BenRG, Rossumcapek, Robbot, Mazin07, Pigsonthewing, Fredrik, Boffy b, Jredmond, Nyh, Iowahwyman, Meelar, Andrew Levine, Michael Snow, Mushroom, Xanzzibar, Dina, Adam78, Exploding Boy, Paul Richter, Ian Maxwell, Gtrmp, Elf, Nifboy, MMBKG, TOttenville8, Brian Kendig, Dissident, Marcika, Ich, Average Earthman, Bkonrad, Gus Polly, Angry candy, Revth, Joe Sewell, Finn-Zoltan, Gracefool, Siroxo, Eequor, Luigi30, Richard cocks, Romero-wiki, Pne, Golbez, Ian Pugh, Thealexfish, ChicXulub, Oklonia, Mackeriv, Utcursch, Mike R, Antandrus, DaveJB, NameofFL Kesac, Jesster79, Secfan, Zerbey, Latitude0116, Bumm13, Kevin B12, Bodnotbod, Spottedowl, Figure, Asbestos, Anirvan, SamSim, Neutrality, Slidewinder, Ukexpat, Jh51681, Walabio, Tjansen, The stuart, Piotras, Corti, Videogamer, Mike Rosoft, Freakofnurture, Monkeyman, Brevity, Gerardvschip, PKFC, Discospinster, Rich Farmbrough, Guanabot, Themusicking, Pak21, D-Notice, Sbo, Night Gyr, Indrian, Prak, Bende-235, ESkog, Slungfish, Violetriga, Fenice, Sgeo, Evice, DrakeCaiman, Ylee, CanisRufus, Chungy, Robert P. O'Shea, Pinzo, Zenohockey, Hayabusa future, Ruyn, LordRM, Matteh, AshSert, Dgpop, TMC1982, Deathawk, Longhair, Smalljim, Viriditas, Cmdrjameson, Clarkbhm, Kevin Myers, Pikawil, DaveGorman, IDX, KBi, Tyan23, Rajah, WikiLeon, DCEdwards1966, Eje211, Jason One, Knucmo2, Frodet, Wendell, Baka toroi, Alansohn, Gary, Gargaj, Jamyskis, Interiot, Arthena, CyberSkull, Diego Moya, WTGDMan1986, Jlascar, Jtalledo, Guga.emc, Goldom, Water Bottle, Redfarmer, Flata, InShaneee, Tuneguru, Fatalerrorrage, Mysdaao, Canek, Bart133, DreamGuy, Wtmitchell, SidP, Here, ProhibitOnions, Rick Sidwell, ReyBrujo, Mattsday, Suruena, Aaarrrggh, Drat, LFaraone, Ianblair23, Gmelfi, Eric Herboso, Kouban, TheCoffee, New Age Retro Hippie, Dismas, Y0u, Tariqabjotu, Mananga, A D Monroe III, Dopefish, Daveydweeb, Soultaco, Thryduulf, Kelly Martin, Simetrical, Firsfron, Starblind, Woohookitty, JarlaxleArtemis, ScottDavis, LOL, Thorpe, Aveilleux, Poiuyt Man, BillC, Gordeonbleu, Urbster1, MattGiuca, Commander Keane, JeremyA, Tckma, MrDarcy, Hbdragon88, Ppk01, Astanhope, Midnightblaze, GregorB, Scm83x, Waldir, Zzyzx11, Combination, Radiant!, Dysepsion, Slgrandson, KyuuA4, Taestell, Patrick2480, FreplySpang, Raymond Hill, Plau, RxS, CoderGnome, Icey, Spot Color Process, Search4Lancer, Sjö, Coneslayer, Sjakkalle, Rjwilmsi, Nightscream, Koavf, George Burgess, Panoptical, Ian Lancaster, Tearsinrain, Quiddity, Tangotango, Jb-adder, NeonMerlin, LjL, Darksasami, Lairor, The wub, MarnetteD, Remurmur, FuriousFreddy, Flatypus222, Yamamoto Ichiro, A Man In Black, Bash, StuartBrady, FlaBot, Mirror Vax, SchuminWeb, TokyoJunkie, CR85747, Stoph, PinkDeoxys, Master Thief Garrett, El Cid, Nihiltres, Crazycomputers, JiFish, Brianreading, SuperDude115, RexNL, Gurch, Mitsukai, Briguy52748, Jordan Elder, Stormwatch, D.brodale, Hibana, Manufacture, Psantora, Chobot, Metropolitan90, DVdm, Bgwhite, Digitalme, Eric B, Mr.Do!, Yoshi348, George Leung, The Rambling Man, Barrettmagic, Subwayguy, YurikBot, Whcisjohngalt, Atoon, RobotE, Tommyt, Crotalus horridus, Butsuri, Jcam, Rob T Firefly, Rtkat3, Kafziel, RussBot, Herzog, Icarus3, Severa, KamuiShirou, Foxxygirltamara, Kerowren, Chensiyuan, Stephenb, Gaius Cornelius, Wgungfu, Kyorosuke, Bisqwit, Wimt, Randall Brackett, RadioKirk, MarcK, EnakoNosaj, Rhindle The Red, MidnightWolf, EngineerScotty, Shanel, NawlinWiki, Pagrashtak, Bachrach44, Captain Yesterday, Shaun F, Schrei, Apokryltaros, Aaron Brenneman, Sted, CecilWard, Matticus78, Larsinio, PhilipO, Kelvingreen, TheMonkofDestiny, RUL3R, Froth, Misza13, Killdevil, PonyToast, Alex43223, Nate1481, Qwertyuiopasdfghjklzxcvbnm~enwiki, MakeChooChooGoNow, CrazyLegsKC, Todfox, EmiOfBrie, Rwalker, Wangi, Vlad, DeadEyeArrow, Strolls, Pelladon, Brisvegas, Typer 525, N. Harmonik, Richardcavell, FF2010, Sandstein, 21655, Zzuuzz, Schavira, Closedmouth, Mike Selinker, Alakazam, Willirennen, JQF, Jedidrunkenllama, TOkKa, ZoFreX, Neilka, Toodiesel, KaHOnas, Shawnc, Kocharsk-, Wainstead, Kevin, HereToHelp, TITROTU, Ghetteaux, David Biddulph, Hmartin, Bdve, RG2, DaProx, NeilN, Zeikcied, Jeff Silvers, DVD R W, Knowledgeum, That Guy, From That Show!, Hartmacnut, Robertd, Locke Cole, Veinor, Crystallina, Mustard~enwiki, Flipkin, SmackBot, Jon Rob, Mattarata, EvilCouch, Khfan93, DuoDeathscyther 02, MarjorieCook, Tarret, Slashme, InverseHypercube, KnowledgeOfSelf, Royalguard11, McGeddon, David.Mestel, Fgk, Sraan, Kilo-Lima, KocjoBot~enwiki, Jagged 85, Sillygostly, Dxco, Fitch, John9275, Delldot, Bragador, SkiBumMSP, PJM, Gjs238, Edgar181, HalfShadow, Cazort, Misterdan, Moralis, Yamaguchi 先生, PeterSymonds, Gilliam, Evan1975, Ohnoitsjamie, Folajimi, Oscarthecat, Skizzik, Twinturbo2, Plotor, Endroit, Bluebot, Soupnyc807, Ottawakismet, Persian Poet Gal, Stimpy9337, DT Strain, Jnelson09, Mokwella, Sirex98, Thumperward, Oli Filth, Raymondluxuryacht, Fplay, Logarithmic, Dschroder, Mike1, Mdwh, Hibernian, Kingfiogojr, The Rogue Penguin, Baa, IMFromKathlene, Gracenotes, Nintendude, KieferSkunk, Stewts5, Lenin and McCarthy, ClaudiaM, Tsca.bot, Can't sleep, clown will eat me, Allemannster, OrphanBot, Talmage, Shunpiker, Rrburke, Gladrius, BarryTheUnicorn, Cantthinkofausername, Esr, Gingerfield rocks, Konzack, SanderK, Chcknwnm, Aresef, GVnayR, COMPFUNK2, Jmlk17, Flyguy649, Pool ninja, PrometheusX303, Шизомби, SevereTireDamage, Nakon, Savidan, JonasRH, Valenciano, Oanabay04, MichaelBillington, EVula, Culturechampion, Dream out loud, MartinRe, NES Boy, Msp0, Sljaxon, SaiRyopk, IanC, Govvy, Salamurai, Luigi.a.cruz, FelisLeo, Drunken Pirate, Sarfa, Ohconfucius, Deepred6502, Lambiam, Poobslag, TheSquirrel, Nitt, Salty!, Rklawton, Kuru, Dreslough, Zephraud, Bando26, Sabalon, Gobonobo, Spell4yr, Ponakis, Automatic Jack, Tktktk, Jacquismo, Rizzoandz, Cyberlink420, Elil, IronGargoyle, Jakehelliwell, Wanna Know My Name? Later, Melody Concerto, MarkSutton, Andypandy.UK, Slakr, Werdan7, Stwalkerster, Taipion, Resident Lune, Avs5221, Mr Stephen, Andyroo316, Waggers, Descubes, RememberMe?, NJA, Ryulong, Manifestation, Goldrushcavi, FaWzY, Animedude360, Mgabrys, Eliashc, TJ Spyke, DabMachine, Walter Day, Hetar, OnBeyondZebrax, Iridescent, 293.xx.xxx.xx, Asmpgmr, Applemask, JoeBot, Snornex, Peterinns, P tasso, Cbrown1023, Hynca-Hooley, Cheesechimp, Anger22, Tawkerbot2, Nsfreeman, George100, Defrag, Dans1120, Signinstranger, Zotdragon, J Milburn, JForget, Friendly Neighbour, *Chosen One*, CmdrObot, Pugs Malone, Waffle88, Dycedarg, Cyrus XIII, Sfgiants320, Rawling, Smiloid, Mika1h, TrapStilton, FredWallace18@yahoo.com, Drinibot, NickW557, FlyingToaster, TheTito, GobeBluth, Untilzero, Preacherdoc, Fro0ty, Touth, El3ment09, Dogman15, Cydebot, Farine, The Ultimate Koopa, Steel, Gogo Dodo, Uker, Yunafan, PembrokeWKorgi, Dancter, DumbBOT, Plasticbadge, Inkington, Guyinblack25, Briantw, Omicronpersei8, Vanished User jdksfajlasd, Legotech, FrancoGG, BetacommandBot, JAF1970, Epbr123, Luminifer, Nolamatic, Hacky, Opheicus, Qwyrxian, Bytebear, Boingo the Clown, Kablammo, 0dd1, N5iln, Andyjsmith, Stonie, Superstooge, Miz Hydee, Headbomb, Louis Waweru, Marek69, John254, Kathovo, Electron9, James086, Rattis1, X201, Mafmafmaf, Sgsilver, Aericanwizard, Lowercase, RFerreira, Dekt-Scrub, Grayshi, Gpollock, AbcXyz, I love me more than you, Michas pi, Scottandrewhutchins, Escarbot, INFOBIZ, Xzema, Mentifisto, EdJogg, Sabeen557, AntiVandalBot, Yuanchosaan, Sg-binbin, Majorly, Yonatan, Luna Santin, Clarenceville Trojan, Seaphoto, Seldane, Ricnun, Opelio, CobraWiki, ASDFGHJKL, CatMan, BigNate37, Pwhitwor, Tmopkisn, Scepia, The Phantom Elot, TTN, Zanic, Cleeseinator, JJarvis, Pbhales, Vinni999, RIPLIUKANG, Bkoppal, Mrath, Pokemon X, Canadian-Bacon, Student Driver, Golgofrinchian, Kariteh, BrowndRemastered, JAnDbot, Holi0023, MER-C, 1 Cent In Mind, Tohru Honda13, Weird Bird, Blood Red Sandman, M.Neko, Smiddle, Atomicfiction, Andonic, Xeno, Panfried, PhilKnight, Joecool94, Kerotan, Repku, Zenpacman, Wikipacmania, Mizipzor, Magioladitis, Person132, Celithemis, Bongwarrior, VoABot II, Kilroyc, JNW, CattleGirl, Janadore, Onlydeathawaits, BudMann9, SineWave, CTF83!, TimothyBodinnar, Vectorgen, Ecksemmess, Jatkins, Kdrx, WikkanWitch, Magasin, DXRAW, Catgut, Wxyzdetroit, Cyktsui, MetsBot, Torchiest, Adrian J. Hunter, 28421u2232nfenfcenc, Allstarecho, XMog, Chivista~enwiki, Groudon199, JMyrleFuller, Thibbs, DerHexer, Edward321, Onigame, Esanchez7587, Lacma50, Jackx3, Patstuart, Hans404, FredMSloniker, Dboocock, Oroso, FisherQueen, Hdt83, MartinBot, Scoobyben, Quatreryukami, John Doe or Jane Doe, Arjun01, Poeloq, Axlq, Mariobro eh, Bealzi, Rettetast, InnocuousPseudonym, R'n'B, CommonsDelinker, Dphower, Tulkolahten, Chaz1dave, WET3, Gatebuster64, RockMFR, J.delanoy, Benji13, Bongomatic, Calamity-Ace, Lg16spears, Bogey97, Cyanolinguophile, Klaus Kratchet, K1darkknight, SemDem, Psiegenthaler, Cocoaguy, AvatarMN, Jickyincognito, Acalamari, Koyanagi, Dispenser, McSly, Port56, The Glory Boy, Namcorules, Deacs777, Floaterfluss, Pacmaniawiki, WillipsBrighton, Gothnic, Kraftlos, Rekiwi, Frogacuda, Safari companion, Iloveaustinferrera, Fjbfour, PapaZeppxxx, Cmichael, Juliancolton, Supersonic94, WJBscribe, FuegoFish, MilosIvanovic, C. Foultz, Jamesofur, Casper10, MrDrake, HiEv, Halo2man77, Scott Illini, Pdcook, Dorftrottel, Useight, Varnent, Themasterofwiki, Xiahou, Idioma-bot, Red Polar Bear Ranger, Xnuala, Lights, 28bytes, VolkovBot, Myuberfishbob, Dadad, ReddShadoe, Goodapollo666, ABF, Orphic, Derekbd, Pleasantville, Kelapstick, Orthologist, Dusk Knight, Jameslwoodward, Rutherfordjigsaw, Katydidit, Sjones23, WOSlinker, Philip Trueman, D-Hell-pers, TXiKiBoT, Oshwah, Ocherstone, Zidonuke, Pacman444, Cosmic Latte, MeStevo, Zeppelinrules123, Cheesemeister, When the sparrow flies, Kriak, Grendel 66, GDonato, Ann Stouter, Anonymous Dissident, Scott3439, Shindo9Hikaru, TheRealFaceOfBoe, Martin451, Biscuitman56, Nuper, LeaveSleaves, PFrisbie, Andrewrost3241981, Bob f it, Master Bigode, Guest9999, Theviolenthero, DGS43825, Antsterr, Realtalk321, Pockack, Greswik, Kmhkmh, Scotty12, 123abc321cba, Enigmaman, Ace747, Tsumaru, Superjustinbros., Pac-mankiksblargen, Joseph A. Spadaro, Alexandrebo, Pppc88, Pointgrey, Aesahettr, Falcon8765, Cw6165, Mr. DigDug, Drutt, Cucumberlava, Phantomdj, Shadowb75, Ashum Besher, Fire woman 11, Magiclite, Jake73, Entirelybs, Brianga, HiDrNick, Freeewilly, BOLONEYBOY51, LuigiManiac, Pengwiin, Jesswuzhere, BigJoeRockHead, Julian2004, Austriacus, Kaleb parham, Wraithdart, Boloneyboy is awsome, Joe4440, Arafel6, Manabunagaoka, God142535544, EJF, Trexfan87, TwiztedSS, SieBot, MuzikJunky, Zenlax, Tommynewman, Dander~enwiki, Trent613, Efflux19, Jauerback, Winchelsea, Gerakibot, Dawn Bard, Skidooshake2021, RJaguar3, Triwbe, 3sides, Wageslave, Vanished user 82345ijgeke4tg, Grundle2600, Merotoker1, Jerryobject, Namcorules's Temporary Account, Fishtron, Keilana, Aillema, Quest for Truth, Flyer22 Reborn, Erushford, Radon210, Elliwood82, AlexWaelde, JSpung, Oxymoron83, Chfriend, Faradayplank, Swimboy31, Divadc, Spock2266, KoshVorlon, Lightmouse, Mr. Kite~enwiki, Techman224, Gunmetal Angel, Apsrobillo, Mhambrook, Spitfire19, CJMiller, Xot, StaticGull, Vyadh, Anchor Link Bot, Tesi1700, Mygerardromance, Geoff Plourde, Superbeecat, PlantTrees, Lord Opeth, Wonchop, Jons63, Escape Orbit, Wjmummert, Kanonkas, Whirlwindlover, Asher196, JadeBoco, Explicit, ImageRemovalBot, Faithlessthewonderboy, Martarius, Sfan00 IMG, Elassint, ClueBot, Kristensson, GorillaWarfare, Artichoker, Lemnar, Foxj, The Thing That Should Not Be, Matdrodes, Pakopako, Rjd0060, Ndenison, Cochonfou, Frmorrison, Boing! said Zebedee, RobertJWalker, CounterVandalismBot, Lampak, Blanchardb, Billyfutile, Piledhigheranddeeper, Trivialist, Tfullwood, Jswd, Dead velvet elvis, Ktr101, Excirial, Jakeownsall, IdeologicalArrest, Jusdafax, Andy pyro, Gamer HK, Matthias 09, Leonard^Bloom, Vivio Testarossa, BeeZspeeZ, Rhododendrites, Iamsoboreeeeed, NuclearWarfare, Cenarium, AndyFielding, Nanakon, Fairseeder, Tnxman307, Apl2007, Dmx13134, Razorflame, Soccerfan583, Jimtsop, JasonAQuest, Xray340, Snoozy01, Anakin4me, Thingg, Project FMF, Versus22, Pzoxicuvybtnrm, MelonBot, Kruusamägi, Andross52, LetsDoThisRight, Seallion, DumZiBoT, Kfedpwnesass, BarretB, Comandercool07, Badmachine, XLinkBot, Yanksfanatic2000, EddieT10, GameLegend, Ost316, Asfh, Little Mountain 5, Avoided, Kotakkasut, SilvonenBot, Mifter, Jasynnash2, Manometer, Stevenh123, Good Olfactory, BrucePodger, Deffydeffydeffy, SimonKSK, The R, ERK, Proofreader77, Jmdbcool, Nolret33, Punkyac, Willking1979, Rcko1968, Ossie7~enwiki, Kiosk lord, Spartan43543, Non-dropframe, Captain-tucker, Next-Genn-Gamer, Joenamathlalala, Racknak, Stentie, Boomur, Ronhjones, TutterMouse, Movingboxes, Cory1225, Fieldday-sunday, Jacksparrow213, CanadianLinuxUser, Fluffernutter, Avengah, Mikeywhitto1, Carbon1237, PPCInformer, Yurayurateikoku, Losier7, Crupi7, TySoltaur, Scottyferguson, Ferroequus, Tpbsp, Spittlespat, Chzz, CUSENZA Mario, LinkFA-Bot, 5 albert square, Bbg070, Mdnavman, Numbo3-bot, Kiril Simeonovski, R013, Oldgregg666, 石, Gail, AdamXtreme18, LuK3, Megaman en m, Yoda3539, Staringupyourskirt, Fishthrowing16, Pacmanfan84, Erg12345, Luckas-bot, Yobot, Tohd8BohaithuGh1, Bull on Steroids, Dandon3.0, StopVandalsNow, KirkCliff2, ZRetro, Eman9405, IW.HG, Typenolies, Eric-Wester, Lloydsayers, Seena is cool, Willd891, AnomieBOT, Gameboy88, A More Perfect Onion, Hairhorn, Qbertmofo, Message From Xenu, Jim1138, ISquishy, Solidsandie, Brerose, Ismashed, Materialscientist, RobertEves92, ImperatorExercitus, Citation bot, Pacmanlover, Felyza, Mechamind90, Jock Boy, An upside down panda, Weirdo66, Hudgeba778, ArthurBot, Lemmys61, Hrishnu, Abram151, Digicaveman, Capricorn42, 4twenty42o, Lex-gc, Gilo1969, Jsharpminor, King of wikisekhon, Swilbz, NFD9001, Arjun101, Pokemonmaster18, Coretheapple, GrouchoBot, Dunovaniscool, Abce2, Resident Mario, Jasonjax94, Wizardist, Sergecross73, Smurdah, Pizza12456, RibotBOT, PacmanMaster, Tufor, Bellerophon, Hastyo, ZoidbergBender, Eugene-elgato, Dinopopsaur, Chaheel Riens, Elemesh, Mobiusone7, A.amitkumar, Willismongoni, Tktru, Pickbothmanlol, Iamzander777, GT5162, FrescoBot, Hej30251, BACONisKING, Logiebo, Rkr1991, Gourami Watcher, DivineAlpha, Citation bot 1, LetsPlayMBP, Kresnas, XxTimberlakexx, I dream of horses, Jonesey95, 2birds1stone, Smuckola, Σ, LaesaMajestas, Necrojesta, Black Squirrel 2, Ffiti, Graniggo, Yunshui, Sytable, Grapesoda22, Lotje, MrX, Nemesis of Reason, Edjca, McpWarrior, Reaper Eternal, Crazy333, Poop6487654, Unrulyevil, Imagepixel, Livestrongest, Lerner888, Lorson, Rarigi, Sirkablaam, Tbhotch, TheGreenMartian, Mkonjibhuvgy, Fivefingerdeathpunch, BenmcD, Davidjess, TheDon215, Haydecha, Onel5969, Cheerleader2, RjwilmsiBot, Noommos, Joseph315, Hrvatskacanadian, DASHBot, Steve03Mills, EmausBot, John of Reading, Orphan Wiki, Andrewz1128, Erjwiki, Eekerz, SHABANIZ, Dewritech, ClapBoy380, G&CP, TuneyLoon, Somebody500, Tommy2010, Wikipelli, K6ka, Finley123, Jonpatterns, H3llBot, Antimatter31, Belrien12, Jackoo75, MaGa, Donner60, Wikiposter0123, HalfElfDragon, Damirgraffiti, Rickraptor707, Carmichael, Ego White Tray, ChuispastonBot, Xyzzyavatar, Petrb, Xanchester, Helpsome, Janburse, Mykmason 001, ClueBot NG, Gareth Griffith-Jones, MrJoshbumstead, MelbourneStar, Satellizer, Shaddim, Name Omitted, Rmwwalker, Link475, Kotleytki, Despatche, Vulpesinculta51, Dragonfirehero, MTG1989, Mclarenf1123, Jason381141, WPSamson, O.Koslowski, REBELSROCK99, Logan The Master, Marechal Ney, Boontje, Auchansa, Widr, Antiqueight, Judsonva, V Debs, Oransky, Supernova888, Helpful Pixie Bot, SoftballPlayer17, Lowercase sigmabot, BG19bot, Vhvbe, Neo Red, Cameron1999, Eric567, Zimmygirl7, MusikAnimal, Dan653, Mark Arsten, Slovak.s, Altaïr, Supernerd11, Floating Boat, SuperSherbet, Ricordisamoa, Snow Blizzard, FatWhite&Nerdi2000, Maxdata2011, Rhinomantis88, Emouldin, Ndtronerud, Anbu121, WebTV3, BattyBot, Oneseventhree, Ajaxfiore, Vanished user lt94ma34le12, KATANAGOD, Cyberbot II, GoShow, Bkat004, Buubarooky, SNAAAAKE!!, Dobie80, Pansybee, Rezonansowy, Dissident93, Hmainsbot1, Agn106, SoledadKabocha, Webclient101, Joey108, Awesomegod112, Salanarara, SuperGamerWiki, Trollollol98, Lugia2453, Leorrra, AldezD, Frosty, BobCortex, ConnollyMiddleSchoolPrincipal, Kekesg101200, Timero Zero, ABdielH, Jokonino, Faizan, Epicgenius, Sbaker100, Bosna Sarajevo, Acetotyce, Medgeorgia, Magicperson6969, NigelBlount, Little Runs With Scissors, Limefrost Spiral, Tentinator, B14709, ArmbrustBot, Kevinfrombk, Eric Corbett, Goodboy456, Ginsuloft, Sandomenico69, PugPal, AJ tobin, StanleyGrohl, Uhbooh, Sam Sailor, Dunkinator, Francois-Pier, Tjenningsadelphi, ScotXW, XXX8906, Booby3127, KyleTVGamer, Epic Failure, Skr15081997, Mal1236, NOTschmied, Comolostone, ThatRusskiiGuy, Monkbot, Soloism, Twerkmachine, Joeleoj123, Elario1242, TheDerpyChip, Cakeinator2000, Bobdiddlybobbagins, Sheldor of Azeroth, MajesticalMan, Balls311, Spiralkamic, Birdman456, Zucat, Pkupku, AliciaHC12, Teddyfazel1, Bearbacchus, Nagifawzy, RyanTQuinn, MarkToader 965, Tcc 613, Iwilsonp, Darryl1970, Maplestrip, Linkslayer1, JimmyWagger, TaqPol, Ducminhdlna, Mediavalia, ToonLucas22, HACKERMAN123, Spiderman2508, Sss04, Pawn X., Perfretstorm, Clyde82, CountryBoy99, Blamheadshot11, Joseesurboss, Entro3.14, Narria43, Coller226, Goatford, EtherealGate, SuperPacMan, Aaronisafish, Hurdur2.912, Endercreeper615, Vijayvaradan, Anon the User and Anonymous: 1624

- **Pirate Adventure** *Source:* https://en.wikipedia.org/wiki/Pirate_Adventure?oldid=679066057 *Contributors:* Alan De Smet, Jonabbey, Ylee, 23skidoo, Frodet, Uucp, D.brodale, Bgwhite, Chris the speller, Bluebot, BOZ, Hu12, MrEoo, Drinibot, BetacommandBot, X201, Salavat, Jay Firestorm, Manderiko, Upholder, Diggerne, Duffaboy, Mr WR, Charliechuck, Retro junkie, WOSlinker, TravelingCat, Wjl2, Uncle Milty, Cityofc, Beast fat, Deltasim, RjwilmsiBot, Andrewdad and Anonymous: 8

- **Pyramid of Doom** *Source:* https://en.wikipedia.org/wiki/Pyramid_of_Doom?oldid=679069231 *Contributors:* Maury Markowitz, Jonabbey, Czolgolz, Bgwhite, BetacommandBot, GermanX, Retro junkie, Philip Trueman, MrKIA11, Mjs1991 and Anonymous: 2

- **Q*bert** *Source:* https://en.wikipedia.org/wiki/Q*bert?oldid=687627559 *Contributors:* Danny, Modemac, Frecklefoot, Lir, Infrogmation, DopefishJustin, Liftarn, Tregoweth, Docu, Schneelocke, Furrykef, Topbanana, Boffy b, Ludraman, Gtrmp, NessSnorlax, Gus Polly, Jdavidb, Gilgamesh~enwiki, Wiki Wikardo, ChicXulub, Bumm13, Spottedowl, Allefant, Tjansen, CALR, Ntm, Discospinster, Will2k, WikiFediaAid, Ylee, Kwamikagami, Aude, Fiveless, Slowco, Dgpop, MoonBeam, DaveGorman, Tyan23, John Fader, A2Kafir, Jasor One, Frodet, Gargaj, CyberSkull, Ashley Pomeroy, Echuck215, Dopefish, Lime~enwiki, Marasmusine, Woohookitty, MrDarcy, DESiegel, Combination, Rad Racer, BD2412, FreplySpang, JIP, Aidepikiw0000, Rjwilmsi, Staecker, FlaBot, Xiopher, Mirror Vax, Master Thief Garrett, Brianreading, Czar, D.brodale, Piniricc65, Bgwhite, Eric B, Check two you, EamonnPKeane, YurikBot, Playstationman, Koveras, Rtkat3, Kirill Lokshin, Gaius Cornelius, Wgungfu, Thiseye, Banes, Larsinio, FlyingPenguins, T-rex, Pelladon, N. Harmonik, Wknight94, SMcCandlish, Jogers, Jedidrunkenllama, GraemeL, 2fort5r, Kevin, Yahnatan, SmackBot, Reedy, Pgk, Jagged 85, AnOddName, Loomer~enwiki, Oscarthecat, Plotor, Bluebot, DarthInsinuate, Emurphy42, Dethme0w, Gldnspud, Pissant, GodEmperorOfHell, Cor anglais 16, Cyberlink420, PaulGS, Asmpgmr, Applemask, Zero sharp, Mikelima~enwiki, DevinCook, Mika1h, Drinibot, Dairhenien, Preachercoc, Digression from a tangent, Cydebot, Mackadet fr, Soetermans, Caejis, Guyinblack25, BetacommandBot, JAF1970, TheYmode, Alianman, Marek69, X201, Zerothis, Jbl1975, Salavat, SpongeSebastian, WinBot, Aruffo, RobJ1981, Yoyo63, Deflective, Areaseven, Rpiredhawk17, Giant Tree, Celithemis, Carlwev, Oskay, Ling.Nut, KJRehberg, Whoop whoop, Fabrictramp, ForestAngel, Keviniskool, Torchiest, Tberla, Thibbs, Dell9300, Brittany Ka, STBot, Sagabot, CommonsDelinker, Brain Rodeo, Lg16spears, Jennbrady, Maproom, Insp nf, Casper10, ACSE, Vranak, WOSlinker, JonnyGURU, Enigmaman, Superjustinbros., Wbdavis29, Sesshomaru, King food, Wraithdart, AlphaPyro, Murmur001, Spock2266, Sfan00 IMG, ClueBot, Hipp999, Jappalang, Haroldinrio, Trivialist, ChardonnayNimeque, Masked Mutant, CowboySpartan, Muro Bot, JasonAQuest, DumZiBoT, Mskeel, Bridies, Feinoha, Ost316, Mm40, Kbdankbot, Addbot, GuyCC75, LaaknorBot, Glane23, Lightbot, Zorrobot, Yobot, AVBr, GamerPro64, Ayrton Prost, AVB, Mqa, Anonymous from the 21th century, AnomieBOT, Tavatar, Qbertmofo, Neptune5000, Piano non troppo, Ismashed, Dada817, Materialscientist, Citation bot, Maxis ftw, LilHelpa, Capricorn42, Nasnema, Pecan314159, Jerzeeshadow, Horath, Peanutcactus, Jhalapch, Derboo, Citation bot 1, Akanbe, I dream of horses, Smuckola, Mjs1991, Dulefni, Eekerz, GoingBatty, Hydao, BookDen, DASHEotAV, ClueBot NG, Darkgod14835, MrJoshbumstead, Zaxxon6, Luvljm, Johnwest1999, Helpful Pixie Bot, BG19bot, Sterling.M.Archer, Jqbeetlee, EdwardH, BattyBot, Dobie80, Guiletheme, Joey108, BDE1982, SolidSnake110, StewieBaby05, RebusDuplex999, Jodosma, Lemaroto, Iseesky, Ugog Nizdast, Mcfaddenskyler, MagicatthemovieS, Crowdsandpowerband, Monkbot, Natsume96, BustaBunny, LightShark1, MiniEstadi1982, Duncan_dontknow, Anarchyte, Markman999, Leo513 and Anonymous: 220

- **Radar Rat Race** *Source:* https://en.wikipedia.org/wiki/Radar_Rat_Race?oldid=682335670 *Contributors:* Timvasquez, Quadell, Bumm13, Dgpop, Polluks, Clgonsal, Mirror Vax, D.brodale, Deville, JLaTondre, Jeff Silvers, SmackBot, ML5, CBM, Drinibot, BetacommandBot, Luminifer, X201, Thibbs, DorganBot, WOSlinker, Casablanca2000in, Sfan00 IMG, Rudolphous, Addbot, Erik9bot, FrescoBot, ClueBot NG, Game-Guru999 and Anonymous: 5

- **Rescue at Rigel** *Source:* https://en.wikipedia.org/wiki/Rescue_at_Rigel?oldid=679171583 *Contributors:* Jhbadger, Maury Markowitz, Frecklefoot, Schneelocke, Arcadian, Hackwrench, Elvarg, Radiant!, Mirror Vax, D.brodale, Bgwhite, DanielM, Tubelius, ShakespeareFan00, Mika1h, Drinibot, BetacommandBot, X201, The Fifth Horseman, Mr WR, JohnnyMrNinja, Sfan00 IMG, Lightbot, Xenobot, Yobot, FrescoBot, Deneb in Cygnus and Anonymous: 3

- **River Rescue** *Source:* https://en.wikipedia.org/wiki/River_Rescue?oldid=679172777 *Contributors:* Dgpop, Frodet, D.brodale, Bgwhite, Krótki, WillKemp, Mika1h, X201, Maxim, MrKIA11, Megata Sanshiro, Ubcule, Chaheel Riens, 3lncr, BG19bot and Anonymous: 1

- **Robotron: 2084** *Source:* https://en.wikipedia.org/wiki/Robotron%3A_2084?oldid=680390610 *Contributors:* Frecklefoot, Lezek, Silverfish, Furrykef, Whcernan, Kizor, Brian Kendig, Gus Polly, Mboverload, DNewhall, Bumm13, Spottedowl, Asbestos, Rich Farmbrough, Pak21, Xezbeth, Martpol, Ylee, Dgpop, JHarris, Aquillon, Tyan23, Frodet, Interiot, Jtalledo, Ashley Pomeroy, DreamGuy, ProhibitOnions, TheDotGamer, Alphageeq, Jackel, Tabletop, Evilmoo, ThomasHarte, Combination, Cjelly, Raymond Hill, RadioActive~enwiki, Rjwilmsi, XP1, Lkoziarz, Impunity, Brighterorange, Nandesuka, Guerrilaradio, Mirror Vax, Master Thief Garrett, Czar, Stormwatch, YurikBot, Hyad, Chroniclev, Gaius Cornelius, Wgungfu, Kvn8907, Thiseye, Larsinio, Rrrky87, Pelladon, N. Harmonik, S icing, Clayhalliwell, Romanista, Liyster, WindFish, SmackBot, DuoDeathscyther 02, SamSock, Septegram, Mediaphd, Plotor, Valley2city, Kurt S Koller, DarthInsinuate, Frap, Zagrebo, GVnayR, Nyletak, Nakon, T-borg, Ryan Roos, LanternLight, Ravimakkar, NeedlerFanPudge, Dungeoneer, TJ Spyke, Walter Day, Teerdnet, Applemask, Paolo999~enwiki, SkyWalker, Mika1h, Drinibot, Cydebot, Steel, DryCleanOnly, Dr. Pizza, Daniel J. Leivick, Guyinblack25, BetacommandBot, X201, RobotG, Scanbus, Darkuni, Aruffo, Scepia, PresN, Xeno, Dougieb, Daniel thomas, FredMSloniker, USN1977, SemDem, JEB215, SharkD, The Glory Boy, Floaterfluss, C. Foultz, Thesofbulletin82, Varnent, KaL YoshiKa, VolkovBot, WOSlinker, TXiKiBoT, Bill nghurst, Eskovan, VeblenBot, MuzikJunky, Clwizard, Vanished user j3roijqwkskjf5kr, Curtisobal, BlueAzure, Svick, Superbeecat, DeviantMan, Asher196, ImageRemovalBot, Martarius, Sfan00 IMG, Beeblebrox, Badger Drink, Trivialist, InternetMeme, Blackwatch21, Ost316, Power2084, Addbot, Stentie, Noozgroop, MuZemike, Luckas-bot, AnomieBOT, Quebec99, Tc1138, Jamie's Nightmare, FrescoBot, LucienBOT, Citation bot 1, Martin IIIa, John of Reading, H3llBot, Bill Hicks Jr., Targaryen, MrMajestyk33, Gyp C, Danim, Helpful Pixie Bot, BG19bot, Gabriel Yuji Frze, The1337gamer, RudolfRed, Laodah, Khazar2, Hmainsbot1, FegelCineplex, Bluecap98, Monkbot, Landingdude13, Hawkman1970, 6309er and Anonymous: 89

- **Scramble (video game)** *Source:* https://en.wikipedia.org/wiki/Scramble_(video_game)?oldid=686785192 *Contributors:* Maury Markowitz, Frecklefoot, Stan Shebs, Angela, Hike395, Furrykef, GreatWhiteNortherner, ChicXulub, Bodnotbod, Spottedowl, Solkoll~enwiki, Indrian, Dgpop, Polluks, DaveGorman, Tyan23, Jason One, Alansohn, Interiot, CyberSkull, Metron4, Toyoda, KevinOKeeffe, ADeveria, Combination, Nandesuka, Mirror Vax, D.brodale, RussBot, Wgungfu, Larsinio, Rmky87, N. Harmonik, SmackBot, McGeddon, Jagged 85, Oscarthecat, ReyVGM, GVnayR, NES Boy, Ravimakkar, Asmpgmr, MarkAlldridge, Mika1h, Drinibot, Cydebot, Quetzalcoatl45, PamD, BetacommandBot, JAF1970, Stonic, Humble Scribe, X201, Scottandrewhutchins, Kung Fu Man, 100110100, Magus05, Avicennasis, Thibbs, The Glory Boy, Frogacuda, WOSlinker, Frenzicus, ImageRemovalBot, Martarius, Hippo99, Trivialist, PixelBot, The Dork Knight, BOTarate, Blackwatch21, Addbot, WuBot, AnomieBOT, Citation bot, Mattg82, Chaheel Riens, FrescoBot, RjwilmsiBot, John of Reading, ZéroBot, SlimGoodbody, Samwalton9, Bnelson112, Magicperson6969, SuperMorioBros, DangerousJXD and Anonymous: 41

- **Sea Wolf (video game)** *Source:* https://en.wikipedia.org/wiki/Sea_Wolf_(video_game)?oldid=662833851 *Contributors:* Samw, Indrian, Dgpop, Danhash, Woohookitty, Wgungfu, PTSE, 2fort5r, Krótki, SmackBot, Jagged 85, Cydebot, Legotech, BetacommandBot, Thijs!bot, Aruffo, WANAX, Sophie means wisdom, Geniac, J.delanoy, WOSlinker, JohnnyMrNinja, Ost316, MatthewVanitas, Addbot, Lightbot, Oaoii, Eekerz, Bingbongboy, Helpful Pixie Bot, Ookpik72, Samwalton9, BattyBot, ChrisGualtieri and Anonymous: 7

- **The Secret of Bastow Manor** *Source:* https://en.wikipedia.org/wiki/The_Secret_of_Bastow_Manor?oldid=679328458 *Contributors:* Paul A, Auric, Longhair, Davidmwilliams, Bgwhite, SmackBot, X201, MrKIA11, Addbot and MuZemike

- **Serpentine (video game)** *Source:* https://en.wikipedia.org/wiki/Serpentine_(video_game)?oldid=684824291 *Contributors:* Ylee, Dgpop, Jeff3000, Bgwhite, GVnayR, HDCase, Mika1h, Aspects, MrKIA11, OlEnglish, MuZemike, Daschnarr, Materialscientist, Citation bot, Deltasim, Trappist the monk, EricSnider, The1337gamer, BattyBot, AntonTchekhov, Goslajuavi, Jianhui67 and Anonymous: 3

- **Shamus (video game)** *Source:* https://en.wikipedia.org/wiki/Shamus_(video_game)?oldid=683268803 *Contributors:* Maury Markowitz, Xezbeth, Ylee, Dgpop, Longhair, Shlomital, Jason One, Adam Field, JIP, Mirror Vax, Bgwhite, Pelladon, N. Harmonik, Th1rt3en, Jagged 85, Marktreut, Thismarty, Ravimakkar, Guroadrunner, Chabanais, Ronaldvd, Mika1h, Drinibot, AndrewHowse, Valodzka, Slordak, X201, Aruffo, Xeno, Avicennasis, Jeff G., Djmips, ImageRemovalBot, Gilbertine goldmark, Ost316, Tricob, Addbot, Lightbot, Yobot, AnomieBOT, Ubcule, Lorson, Shaddim, The1337gamer, BattyBot, Dobie80 and Anonymous: 12

- **Sky Hop** *Source:* https://en.wikipedia.org/wiki/Sky_Hop?oldid=625464138 *Contributors:* Toddst1, MuZemike, Lenez, Deltasim, BG19bot, Gorobay and Anonymous: 1

- **Snooker (video game)** *Source:* https://en.wikipedia.org/wiki/Snooker_(video_game)?oldid=679332411 *Contributors:* Postdlf, Rich Farmbrough, Bgwhite, Cpc464, SMcCandlish, Ww2censor, GVnayR, Hindleyite, CommonsDelinker, Retro junkie, ImageRemovalBot, MrKIA11, Materialscientist, Armbrust, Noggin143 and Anonymous: 1

- **Squish 'em** *Source:* https://en.wikipedia.org/wiki/Squish_'em?oldid=679470753 *Contributors:* Psychonaut, Bgwhite, RussBot, N. Harmonik, Odie5533, Magioladitis, Someone another, Ost316, The Bushranger, Phoenix B 1of3, Sergecross73, Friol, Hydao, Shaddim, Helpful Pixie Bot and Anonymous: 1

- **Star Trek (arcade game)** *Source:* https://en.wikipedia.org/wiki/Star_Trek_(arcade_game)?oldid=684252199 *Contributors:* Invalidname, Pixel8, Indrian, TMC1982, CyberSkull, Woohookitty, Jeff3000, MrLeo, Icey, TnS, Wgungfu, Ritchy, Bobak, AlexCharyna, Krótki, Jagged 85, Propound, GVnayR, S@bre, Peyre, CmdrObot, Mika1h, Cydebot, BetacommandBot, Lord Hawk, X201, Sm8900, RandomFreshInk, Rolod~enwiki, Richard D. LeCour, Aatrek, WOSlinker, Wiendietry~enwiki, Ham Pastrami, Martarius, Trivialist, Gerald G-Money, JasonAQuest, Addbot, Dawynn, Colt9033, Yobot, Miyagawa, FrescoBot, Entilzha37, Lorson, ZéroBot, Braincricket, Mjenison, Monkbot and Anonymous: 16

- **Strange Odyssey** *Source:* https://en.wikipedia.org/wiki/Strange_Odyssey?oldid=679473181 *Contributors:* Jonabbey, Ylee, Czolgolz, SchuminWeb, D.brodale, Bgwhite, Rogermw, BOZ, BetacommandBot, X201, Salavat, R'n'B, Retro junkie, Prhartcom, MrKIA11, Cityofc, RjwilmsiBot and Anonymous: 2

- **Sword of Fargoal** *Source:* https://en.wikipedia.org/wiki/Sword_of_Fargoal?oldid=675868829 *Contributors:* Frecklefoot, Zoicon5, Pakaran, Cyberia23, Vadmium, ChicXulub, DNewhall, Bumm13, Sysy, KillerChihuahua, YUL89YYZ, LeeHunter, RJHall, Ylee, Hu, GregorB, ADeveria, Master Thief Garrett, TnS, N. Harmonik, SmackBot, Felicity4711, Sigma 7, Harryboyles, Iridescent, AA Pilot16, Mika1h, Drinibot, Mathsgeek, Alaibot, X201, Jaysweet, Jay Gatsby, Gwern, Coin945, Freelancelot, Bion2u2, Mild Bill Hiccup, 718 Bot, Sun Creator, JasonAQuest, Ost316, Addbot, Dawynn, Lightbot, AnomieBOT, 78.26, Sławomir Biały, Lorson, Alcazar84, Shaddim, The1337gamer, BattyBot, SNAAAAKE!!, AdrianGamer and Anonymous: 27

- **The Perils of Willy** *Source:* https://en.wikipedia.org/wiki/Miner_Willy?oldid=678579787 *Contributors:* Tarquin, Liftarn, Malcohol, Headlessness, Rls, Shantavira, Neilc, Mike Rosoft, Deadlock, Xezbeth, Richard W.M. Jones, Pearle, Halsteadk, Hoary, Marasmusine, The wub, Stx, Luk, SmackBot, Oscarthecat, Zagrebo, Jimd, Mika1h, Cydebot, Ebyabe, Mentifisto, RobotG, TTN, Kung Fu Man, Retro junkie, Lightbot, Yobot, Chaheel Riens, Erik9bot, Fortdj33, Skull33, Dgoop, NeoGenPT, Shaddim, Cyberbot II, SNAAAAKE!!, Mogism and Anonymous: 13

- **Traxx (video game)** *Source:* https://en.wikipedia.org/wiki/Traxx_(video_game)?oldid=684417195 *Contributors:* Dgpop, Bgwhite, Hahnchen, 2fort5r, X201, Trivialist, MuZemike and Anonymous: 1

- **Ultima: Escape from Mt. Drash** *Source:* https://en.wikipedia.org/wiki/Ultima%3A_Escape_from_Mt._Drash?oldid=665936915 *Contributors:* Wwwwolf, Deewiant, Ylee, Art LaPella, Mindmatrix, Sega381, Remurmur, Master Thief Garrett, Koveras, N. Harmonik, Ravimakkar, Klytos, Mika1h, Drinibot, Cydebot, AniMate, Xeno, Dream Focus, Tokyogirl79, WOSlinker, TXiKiBoT, JohnnyMrNinja, Muro Bot, Addbot, Lightbot, OlEnglish, BlueBallz, Lorson, John of Reading, KasperFauerby, Helpful Pixie Bot, The1337gamer, AdrianGamer, Spiderjerky and Anonymous: 18

- **Voodoo Castle** *Source:* https://en.wikipedia.org/wiki/Voodoo_Castle?oldid=680285421 *Contributors:* Alan De Smet, Jonabbey, Klemen Kocjancic, 23skidoo, Czolgolz, JIP, Brighterorange, D.brodale, Bgwhite, N. Harmonik, SmackBot, Bluebot, Hu12, Drinibot, Alaibot, BetacommandBot, X201, RobJ1981, Diggernet, Duffaboy, Gwern, Mr WR, Sydric, Retro junkie, Lightbot, OMPIRE and Anonymous: 7

- **Wacky Waiters** *Source:* https://en.wikipedia.org/wiki/Wacky_Waiters?oldid=680285790 *Contributors:* Rich Farmbrough, CyberSkull, Mailer diablo, GregorB, Bgwhite, Mikeblas, N. Harmonik, Canley, SmackBot, Oscarthecat, Ck lostsword, Amalas, Mika1h, Drinibot, BetacommandBot, X201, RobJ1981, MegX, Duffaboy, Gwern, Mr WR, Y, Krano, Deltasim and Anonymous: 5

5.2 Images

- **File:3D_silicon_fish_commodore_vic20_box.png** *Source:* https://upload.wikimedia.org/wikipedia/en/f/f7/3D_silicon_fish_commodore_vic20_box.png *License:* Fair use *Contributors:*
  http://www.retrogames.co.uk/031784/Commodore/3D-Silicon-Fish-by-Thor *Original artist:* ?

- **File:4860_-_VIC-20_Mainboard.JPG** *Source:* https://upload.wikimedia.org/wikipedia/commons/5/5b/4860_-_VIC-20_Mainboard.JPG *License:* Public domain *Contributors:* Own work *Original artist:* Sven.petersen

- **File:Adventureland_Cover.png** *Source:* https://upload.wikimedia.org/wikipedia/en/6/62/Adventureland_Cover.png *License:* Fair use *Contributors:*
  http://image.com.com/gamespot/images/bigboxshots/9/564079_29257_front.jpg *Original artist:* ?

- **File:Ambox_current_red.svg** *Source:* https://upload.wikimedia.org/wikipedia/commons/9/98/Ambox_current_red.svg *License:* CC0 *Contributors:* self-made, inspired by Gnome globe current event.svg, using Information icon3.svg and Earth clip art.svg *Original artist:* Vipersnake151, penubag, Tkgd2007 (clock)

- **File:Ambox_important.svg** *Source:* https://upload.wikimedia.org/wikipedia/commons/b/b4/Ambox_important.svg *License:* Public domain *Contributors:* Own work, based off of Image:Ambox scales.svg *Original artist:* Dsmurat (talk · contribs)

- **File:Apple_Panic.png** *Source:* https://upload.wikimedia.org/wikipedia/en/b/b1/Apple_Panic.png *License:* ? *Contributors:*
  Screenshot from the game, captured by Huwmanbeing *Original artist:*
  Ben Serki / Brøderbund

- **File:Apple_panic.jpg** *Source:* https://upload.wikimedia.org/wikipedia/en/1/17/Apple_panic.jpg *License:* Fair use *Contributors:*
  [1] *Original artist:* ?

- **File:Arcadia_cassette_inlay.jpg** *Source:* https://upload.wikimedia.org/wikipedia/en/d/d3/Arcadia_cassette_inlay.jpg *License:* Fair use *Contributors:*
  scanned image
  Original artist: ?

- **File:Artiller_Duel_Atari_2600_screenshot1.png** *Source:* https://upload.wikimedia.org/wikipedia/en/f/f5/Artiller_Duel_Atari_2600_screenshot1.png *License:* Fair use *Contributors:*
  I created this image file.
  Original artist: ?

- **File:Atari-2600-Light-Sixer-FL.jpg** *Source:* https://upload.wikimedia.org/wikipedia/commons/7/79/Atari-2600-Light-Sixer-FL.jpg *License:* Public domain *Contributors:* Own work *Original artist:* Evan-Amos

- **File:Atari-bz-arcade.jpg** *Source:* https://upload.wikimedia.org/wikipedia/commons/b/b9/Atari-bz-arcade.jpg *License:* CC BY-SA 2.5 *Contributors:* Transferred from en.wikipedia to Commons. *Original artist:* The original uploader was Rees11 at English Wikipedia

- **File:Atlantis_Atari_2600_screenshot1a.png** *Source:* https://upload.wikimedia.org/wikipedia/en/9/9d/Atlantis_Atari_2600_screenshot1a.png *License:* Fair use *Contributors:*
  I created this image file myself.
  Original artist: ?

- **File:Battlezone(Poster).jpg** *Source:* https://upload.wikimedia.org/wikipedia/en/8/8c/Battlezone%28Poster%29.jpg *License:* Fair use *Contributors:*
  http://www.arcadeflyers.com/?page=thumbs&db=videodb&id=116 *Original artist:* ?

- **File:BlitzVic20.jpg** *Source:* https://upload.wikimedia.org/wikipedia/en/a/a3/BlitzVic20.jpg *License:* Fair use *Contributors:*
  photograph by User:Phooto *Original artist:* ?

- **File:Bradley_Trainer_screenshot.png** *Source:* https://upload.wikimedia.org/wikipedia/commons/8/8a/Bradley_Trainer_screenshot.png *License:* Public domain *Contributors:* http://www.atariage.com/news/Bradley/ *Original artist:* US Army

- **File:C2n_waveform.png** *Source:* https://upload.wikimedia.org/wikipedia/commons/c/c5/C2n_waveform.png *License:* CC BY-SA 3.0 *Contributors:* http://en.wikipedia.org/wiki/File:C2n_waveform.png *Original artist:* Electron9 on english Wikipedia

- **File:C64_Petscii_Charts.png** *Source:* https://upload.wikimedia.org/wikipedia/commons/c/c4/C64_Petscii_Charts.png *License:* Public domain *Contributors:* ? *Original artist:* ?

- **File:CBM64CartridgeRadarRatRace.JPG** *Source:* https://upload.wikimedia.org/wikipedia/commons/4/4d/CBM64CartridgeRadarRatRace.JPG *License:* Public domain *Contributors:* Own work *Original artist:* ML5

- **File:Cbmcharset-modes.png** *Source:* https://upload.wikimedia.org/wikipedia/en/f/f5/Cbmcharset-modes.png *License:* Cc-by-sa-3.0 *Contributors:* ? *Original artist:* ?

- **File:Chalk_stub.png** *Source:* https://upload.wikimedia.org/wikipedia/commons/f/fc/Chalk_stub.png *License:* CC-BY-SA-3.0 *Contributors:* Own work *Original artist:* Myself

- **File:Choplifter.png** *Source:* https://upload.wikimedia.org/wikipedia/en/d/d1/Choplifter.png *License:* ? *Contributors:*
  Screenshot from the game, captured by Huwmanbeing. *Original artist:*
  Dan Gorlin/Brøderbund

- **File:Choplifter_title.png** *Source:* https://upload.wikimedia.org/wikipedia/en/7/7f/Choplifter_title.png *License:* ? *Contributors:* Screenshot from the game, captured by Huwmanbeing. *Original artist:*
Dan Gorlin/Brøderbund

- **File:Commodore-64-Computer.png** *Source:* https://upload.wikimedia.org/wikipedia/commons/3/34/Commodore-64-Computer.png *License:* Public domain *Contributors:* Own work *Original artist:* Evan-Amos

- **File:Commodore-64-back_cassette_port.jpg** *Source:* https://upload.wikimedia.org/wikipedia/commons/f/ff/Commodore-64-back_cassette_port.jpg *License:* Public domain *Contributors:* https://commons.wikimedia.org/wiki/File:Commodore-64-Back.jpg *Original artist:* Evan-Amos

- **File:Commodore-VIC-20-FL.jpg** *Source:* https://upload.wikimedia.org/wikipedia/commons/f/f1/Commodore-VIC-20-FL.jpg *License:* Public domain *Contributors:* Own work *Original artist:* Evan-Amos

- **File:Commodore_1530_Datasette_VIC-20.jpg** *Source:* https://upload.wikimedia.org/wikipedia/commons/8/86/Commodore_1530_Datasette_VIC-20.jpg *License:* CC BY 3.0 *Contributors:* Own work *Original artist:* Bilby

- **File:Commodore_Basic_screenshot.jpg** *Source:* https://upload.wikimedia.org/wikipedia/commons/5/52/Commodore_Basic_screenshot.jpg *License:* CC BY 2.0 *Contributors:* originally posted to **Flickr** as Playing Space Invader... *Original artist:* Sven

- **File:Commons-logo.svg** *Source:* https://upload.wikimedia.org/wikipedia/en/4/4a/Commons-logo.svg *License:* ? *Contributors:* ? *Original artist:* ?

- **File:Computer-aj_aj_ashton_01.svg** *Source:* https://upload.wikimedia.org/wikipedia/commons/d/d7/Desktop_computer_clipart_-_Yellow_theme.svg *License:* CC0 *Contributors:* https://openclipart.org/detail/105871/computeraj-aj-ashton-01 *Original artist:* AJ from openclipart.org

- **File:Contiki-C64.png** *Source:* https://upload.wikimedia.org/wikipedia/commons/5/5c/Contiki-C64.png *License:* BSD *Contributors:* Transferred from en.wikipedia to Commons by IngerAlHaosului using CommonsHelper. *Original artist:* The original uploader was Rrelf at English Wikipedia

- **File:Contiki-avr.png** *Source:* https://upload.wikimedia.org/wikipedia/commons/9/93/Contiki-avr.png *License:* BSD *Contributors:* Transferred from en.wikipedia to Commons by IngerAlHaosului using CommonsHelper. *Original artist:* The original uploader was Rrelf at English Wikipedia

- **File:Contiki-ipv6-rpl-cooja-simulation.png** *Source:* https://upload.wikimedia.org/wikipedia/commons/4/45/Contiki-ipv6-rpl-cooja-simulation.png *License:* CC BY-SA 3.0 *Contributors:* Own work *Original artist:* Adnk

- **File:CopsNRobbersAtari.gif** *Source:* https://upload.wikimedia.org/wikipedia/en/c/c7/CopsNRobbersAtari.gif *License:* Fair use *Contributors:* http://www.atarimania.com/detail_soft.php?MENU=8&VERSION_ID=1340 *Original artist:* ?

- **File:CopsNRobbersC64.png** *Source:* https://upload.wikimedia.org/wikipedia/en/f/fe/CopsNRobbersC64.png *License:* Fair use *Contributors:* http://www.mobygames.com/game/c64/cops-and-robbers/screenshots/gameShotId,140820/ *Original artist:* ?

- **File:CopsNRobbersElectron.gif** *Source:* https://upload.wikimedia.org/wikipedia/en/3/32/CopsNRobbersElectron.gif *License:* Fair use *Contributors:*
ElectrEm emulator
Original artist: ?

- **File:Cops_'n'_Robbers_Cover.jpg** *Source:* https://upload.wikimedia.org/wikipedia/en/e/e6/Cops_%27n%27_Robbers_Cover.jpg *License:* Fair use *Contributors:*
http://image.com.com/gamespot/images/bigboxshots/0/569490_1712_front.jpg *Original artist:* ?

- **File:Crystal_Clear_app_Community_Help.png** *Source:* https://upload.wikimedia.org/wikipedia/commons/6/64/Crystal_Clear_app_Community_Help.png *License:* LGPL *Contributors:* All Crystal Clear icons were posted by the author as LGPL on kde-look; *Original artist:* Everaldo Coelho and YellowIcon;

- **File:Crystal_Clear_app_kedit.svg** *Source:* https://upload.wikimedia.org/wikipedia/commons/e/e8/Crystal_Clear_app_kedit.svg *License:* LGPL *Contributors:* Sabine MINICONI *Original artist:* Sabine MINICONI

- **File:Datasette.JPG** *Source:* https://upload.wikimedia.org/wikipedia/commons/d/d3/Datasette.JPG *License:* Public domain *Contributors:* Own work *Original artist:* Joho345

- **File:Datasette.ogg** *Source:* https://upload.wikimedia.org/wikipedia/commons/a/ad/Datasette.ogg *License:* Public domain *Contributors:* Own work *Original artist:* Rolypolyman

- **File:Datasette_c2n.jpg** *Source:* https://upload.wikimedia.org/wikipedia/commons/c/c6/Datasette_c2n.jpg *License:* CC-BY-SA-3.0 *Contributors:* Own work *Original artist:* boffy_b

- **File:Deadly_Duck_Cover.jpg** *Source:* https://upload.wikimedia.org/wikipedia/en/1/1e/Deadly_Duck_Cover.jpg *License:* Fair use *Contributors:*
http://image.com.com/gamespot/images/bigboxshots/1/584671_37631_front.jpg *Original artist:* ?

- **File:Demon_Attack_box_art.jpg** *Source:* https://upload.wikimedia.org/wikipedia/en/6/67/Demon_Attack_box_art.jpg *License:* Fair use *Contributors:*
Atari Age [1] *Original artist:* ?

- **File:Demonattack.png** *Source:* https://upload.wikimedia.org/wikipedia/en/1/13/Demonattack.png *License:* Fair use *Contributors:*
Computer game itself
Original artist: ?

- **File:Dig_Dug_Famicom_cartridge.jpg** *Source:* https://upload.wikimedia.org/wikipedia/commons/a/a8/Dig_Dug_Famicom_cartridge.jpg *License:* CC BY 2.0 *Contributors:* https://flic.kr/p/9abcSx *Original artist:* Bryan Ochalla

- **File:Dig_Dug_Flyer.png** *Source:* https://upload.wikimedia.org/wikipedia/en/9/95/Dig_Dug_Flyer.png *License:* Fair use *Contributors:* http://www.arcadeflyers.com/?page=flyer&db=videodb&id=302&image=1 *Original artist:* ?

- **File:Digdug.png** *Source:* https://upload.wikimedia.org/wikipedia/en/a/a0/Digdug.png *License:* Fair use *Contributors:* http://www.eightyeightynine.com/games/dig-dug.html *Original artist:* ?

- **File:Diskette_SEGA_Frogger.jpg** *Source:* https://upload.wikimedia.org/wikipedia/commons/2/27/Diskette_SEGA_Frogger.jpg *License:* CC BY-SA 3.0 *Contributors:* Own work *Original artist:* S.J. de Waard

- **File:Dkong_end.png** *Source:* https://upload.wikimedia.org/wikipedia/en/8/8f/Dkong_end.png *License:* Fair use *Contributors:* http://www010.upp.so-net.ne.jp/muu-word/acdkong.html *Original artist:* ?

- **File:Donkey_Kong_Gameplay.png** *Source:* https://upload.wikimedia.org/wikipedia/en/f/fd/Donkey_Kong_Gameplay.png *License:* Fair use *Contributors:* http://web.archive.org/web/20050824043108/http://www010.upp.so-net.ne.jp/muu-word/acdkong.html *Original artist:* Nintendo

- **File:Donkey_Kong_arcade.jpg** *Source:* https://upload.wikimedia.org/wikipedia/commons/8/88/Donkey_Kong_arcade.jpg *License:* CC BY-SA 2.0 *Contributors:* base on http://www.flickr.com/photos/zapwizard/34102189/ (ZapWizard, cc-by-sa-2.0) and work of bayo *Original artist:* Joshua Driggs (ZapWizard) (take photo), bayo (remove the background)

- **File:Donkey_Kong_flier.jpg** *Source:* https://upload.wikimedia.org/wikipedia/en/1/14/Donkey_Kong_flier.jpg *License:* ? *Contributors:* ? *Original artist:* ?

- **File:DragonfireBoxShotAtari2600.jpg** *Source:* https://upload.wikimedia.org/wikipedia/en/6/6b/DragonfireBoxShotAtari2600.jpg *License:* Fair use *Contributors:*
  http://www.gamefaqs.com/atari2600/584694-dragonfire/images *Original artist:* ?

- **File:Edit-clear.svg** *Source:* https://upload.wikimedia.org/wikipedia/en/f/f2/Edit-clear.svg *License:* Public domain *Contributors:* The Tango! Desktop Project. *Original artist:*
  The people from the Tango! project. And according to the meta-data in the file, specifically: "Andreas Nilsson, and Jakub Steiner (although minimally)."

- **File:Emoji_u1f4be.svg** *Source:* https://upload.wikimedia.org/wikipedia/commons/f/fb/Emoji_u1f4be.svg *License:* Apache License 2.0 *Contributors:* https://code.google.com/p/noto/ *Original artist:* Google

- **File:Festival_du_Jeu_Vidéo_-_2010-09-11_-_Game_&_Watch_Donkey_Kong.jpg** *Source:* https://upload.wikimedia.org/wikipedia/commons/1/1a/Festival_du_Jeu_Vid%C3%A9o_-_2010-09-11_-_Game_%26_Watch_Donkey_Kong.jpg *License:* CC BY-SA 3.0 *Contributors:* Own work *Original artist:* Jesmar

- **File:Flag_of_Japan.svg** *Source:* https://upload.wikimedia.org/wikipedia/en/9/9e/Flag_of_Japan.svg *License:* PD *Contributors:* ? *Original artist:* ?

- **File:Folder_Hexagonal_Icon.svg** *Source:* https://upload.wikimedia.org/wikipedia/en/4/48/Folder_Hexagonal_Icon.svg *License:* Cc-by-sa-3.0 *Contributors:* ? *Original artist:* ?

- **File:Frogger_1_xbla_cover.jpg** *Source:* https://upload.wikimedia.org/wikipedia/en/2/2./Frogger_1_xbla_cover.jpg *License:* Fair use *Contributors:*
  Official site *Original artist:* ?

- **File:Frogger_game_arcade.png** *Source:* https://upload.wikimedia.org/wikipedia/en/c/cc/Frogger_game_arcade.png *License:* ? *Contributors:* ? *Original artist:* ?

- **File:Galaxian.png** *Source:* https://upload.wikimedia.org/wikipedia/en/0/09/Galaxian.png *License:* ? *Contributors:*
  Video-game or computer emulator
  Original artist: ?

- **File:Galaxian_-_mnactec.JPG** *Source:* https://upload.wikimedia.org/wikipedia/commons/d/d8/Galaxian_-_mnactec.JPG *License:* CC BY-SA 3.0 *Contributors:* Own work *Original artist:* Jordiferrer

- **File:Galaxian_flyer.jpg** *Source:* https://upload.wikimedia.org/wikipedia/en/5/5f/Galaxian_flyer.jpg *License:* Fair use *Contributors:*
  May be found at the following website: The Arcade Flyer Archive *Original artist:* ?

- **File:Gamepad.svg** *Source:* https://upload.wikimedia.org/wikipedia/en/b/be/Gamepad.svg *License:* ? *Contributors:* ? *Original artist:* ?

- **File:Gnome-joystick.svg** *Source:* https://upload.wikimedia.org/wikipedia/commons/b/b6/Gnome-joystick.svg *License:* LGPL *Contributors:* http://ftp.gnome.org/pub/GNOME/sources/gnome-themes-extras/0.9/gnome-themes-extras-0.9.0.tar.gz *Original artist:* David Vignoni

- **File:Gnome-mime-sound-openclipart.svg** *Source:* https://upload.wikimedia.org/wikipedia/commons/8/87/Gnome-mime-sound-openclipart.svg *License:* Public domain *Contributors:* Own work. Based on File:Gnome-mime-audio-openclipart.svg, which is public domain. *Original artist:* User:Eubulides

- **File:Gorftitlescreen.png** *Source:* https://upload.wikimedia.org/wikipedia/en/1/1b/Gorftitlescreen.png *License:* Fair use *Contributors:*
  Video-game or computer emulator
  Original artist: ?

- **File:HareraiserZX.gif** *Source:* https://upload.wikimedia.org/wikipedia/en/0/0a/HareraiserZX.gif *License:* Fair use *Contributors:* http://www.adamdawes.com/retrogaming/rg_17_playtowin.html *Original artist:* ?

- **File:Q*Bert_concept_sketch.jpg** *Source:* https://upload.wikimedia.org/wikipedia/en/9/9b/Q%2ABert_concept_sketch.jpg *License:* Fair use *Contributors:* **Original publication**: EDGE Online

  Immediate source: http://www.edge-online.com/features/making-qbert/2/ *Original artist:* Jeff Lee

- **File:Q*bert_Atari_2600_Screenshot.png** *Source:* https://upload.wikimedia.org/wikipedia/en/5/54/Q%2Abert_Atari_2600_Screenshot.png *License:* Fair use *Contributors:* Screenshot taken directly from the game. *Original artist:* Parker Brothers / Gottlieb

- **File:Q*bert_arcade_cabinet.jpg** *Source:* https://upload.wikimedia.org/wikipedia/en/7/7c/Q%2Abert_arcade_cabinet.jpg *License:* Fair use *Contributors:* http://www.arcade-history.com/?n=qbert-model-gv-103&page=detail&id=2094 *Original artist:* Gottlieb

- **File:Q*bert_merchandise_advertisement_flyer.jpg** *Source:* https://upload.wikimedia.org/wikipedia/en/c/c8/Q%2Abert_merchandise_advertisement_flyer.jpg *License:* Fair use *Contributors:* http://flyers.arcade-museum.com/?page=thumbs&db=videodb&id=2556 *Original artist:* Gottlieb

- **File:Qbert.png** *Source:* https://upload.wikimedia.org/wikipedia/en/5/5e/Qbert.png *License:* Fair use *Contributors:* Screenshot taken by User:Spottedowl *Original artist:* ?

- **File:Qbertsqubes.png** *Source:* https://upload.wikimedia.org/wikipedia/en/8/86/Qbertsqubes.png *License:* Fair use *Contributors:* Screenshot taken by User:Spottedowl *Original artist:* ?

- **File:Question_book-new.svg** *Source:* https://upload.wikimedia.org/wikipedia/en/9/99/Question_book-new.svg *License:* Cc-by-sa-3 0 *Contributors:* Created from scratch in Adobe Illustrator. Based on Image:Question book.png created by User:Equazcion *Original artist:* Tkgd2007

- **File:River_Rescue-screenshot.png** *Source:* https://upload.wikimedia.org/wikipedia/en/e/2a/River_Rescue-screenshot.png *License:* Fair use *Contributors:* Screenshot from game running on VICE Vic-20 emulator *Original artist:* ?

- **File:Robotron:_2084.png** *Source:* https://upload.wikimedia.org/wikipedia/en/7/79/Robotron%3A_2084.png *License:* Fair use *Contributors:* Screenshot taken by User:Tyan23 *Original artist:* ?

- **File:RobotronX-gameplay.jpg** *Source:* https://upload.wikimedia.org/wikipedia/en/f/fa/RobotronX-gameplay.jpg *License:* Fair use *Contributors:* http://reviews.cnet.com/legacy-game-platforms/robotron-x-playstation/4528-9882_7-30965615-3.html?tag=mncol *Original artist:* ?

- **File:Robotron_flyer.png** *Source:* https://upload.wikimedia.org/wikipedia/en/e/ee/Robotron_flyer.png *License:* Fair use *Contributors:* May be found at the following website: The Arcade Flyer Archive *Original artist:* ?

- **File:Rubik'{}s_cube_v3.svg** *Source:* https://upload.wikimedia.org/wikipedia/commons/b/b6/Rubik%27s_cube_v3.svg *License:* CC-BY-SA-3.0 *Contributors:* Image:Rubik'{}s cube v2.svg *Original artist:* User:Booyabazooka, User:Meph666 modified by User:Niabot

- **File:Scramble_arcade_flyer.jpg** *Source:* https://upload.wikimedia.org/wikipedia/en/1/15/Scramble_arcade_flyer.jpg *License:* Fair use *Contributors:* The Arcade Flyer Archive *Original artist:* ?

- **File:Screen_color_test_CommodoreVIC20_Multicolor.png** *Source:* https://upload.wikimedia.org/wikipedia/commons/c/cb/Screen_color_test_CommodoreVIC20_Multicolor.png *License:* Public domain *Contributors:* Own work (Original text: *self-made*) *Original artist:* Ricardo Cancho Niemietz (talk)

- **File:Sea_wolf_arcade_midway_flyer.jpg** *Source:* https://upload.wikimedia.org/wikipedia/en/1/19/Sea_wolf_arcade_midway_flyer.jpg *License:* Fair use *Contributors:* May be found at the following website: The Arcade Flyer Archive *Original artist:* ?

- **File:Secret_of_bastow_manor_01.gif** *Source:* https://upload.wikimedia.org/wikipedia/en/3/36/Secret_of_bastow_manor_01.gif *License:* Fair use *Contributors:* From Lemon. *Original artist:* ?

- **File:Simons_Basic_Splash_Screen.gif** *Source:* https://upload.wikimedia.org/wikipedia/en/9/96/Simons_Basic_Splash_Screen.gif *License:* PD *Contributors:* ? *Original artist:* ?

- **File:Snooker_balls_triangled.png** *Source:* https://upload.wikimedia.org/wikipedia/commons/3/3a/Snooker_balls_triangled.png *License:* CC BY-SA 2.5 *Contributors:* Own work *Original artist:* Myself

- **File:Space_Invaders_Paris.JPG** *Source:* https://upload.wikimedia.org/wikipedia/commons/1/18/Space_Invaders_Paris.JPG *License:* Public domain *Contributors:* Own work *Original artist:* KoS

- **File:SpeedScript_128_in_action.png** *Source:* https://upload.wikimedia.org/wikipedia/commons/f/fe/SpeedScript_128_in_action.png *License:* Public domain *Contributors:* ? *Original artist:* ?

- **File:SpeedScript_3.0.png** *Source:* https://upload.wikimedia.org/wikipedia/en/6/60/SpeedScript_3.0.png *License:* Fair use *Contributors:* COMPUTE! Publications *Original artist:* ?

- **File:Speedscript_3.2_for_Commodore_64.png** *Source:* https://upload.wikimedia.org/wikipedia/en/0/0b/Speedscript_3.2_for_ Commodore_64.png *License:* Cc-by-sa-3.0 *Contributors:*
  Screenshot from VICE by coldacid *Original artist:*
  SpeedScript user interface by Charles Brannon. Text by various Wikipedia editors and authors (see below). Screenshot by coldacid.

- **File:SquishemC64screenshot.png** *Source:* https://upload.wikimedia.org/wikipedia/en/a/af/SquishemC64screenshot.png *License:* Fair use *Contributors:*
  VICE emulator
  Original artist: ?

- **File:Star_empty.svg** *Source:* https://upload.wikimedia.org/wikipedia/commons/4/49/Star_empty.svg *License:* CC BY-SA 2.5 *Contributors:* Made with Inkscape from Stars615.svg: <a href='//commons.wikimedia.org/wiki/File:Stars615.svg' class='image'><img alt='Stars615.svg' src='https://upload.wikimedia.org/wikipedia/commons/thumb/7/72/Stars615.svg/96px-Stars615.svg.png' width='96' height='17' srcset='https: //upload.wikimedia.org/wikipedia/commons/thumb/7/72/Stars615.svg/144px-Stars615.svg.png 1.5x, https://upload.wikimedia.org/wikipedia/ commons/thumb/7/72/Stars615.svg/192px-Stars615.svg.png 2x' data-file-width='640' data-file-height='110' /></a>. *Original artist:* This vector image was created with Inkscape by Conti from the original images by RedHotHeat, and then manually edited.

- **File:Star_full.svg** *Source:* https://upload.wikimedia.org/wikipedia/commons/5/51/Star_full.svg *License:* Public domain *Contributors:* Made with Inkscape from Image:Stars615.svg. *Original artist:* User:Conti from the original images by User:RedHotHeat

- **File:Sword_map.jpg** *Source:* https://upload.wikimedia.org/wikipedia/en/9/91/Sword_map.jpg *License:* ? *Contributors:* ? *Original artist:* ?

- **File:Sword_map1.png** *Source:* https://upload.wikimedia.org/wikipedia/en/1/17/Sword_map1.png *License:* ? *Contributors:* ? *Original artist:* ?

- **File:Symbol_book_class2.svg** *Source:* https://upload.wikimedia.org/wikipedia/commons/8/89/Symbol_book_class2.svg *License:* CC BY-SA 2.5 *Contributors:* Mad by Lokal_Profil by combining: *Original artist:* Lokal_Profil

- **File:Symbol_list_class.svg** *Source:* https://upload.wikimedia.org/wikipedia/en/d/db/Symbol_list_class.svg *License:* Public domain *Contributors:* ? *Original artist:* ?

- **File:Synaptic.png** *Source:* https://upload.wikimedia.org/wikipedia/commons/0/05/Synaptic.png *License:* GPL *Contributors:* [1] *Original artist:* en:User:Burgundavia

- **File:Text_document_with_red_question_mark.svg** *Source:* https://upload.wikimedia.org/wikipedia/commons/a/a4/Text_document_with_ red_question_mark.svg *License:* Public domain *Contributors:* Created by bdesham with Inkscape; based upon Text-x-generic.svg from the Tango project. *Original artist:* Benjamin D. Esham (bdesham)

- **File:Tudengite_Kevadpäevad_2009,_kostümeeritud_teatejooks_09.JPG** *Source:* https://upload.wikimedia.org/wikipedia/commons/4/4b/ Tudengite_Kevadp%C3%A4evad_2009%2C_kost%C3%BCmeeritud_teatejooks_09.JPG *License:* CC BY-SA 3.0 *Contributors:* Own work *Original artist:* Ivo Kruusamägi

- **File:Ultima_Escape_from_Mt_Drash_cover.png** *Source:* https://upload.wikimedia.org/wikipedia/en/1/13/Ultima_Escape_from_Mt_ Drash_cover.png *License:* Fair use *Contributors:*
  May be found at the following website: [1] *Original artist:* ?

- **File:VIC-20_SpeedScript.gif** *Source:* https://upload.wikimedia.org/wikipedia/en/7/7c/VIC-20_SpeedScript.gif *License:* Cc-by-sa-3.0 *Contributors:*
  Screen capture of VIC20 version in operation.
  Original artist: ?

- **File:VICE.png** *Source:* https://upload.wikimedia.org/wikipedia/commons/3/3d/VICE.png *License:* GPL *Contributors:* ? *Original artist:* ?

- **File:ViC20_Cartridge.jpg** *Source:* https://upload.wikimedia.org/wikipedia/commons/e/ef/ViC20_Cartridge.jpg *License:* Public domain *Contributors:* Own work *Original artist:* ?

- **File:Vic20_16k.jpg** *Source:* https://upload.wikimedia.org/wikipedia/commons/5/5c/Vic20_16k.jpg *License:* CC BY-SA 3.0 *Contributors:* Own work *Original artist:* Alex Lozupone

- **File:Vice-c64.png** *Source:* https://upload.wikimedia.org/wikipedia/commons/b/b2/Vice-c64.png *License:* Public domain *Contributors:* Own work *Original artist:* Original uploader was DASHBot at en.wikipedia

- **File:Wiki_letter_w_cropped.svg** *Source:* https://upload.wikimedia.org/wikipedia/commons/1/1c/Wiki_letter_w_cropped.svg *License:* CC-BY-SA-3.0 *Contributors:*
- Wiki_letter_w.svg *Original artist:* Wiki_letter_w.svg: Jarkko Piiroinen

- **File:Zx_arcadia_screenshot.png** *Source:* https://upload.wikimedia.org/wikipedia/en/b/b9/Zx_arcadia_screenshot.png *License:* Fair use *Contributors:*
  Emulator
  Original artist: ?

5.3 Content license